ADVANCES IN
HUMAN GENETICS 11

CONTRIBUTORS TO THIS VOLUME

Diane Wilson Cox
University of Toronto
Toronto, Canada

Robert J. Desnick
Mount Sinai School of Medicine
New York, New York

R. C. Elston
Louisiana State University Medical Center
New Orleans, Louisiana

Magne K. Fagerhol
Ullevaal Hospital
Oslo, Norway

Gregory A. Grabowski
Mount Sinai School of Medicine
New York, New York

Attallah Kappas
The Rockefeller University Hospital
New York, New York

David G. Nathan
Harvard Medical School
Boston, Massachusetts

Stuart H. Orkin
Harvard Medical School
Boston, Massachusetts

Shigeru Sassa
The Rockefeller University Hospital
New York, New York

A Continuation Order Plan is available for this series. A continuation order will bring delivery of each new volume immediately upon publication. Volumes are billed only upon actual shipment. For further information please contact the publisher.

ADVANCES IN HUMAN GENETICS 11

Edited by

Harry Harris

Harnwell Professor of Human Genetics
University of Pennsylvania, Philadelphia

and

Kurt Hirschhorn

Herbert H. Lehman Professor and Chairman of Pediatrics
Mount Sinai School of Medicine of The City University of New York

SPRINGER SCIENCE+BUSINESS MEDIA, LLC

The Library of Congress cataloged the first volume of this title as follows:

Advances in human genetics. 1-
    New York, Plenum Press, 1970-

        (1) v. illus. 24-cm.

        Editors: v. 1- H. Harris and K. Hirschhorn.

        1. Human genetics—Collected works. I. Harris, Harry, ed. II. Hirschhorn, Kurt,
1926-   joint ed.
        QH431.A1A32   573.2'1   77-84583

Library of Congress Card Catalog Number 77-84583
ISBN 978-1-4615-8305-9   ISBN 978-1-4615-8303-5 (eBook)
DOI 10.1007/978-1-4615-8303-5

© 1981 Springer Science+Business Media New York
Originally published by Plenum Press, New York in 1981
Softcover reprint of the hardcover 1st edition 1981

ARTICLES PLANNED FOR FUTURE VOLUMES:

CONTENTS OF EARLIER VOLUMES:

Preface to Volume 1

During the last few years the science of human genetics has been expanding almost explosively. Original papers dealing with different aspects of the subject are appearing at an increasingly rapid rate in a very wide range of journals, and it becomes more and more difficult for the geneticist and virtually impossible for the nongeneticist to keep track of the developments. Furthermore, new observations and discoveries relevant to an overall understanding of the subject result from investigations using very diverse techniques and methodologies and originating in a variety of different disciplines. Thus, investigations in such various fields as enzymology, immunology, protein chemistry, cytology, pediatrics, neurology, internal medicine, anthropology, and mathematical and statistical genetics, to name but a few, have each contributed results and ideas of general significance to the study of human genetics. Not surprisingly it is often difficult for workers in one branch of the subject to assess and assimilate findings made in another. This can be a serious limiting factor on the rate of progress.

Thus, there appears to be a real need for critical review articles which summarize the positions reached in different areas, and it is hoped that *Advances in Human Genetics* will help to meet this requirement.

Each of the contributors has been asked to write an account of the position that has been reached in the investigations of a specific topic in one of the branches of human genetics. The reviews are intended to be critical and to deal with the topic in depth from the writer's own point of view. It is hoped that the articles will provide workers in other branches of the subject, and in related disciplines, with a detailed account of the results so far obtained in the particular area, and help them to assess the relevance of these discoveries to aspects of their own work, as well as to the science as a whole. The reviews are also intended to give the reader some idea of the nature of the technical and methodological problems involved, and to indicate new directions stemming from recent advances.

The contributors have not been restricted in the arrangement or organization of their material or in the manner of its presentation, so that the reader should be able to appreciate something of the individuality of approach which goes to make up the subject of human genetics, and which, indeed, gives it much of its fascination.

HARRY HARRIS
The Galton Laboratory
University College London

KURT HIRSCHHORN
Division of Medical Genetics
Department of Pediatrics
Mount Sinai School of Medicine

Preface to Volume 10

This is the tenth volume of *Advances in Human Genetics* and some fifty different reviews covering a very wide range of topics have now appeared. Many of the earlier articles still stand as valuable sources of reference. But the subject continues to move forward at an increasing speed and its vitality is indicated by its remarkable recruitment of young investigators. New areas of research which could hardly have been envisaged only a few years ago have emerged, and quite unexpected discoveries have been made in parts of the subject which only recently had come to be thought of as fully explored. So there continues to be a need for authoritative and critical reviews intended to keep workers in the various branches of this seemingly ever-expanding subject fully informed about the progress that is being made and also, of course, to provide a ready and accessible account of new developments in human genetics for those whose primary interests are in other fields of biological and medical research.

We see no reason to alter the general policy which was outlined in the preface to the first volume. We believe that it has served our readers well. The subject seems to us to be just as exciting and intellectually stimulating and rewarding as it did when this series was first started. We expect the next decade of research in human genetics to be as innovative and productive as the last and our aim is to record its progress in *Advances in Human Genetics*.

HARRY HARRIS
University of Pennsylvania, Philadelphia

KURT HIRSCHHORN
Mount Sinai School of Medicine
of the City University of New York

xi

NOTE ABOUT ADDENDA

To make the volume as up-to-date as possible, each author was given the opportunity to write a short Addendum at the time he or she received the page proofs of that particular chapter. This allows for any important new material to be presented at the latest possible time in the publication process. The Addenda are all presented together at the end of the book, beginning on page 371.

Contents

Chapter 2

Segregation Analysis

R. C. Elston

Chapter 3

Genetic, Metabolic, and Biochemical Aspects of the Porphyrias

Shigeru Sassa and Attallah Kappas

Chapter 4

The Molecular Genetics of Thalassemia

Stuart H. Orkin and David G. Nathan

Chapter 5

Advances in the Treatment of Inherited Metabolic Diseases

Robert J. Desnick and Gregory A. Grabowski

Chapter 1

The Pi Polymorphism
Genetic, Biochemical, and Clinical Aspects of Human α_1-Antitrypsin

Magne K. Fagerhol

Ullevaal Hospital
Oslo, Norway

Diane Wilson Cox

Research Institute, The Hospital for Sick Children and
Departments of Paediatrics, Medical Genetics, and Medical Biophysics
University of Toronto
Toronto, Canada

INTRODUCTION

Few polymorphic plasma proteins have attracted more interest among scientists during the last ten years than the Pi (protease inhibitor) system of human α_1-antitrypsin (α_1AT). There are several reasons for this interest. The Pi system was, from the beginning, associated with disease, which heralded the possibility of studying the pathogenetic mechanism leading to tissue damage. With the development of high-resolution techniques, the Pi system was soon shown to comprise more than 30 codominant alleles. Several of these alleles are polymorphic in many ethnic groups, thus providing a useful marker for population geneticists. The finding of linkage between the Pi system and the Gm system of human IgG has contributed to chromosome mapping and to studies on crossing-over processes. The Pi system may provide insight into selective forces.

Scientists in many countries are presently working out the detailed structure of the α_1AT molecule, including the amino acid sequence and the carbohydrate side chains. These studies will provide important information on the accumulation of the Pi^Z allele product in hepatocytes (associated with cirrhosis in many cases), on environmental factors that can block the physiological function of $\alpha_1 AT$, on the mechanism of its interaction with proteases, and on the mechanisms for the synthesis and release of glycoproteins.

The study of proteases and their inhibitors is becoming of prime interest, not only for biochemists and biologists, but also for clinicians, since such substances play major roles in preserving homeostasis in the body. For example, the delicate balance between hemorrhage and thrombosis is maintained by a proper activation of proteases which must be counterbalanced by protease inhibitors. The same types of molecules are important as mediators in the inflammatory response, in protection against tissue damage, in removal of dead cells and tissue, and in healing processes.

With increasing knowledge of the genetic constitution of many individuals in various populations, it becomes evident that diseases often are caused by one or more environmental factors acting upon individuals with a certain inherited predisposition. The term *ecogenetics* has been used for this field. The Pi system offers useful examples and models for further studies. Individuals who are homozygous for the Pi^Z gene have a high risk of developing childhood cirrhosis or early adulthood emphysema, but some escape both diseases. Characteristically, ZZ emphysema patients are males that have been heavy smokers for many years, and the smoking seems to subtract an average of a least ten years from their life-span. Generally speaking, though, little is yet known of the additional genetic and/or environmental factors that lead to disease in some Pi type ZZ individuals.

In recent years many reports have suggested an increased frequency of some heterozygous Pi types among patients with various kinds of disease such as rheumatoid arthritis, nephritis, asthma, and lung cancer. This suggests that the level or type of $\alpha_1 AT$ in the body may influence the pathophysiological processes and thereby be one of the factors that determine the occurrence, course, and severity of clinical disease.

HISTORICAL REVIEW

The name α_1-antitrypsin was suggested by Schultze *et al.* (1962) for the protein they had described as α_1-3,5-glycoprotein seven years earlier (Schultze *et al.*, 1955). In the meantime, other groups had associated most of the serum trypsin inhibitory activity with the α_1-globulin fraction (Moll *et al.*, 1958; Bundy and Mehl, 1959).

The history of the Pi system began with the observation of Laurell and Eriksson (1963) in Malmo, Sweden, that in sera from some patients, the α_1-globulin band was absent on paper or agarose gel electrophoresis. They found this deficiency associated with pulmonary disease, and shortly after, also discovered a slow variant (later identified as Pi^X), with normal protein concentration and trypsin inhibitory capacity. Axelsson and Laurell (1965) reported studies on a large family suggesting that the genes for α_1AT deficiency and the slowly migrating variant were alleles.

Independently of the Malmo group, Fagerhol and Braend (1965) were working on improvements of the starch-gel electrophoretic technique to look for genetic variation in the prealbumin (Pr) region. Shortly after their report of the Pr system, they suspected that the Pr bands were in fact α_1AT, a suggestion also presented to them in correspondence from Dr. K. Heide at the Behringwerke laboratory.

The identity of the Pr and α_1AT systems was soon proven (Fagerhol and Laurell, 1967). The symbol Pi was chosen for the α_1AT polymorphism, in consultation with Dr. Schultze, since the protein is one of the major protease inhibitors of human plasma. Already at that time few if any believed that trypsin was the major target for inhibition by α_1AT.

Pi typing procedures were not easily established in all laboratories, partly because suitable hydrolyzed starch of high quality was not available and partly because details of the method varied in different laboratories.

The introduction of isoelectric focusing in polyacrylamide gels (PIEF) not only bypassed the starch problem but also revealed new polymorphic and rare variants, so that a significant extension of the Pi system was obtained. By isofocusing, 20–30% of the individuals in many populations have Pi types other than the "normal" MM type or its subtypes.

Pi VARIANTS

Nomenclature of Genetic Variants

A large number of Pi variants have now been described. Some of these appear to be identical but have been given several different names. In order to help clarify the nomenclature for the genetic variants of α_1AT, a group of scientists involved in studies of α_1AT variants met in Rouen in July, 1978, to formulate nomenclature guidelines. Recommendations from consensus at the Rouen Nomenclature Meeting have been published in full (Rouen Report, 1980) and are briefly outlined here.

The name α_1-antitrypsin (α_1AT) has been maintained because of its common usage in the literature since early reports of its purification and characterization (Schultze *et al.*, 1962). Recently, "α_1-protease inhibitor" has been favored in the biochemical literature because of its more accurate description of function. The genetic locus for α_1AT is *Pi*. A Nomenclature Committee of the International Linkage Workshop is currently preparing a unified system for gene loci which will be appropriate for computer application. In this context, the *Pi* locus will be written as *PI*.

Prior to the Rouen Nomenclature Meeting, the Pi variants were given letters according to their relative mobility in acid starch gel. Because of the wide-spread use of isoelectric focusing, it was recommended that subsequent variants should be designated by a letter corresponding to the relative position by isoelectric focusing in polyacrylamide (PIEF). When there is no available letter, the letter of the closest anodal allele may be used, with the addition of a place of origin name, or a numeral for polymorphic alleles. Allele symbols are therefore written as Pi^S, Pi^{M2}, $Pi^{Mmalton}$. Phenotypes would be designated as M2S, MmaltonM3, M1M3, etc. A three-letter abbreviation may be used for names, e.g., Mmal for Mmalton, Pstl for Pstlouis, etc. The M subtypes are designated M1, M3, and M2, beginning with the most anodal. The use of M without a numerical designation indicates that the method of Pi typing has not distinguished subvariants. The allele giving no detectable α_1AT in serum is Pi^{null} (Talamo *et al.*, 1973), represented in genotypes by a dash, e.g., Pi S–. Where a family study has not excluded the possibility of a Pi^{null} allele, homozygous types should be designated by a single allele symbol, e.g., S (phenotype) rather than SS (genotype).

Reference laboratories have been established in Amsterdam, Toronto, and Chapel Hill for the comparison of possible new variants. The requirements for requesting testing at the reference laboratories have been outlined. These laboratories are prepared to compare submitted variants with previously established variants, after criteria for such testing have been met.

When describing, in the literature, variants which have been confirmed as new in the reference laboratories, inclusion of photographs showing a comparison with other close variants by isoelectric focusing, and by acid starch gel and agarose electrophoresis is recommended. The extra bands should be confirmed as α_1AT by an immunological technique, such as crossed immunoelectrophoresis or immunofixation. A family study and quantitation of $\alpha_1 AT$ should also be included. These recommendations should facilitate both population studies and biochemical characterization of the Pi variants. (See Addendum.)

Techniques for Identification

Starch Gel Electrophoresis

Prior to 1976, the Pi variants were identified primarily by acid starch gel electrophoresis, with the exception of a few variants which had been identified by agarose electrophoresis. With an acid starch gel, at about pH 5, $\alpha_1 AT$ migrates more anodally than many of the serum proteins including albumin. At this acid pH, the microheterogeneity of $\alpha_1 AT$ can be observed. In some individuals, haptoglobin can interfere with the Pi pattern but can be removed by adding a small amount of hemoglobin to produce hemoglobin–haptoglobin complex.

The deficient type of $\alpha_1 AT$ (Z) was found to migrate more slowly than normal. The genetic variants were named in order of their relative mobility in the acid starch gel system: F (fast), M (medium), S (slow), and Z (the most cathodal) (Fagerhol, 1967). Subsequently discovered variants were also given alphabetic designations in accordance with their relative mobility in starch.

While isoelectric focusing in polyacrylamide gels has to a large extent replaced starch gel electrophoresis, the starch method remains important for the characterization of genetic variants. Several variants with similar mobility by PIEF can be differentiated by starch.

The discontinuous buffer system used for acid starch gel electrophoresis establishes a pH gradient in a limited part of the gel for a limited period of time. It therefore represents a type of isoelectric focusing. Additional factors are probably also important since some variants behave differently on starch gel than by PIEF.

The technique for starch gel electrophoresis remains as originally described (Fagerhol, 1968). The quality of the hydrolyzed starch is of great importance and the most distinct patterns are obtained when potato starch is hydrolyzed specifically for this purpose. The concentration of starch and the pH of the gel buffer must be adjusted for each starch lot and probably for each laboratory. The starch concentration must be adequate for producing a gel which can be conveniently sliced. This is usually around 12–14%. An increased in the starch concentration decreases the mobility of α_1AT. Increasing the pH of the gel buffer also decreases the anodal mobility of α_1AT. A slight increase (about 10%) in the acidity of the gel buffer has been found to be preferable for distinguishing the variants more anodal to M (Fagerhol, 1967).

Crossed immunoelectrophoresis following electrophoresis in the starch gel is necessary for the recognition of deficient variants and to detect the minor components of the variant patterns (Fagerhol and Laurell, 1967). Immunofixation directly in the starch gel has also been described (Lieberman and Gaidulis, 1976).

Isoelectric Focusing in Polyacrylamide (PIEF)

PIEF offers increased resolution of the Pi variants, better reproducibility between laboratories, and the possibility of handling up to 50 samples on a single gel. Some variants identical by starch gel electrophoresis are different by PIEF. PIEF has been particularly useful in identifying common subtypes of M. The multiple bands of α_1AT focus between pH 4.46 and 4.69. Ampholytes between pH 3.5 and 6 are generally used. Ampholytes tend to show variation from lot to lot and the particular mix of ampholytes (e.g., 3.5–5 and 4–6) can be selected for maximum separation of variants. Several methods for the production of suitable gels have been outlined and differ slightly in running conditions, in stabilizing agent (sucrose, glycerol), and in polymerizing agents (Allen et al., 1974; Arnaud et al., 1974; Lebas et al., 1974; Pierce et al., 1976). Narrow pH-range prepared gels (LKB PAGplates, pH 4–5) are available commercially.

Because of the increased resolution of PIEF, there is a greater tendency to observe distorted band patterns produced by aging of the serum or by complexing of α_1AT with other proteins in certain pathological states. These problems frequently can be overcome by the reduction or all sera or plasma samples, or at least those showing unusual patterns, with a reducing agent such as dithioerythritol or dithiothreitol (Pierce et al., 1976; Pierce and Eradio, 1979). Cysteine reduction is particularly useful for plasma samples and stabilizes the α_1AT for several days at 4°C, a useful feature when repeat analyses are to be made. An excess of heparin, which obscures the Pi region, can be cleared using protamine sulfate (Klasen and de Brij, 1979). Occasionally, aged samples deteriorate to such an extent that even a normal Pi type appears deficient, although in such samples no clear Pi Z bands are seen.

An additional complication in the interpretation of Pi patterns by PIEF is the occurrence of other plasma proteins within the $\alpha_1 AT$ region. These can become particularly obvious in certain disease states where the amount of $\alpha_1 AT$ is very low. In such cases, the use of an immunological technique is important. Crossed immunoelectrophoresis has been described for use with PIEF (Lebas et al., 1974).

This technique is technically difficult and the same purpose can be achieved more easily by immunofixation directly on the gel (Johnson, 1976) or by print "immunofixation" using cellulose acetate (Arnaud et al., 1977d). The high resolution of PIEF is maintained in the latter two methods.

An increase in running time to four or five hours increases separation by distorting the linearity of the pH gradient and increasing the resolution in certain areas of the gel. This same effect can be obtained by the use of N-(2)acetamido)-2-aminomethane sulfonic acid, a sulfonic acid derivative (Frants and Eriksson, 1978; Frants et al., 1978). While increasing the resolution in certain areas and improving discrimination, for example, for M subtypes, the Z band can become indistinct or entirely disappear under these conditions. Carrier ampholytes synthesized with condensing reagents have been used to produce a clear resolution of Pi types, the resolving power being 0.005 pH unit (Charlionet et al., 1979).

Agarose Electrophoresis

A deficiency of $\alpha_1 AT$ can be recognized by agarose electrophoresis (Johanssen, 1972). For the study of Pi variants, agarose electrophoresis is useful to distinguish variants similar by other methods, e.g., S and V.

Good resolution can be obtained with attention to the quality of agarose, application of the sample through a plastic template, and careful attention to running conditions (Jeppsson et al., 1979b).

Characterization of Pi Variants

Pi M, now further divided into subtypes, is the most common allele in all populations tested to date. All alleles appear to fit a model of multiple

TABLE I. Established Genetic Variants

Anodal Pi variant (starch)	Reference	Cathodal Pi variant (starch)	Reference
B	Martin et al. (1973)	Mchapelhill (Mcha)	Johnson (1976)
C	Robinet-Levy and Reunier (1972)	Nhampton (Nham)	Arnaud et al. (1979)
D	Robinet-Levy and Reunier (1972)	N	Cox and Celhoffer (1974)
E	Fagerhol (1972b)	Pstlouis (Pstl)	Rouen Report (1980)[a]
Elemberg (Elem)	Rouen Report (1980)[a]; Cox (1981)	P	Fagerhol and Hauge (1968)
Etokyo (Etok)	Rouen Report (1980)[a]	Pbudapest (Pbud)	Cox (1977, 1981)
F	Fagerhol (1967)	R	Cox (1977; 1981)
Ecincinnati (Ecin)	Rouen Report (1980)[a]	S	Fagerhol (1967)
G	Fagerhol (1972b)	T	Kühnl and Spielmann (1979)
I	Fagerhol (1967)	Pkyoto (Pkyo)	Rouen Report (1980)[a]
L	Vandeville et al. (1974)	V	Fagerhol (1967)
M subvariants	(See text)	Wsalermo (Wsal)	Cook (1975)
(M1, M2, M3)			
Mmalton (Mmal)	Cox (1976)	X	Axelsson and Laurell (1965)
		Ytoronto (Ytor)	Rouen Report (1980)[a]; Cox (1981)
			Cook (1975)
		Ybrighton (Ybri)	Laurell and Eriksson (1963)
		Z	Hug et al. (1979)
Null	Talamo et al. (1973)	Zpratt (Zpra)	

[a] Report of Pi Nomenclature Workshop, Rouen, 1978 (Cox et al., 1980).

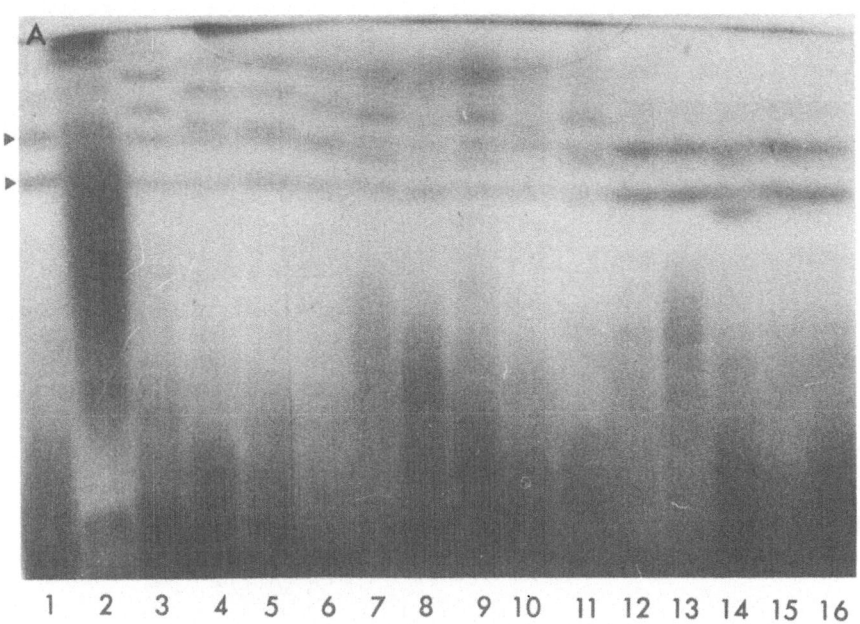

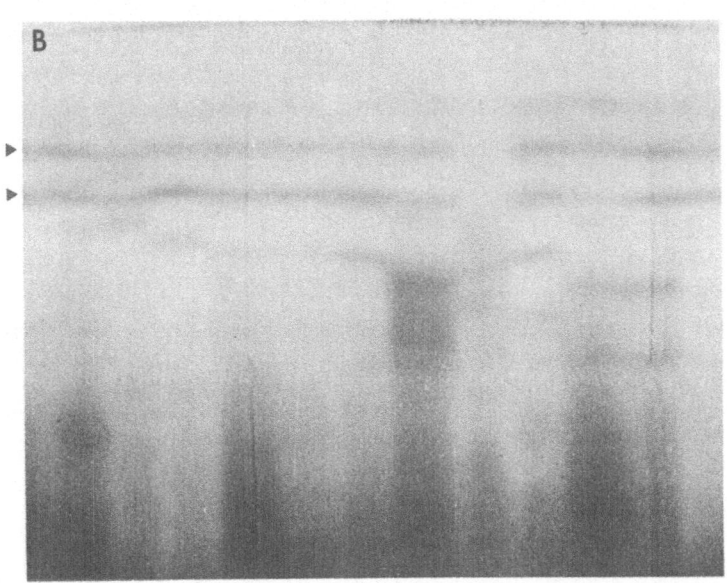

Fig. 1. Pi variants by acid starch gel electrophoresis. Arrows indicate major bands of M. Anode at top. (A) Anodal variants: 1—M, 2—BM, 3—DM, 4—EM2, 5—EM, 6—FM2, 7—EcinM, 8—GM, 9—EcinM, 10—IM, 11—LM, 12—M1M2, 13—M1M3, 14—MN, 15—MNham, 16—M. (B) Cathodal variants: 1—M, 2—MP, 3—MPstl, 4—MR, 5—MS, 6—MT, 7—MV, 8—MWsal, 9—YbriZ, 10—MX, 11—MY, 12—MYtor, 13—MZ.

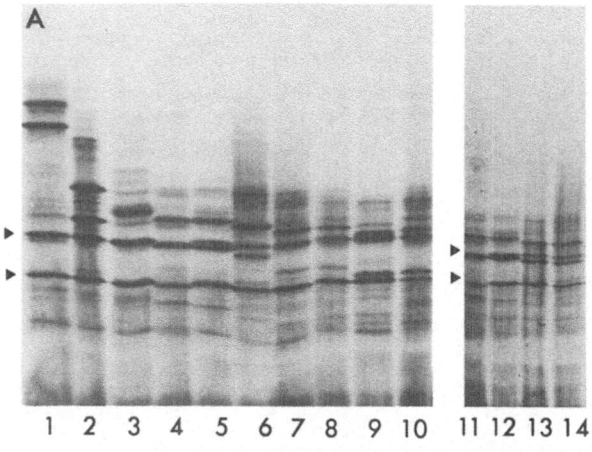

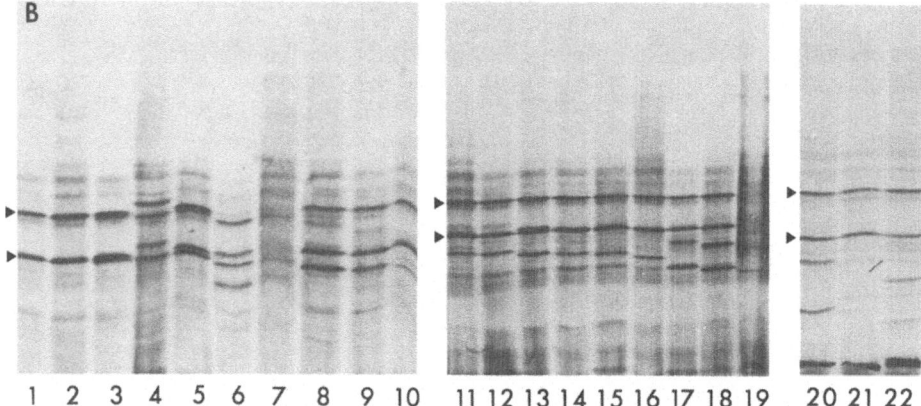

Fig. 2. Pi variants by polyacrylamide isoelectric focusing (LKB PAG plates, pH 4–5). Arrows indicate major bands of M1. Anode at top. (A) Anodal variants: 1—BM1, 2—CM1, 3—DM1, 4—EM2, 5—ElemM1, 6—EcinM3, 7—GM3, 8—IMI, 9—FM1, 10—LM1, 11—EM2, 12—ElemM1, 13—EtokM1, 14—EcinM3. (B) Cathodal variants: 1—M1, 2—M1M3, 3—M1M2, 4—LMcha, 5—M1N, 6—NhamS, 7—Mmal, 8—M1Pbud, 9—M1Pstl, 10—M1P, 11—M1P, 12—M1Pkyo, 13—M1N, 14—M1S, 15—M1T, 16—M1Wsal, 17—M1X, 18—M1Ytor, 19—Z, 20—M1Ytor, 21—M1Z, 22—M1Zpra.

autosomal codominant alleles at one locus, as proposed earlier by Fagerhol and Gedde-Dahl (1969). The anodal variants by starch gel electrophoresis or isoelectric focusing are designated from B to L; variants cathodal to M are designated from N to Z. A list of the genetic variants described to date is listed in Table I. This includes all variants which have been tested in two laboratories in addition to that of the investigator, have occurred in more than one member of a family, and are available for further comparison, as recommended at the Rouen Nomenclature Meeting (Rouen Report, 1980). Studies and comparisons have been incomplete for Mduarte (Lieberman *et al.*, 1976) and M-like (Kueppers *et al.*, 1977), which are probably identical to Mmalton; for W (Fagerhol and Tenfjord, 1968), Wconstantine (Khitri *et al.*, 1977) and Y (Ymilford) (Cook, 1975). The subdivision of the common Pi M type into, first, two, then three polymorphic subtypes using PIEF, occurred simultaneously in several laboratories: M2 was differentiated first (Constans and Viau, 1975; Frants and Eriksson, 1976; Kueppers, 1976; Van den Broek *et al.*, 1976), then the more anodal variant M3 (Genz *et al.*, 1977; Klasen *et al.*, 1977; Frants and Eriksson, 1978). Beginning with the most anodal, these subtypes are M1, M3, and M2. Several names were used prior to the Nomenclature Meeting (Rouen Report, 1980): M1 = Ma or M, M3 = M1 or M2, and M2 = M_N, Mc, M1, or M3.

The appearance of the Pi variants by acid starch gel electrophoresis is shown in Fig. 1. Their appearance by PIEF is shown in Fig. 2. Diagrams of the anodal and cathodal variants by these two methods are shown in Figs. 3 and 4. The band positions of F by PIEF appear to change with aging. I and G may be identical, the apparent differences being due to the accompanying M subvariant (Martin, 1979, personal communication).

Most of the Pi variants described to date have been identifiable by their mobility in electrophoretic systems. Other variants are expected which could be differentiated by their biochemical properties, such as

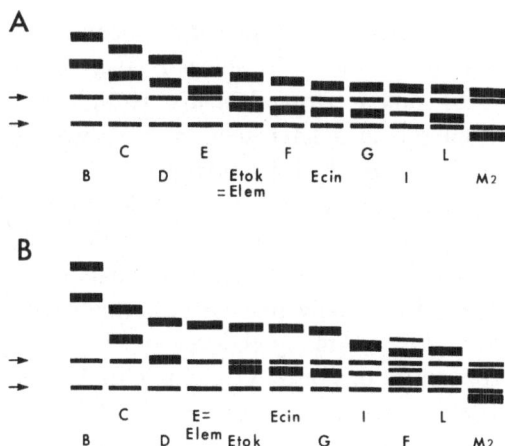

Fig. 3. Diagram of anodal Pi variants. Arrows and corresponding bands indicate position of major bands of M1. Anode at top. (A) By acid starch gel electrophoresis; (B) by isoelectric focusing.

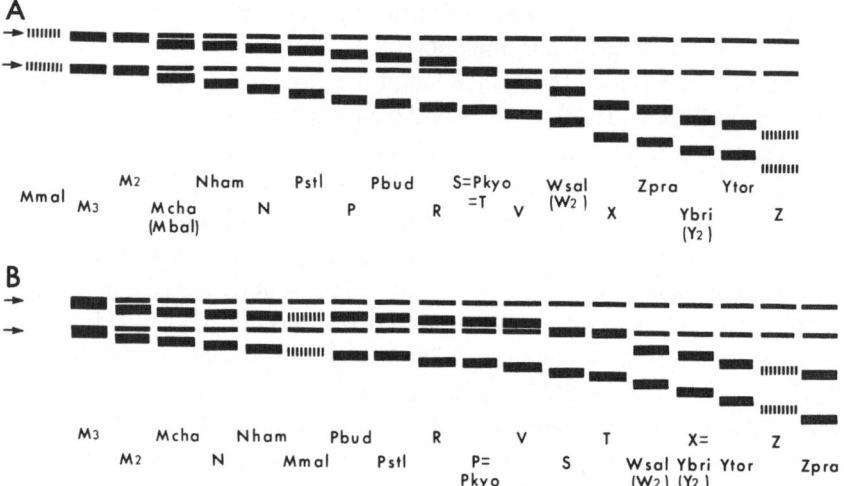

Fig. 4. Diagram of cathodal Pi variants. Arrows and corresponding bands indicate position of major bands of M1. Anode at top. (A) By acid starch gel electrophoresis; (B) by isoelectric focusing.

reduced stability or kinetic differences. One variant with M mobility which shows marked instability at a low pH has been described by Lie-Injo (1976). This pattern is recognized by a marked reduction in intensity of the M pattern by acid starch gel electrophoresis with a normal amount of α_1AT measured immunologically. The appearance by PIEF has not been reported.

The Deficiency Alleles

A slight reduction in the concentration of α_1AT (50–70% of normal) is associated with the alleles *P* (Fagerhol and Hauge, 1969), *S* (Fagerhol, 1969), and *I* (Arnaud *et al.*, 1978). These quantitative differences are discussed in a later section; here we will discuss those alleles which produce a marked deficiency.

The most common of the deficiency alleles is Pi^Z, present in most populations at a frequency of 0.02 or less. In two large series including a total of more than 140 individuals of Pi type Z, the concentration of α_1AT was usually 10–15% of normal (Fagerhol and Laurell, 1970; Kueppers, 1978). Among children of Pi type ZZ who have liver disease, the concentrations of α_1AT can rise to 50% of normal (Moroz *et al.*, 1976*b*). The low serum concentration of α_1AT in Pi Z individuals is due to impaired secretion of the Z protein, which accumulates in inclusions in the rough endoplasmic reticulum of the hepatocytes. The pathological and biochemical aspects will be discussed subsequently.

The Mmalton allele, with a mobility slightly anodal to M by acid starch gel electrophoresis and cathodal to M2 by IEF is considerably more rare than Z. However, it produces as marked a serum deficiency of α_1AT. Two homozygotes have been reported in one family and both have normal tests of lung function (Cox, 1976). However, both were less than 30 years of age and may not have yet reached an age to show lung destruction.

The "M-like" variant (Kueppers et al., 1977) could be identical to Mmalton and to Mduarte (Lieberman et al., 1976). They appear to be similar from the description of their electrophoretic mobilities, but no direct comparisons have yet been possible. The one Mduarte homozygote had severe emphysema at 48 years of age. This homozygote had liver inclusions similar to those found in the livers of Pi Z individuals. Therefore, there may be two variants of α_1AT that have an abnormality of secretion.

Pi^{null} is a rare deficiency allele (Talamo et al., 1973). The proband had severe emphysema by his mid-twenties. Heterozygosity for Pi^{null} (i.e., M–) cannot be detected by Pi typing but can be inferred from the reduced quantities of α_1AT and from pedigree data. It should be kept in mind that the presence of the null allele can produce apparent paternal exclusion, where a Pi Z– father has a Pi M– child. Several other families have been reported in which the null or "silent" allele appears to be present. The original Pi–– homozygote is now deceased and no liver autopsy was possible. The amount of α_1AT in the serum was less than 1/16,000 of the normal [Talamo, 1979 (personal communication)].

A Portuguese child has been reported who also appears to be Pi–– (Feldmann et al., 1975). Her serum had 1/200 the normal amount of α_1AT, higher than in the original Pi–– homozygote (Talamo et al., 1973). Because of the difference in quantity, it is not clear if this is the same variant and has been called Mrouen. The mobility was shown to be identical to that of M by starch gel electrophoresis. No liver inclusions were observed in this homozygote.

Another type of α_1AT deficiency has been described in a 15-year-old homozygote with dyspnea due to restrictive pulmonary disease (Langley et al., 1979). No distinct pattern could be obtained in several electrophoretic systems.

Population Studies of the Pi Alleles

Frequencies of the Pi alleles have been obtained for numerous populations. In all populations tested, Pi^M is the most common allele. Pi^S and Pi^Z are frequently polymorphic.

For the earliest population studies, starch gel electrophoresis was used. These population studies have been previously summarized (Fagerhol and Laurell, 1970; Kellermann and Walter, 1970). The major population differences were a low frequency of variants in Finns and Lapps (Fagerhol *et al.*, 1969) and in Asian populations (Fagerhol and Tenfjord, 1968; Kellermann and Walter, 1970), and a high frequency of Pi^S in Spaniards and Portuguese (Fagerhol, 1968).

Table II includes more recent studies, in which typing has been carried out by acid starch gel electrophoresis followed by crossed immunoelectrophoresis on each sample unless otherwise indicated, or by isoelectric focusing. Results listed in previous summary tables are not included.

Certain technical points should be kept in mind when interpreting population data. In studies in which acid starch gel electrophoresis has been used and not all samples have been studied by crossed immunoelectrophoresis, the frequency of the Pi^Z allele could be underestimated. Aged samples could be mistaken for Pi type FM by starch gel electrophoresis, as discussed by Fagerhol (1972a), or for E by isoelectric focusing. Aged samples or those with a large excess of heparin sometimes have no $\alpha_1 AT$ bands, suggestive of, but not the same as Pi^Z. The rare deficiency alleles are not likely to be recognized in the presence of M. Variants which have not been compared with recognized standards by starch gel electrophoresis and by isoelectric focusing may not actually be identical in mobility to variants of the same name in different studies. Of course, those variants identical in all electrophoretic systems used do not necessarily have the same amino acid substitution.

Several general trends can be seen from Table II. Variant alleles, particularly the Z allele, are relatively infrequent in northern populations and in blacks. The highest frequency of Pi^Z has been reported in the New Zealand Maoris, although starch gel electrophoresis was used without crossed immunoelectrophoresis in that study. In other populations, the frequency of Pi^Z is relatively similar. The frequency of Pi^S can vary widely. A high frequency is found in Spain and Portugal. The frequency of S is higher in France and Southern Italy than in most European and

TABLE II. Allele Frequencies in Selected Populations

Population	N	Pi alleles					Reference[d]
		M	S	F	Z	Other	
Europe							
Finland	548	0.9662	0.0173	0.0018	0.0137	0.0009	1
Ireland	1000	0.9365	0.0385	0.0015	0.0200	0.0035	2
England	926	0.9244	0.0475	0.0022	0.0221	0.0038	3
Netherlands	1474	0.9558	0.0160	0.0018	0.0123	0.0141	4
France	1653	0.9019	0.0714	0.0036	0.0142	0.0090	5
Italy	202	0.9505	0.0297	0.0074	0.0099	0.0025	6
Italy	500	0.9110	0.0670	0.0040	0.0150	0.0030	7
Spain[a]	576	0.8810	0.1291	—	0.0052	—	8
Portugal	330	0.8651	0.1152	—	0.0182	0.0015	9
North America							
U.S. white	1933	0.9483	0.0360	b	0.0116	b	10
U.S. white	283[c]	0.9682	0.0265	—	0.0053	—	11
U.S. black	204	0.9804	0.0098	b	0.0049	b	10
U.S. black	549[c]	0.9817	0.0146	—	0.0037	—	11
Eskimo (Northwest Territories)	170	0.9941	0.0059	—	—	—	12
Other							
Somalia Bantu	347[c]	0.9668	0.0144	0.0043	0.0072	—	13
New Zealand Maori[a]	487	0.9589	0.0051	0.0010	0.0349	—	14

[a] Not all samples by crossed, immunoelectrophoresis after starch, or by IEF.
[b] Figures not given.
[c] In newborns.
[d] (1) Arnaud et al., 1977b; (2) Blundell et al., 1975; (3) Arnaud et al., 1979; (4) Hoffman and van den Broek, 1976; (5) Arnaud et al., 1977a; (6) Klasen et al., 1978; (7) Piantelli et al., 1978; (8) Goedde et al., 1973; (9) Martin et al., 1976; (10) Pierce et al., 1975; (11) Evans et al., 1977; (12) Cox et al., 1978; (13) Massi and Vecchio, 1977; (14) Janus et al., 1975.

TABLE III. Relative Frequencies of M Subtypes

Population	N	Pi^M Frequency	Relative frequency of subtype			Reference[c]
			M1	M2	M3	
Finland	136	0.99	0.80	0.08	0.12	1
Netherlands	708	0.98	0.83	0.14	0.03[a]	2
	131	0.93	0.81	0.14	0.05	1
England	926	0.92	0.81	0.13	0.06	3
Germany	538	0.96	0.78	0.15	0.07	4
Canada	68[b]	0.94[b]	0.66	0.27	0.07	5
U.S. whites		0.94	0.68	0.20	0.12	6
U.S. blacks		0.98	0.92	0.03	0.005	6
West Africa—Bozo	102	0.97	0.96	0.02	0.04	1

[a] Estimate.

[b] Pi^M frequency from 1027 normal adults and children; subtypes based on 68 individuals of Pi type MS.

[c] (1) Frants and Eriksson, 1978; (2) Klasen et al., 1977; (3) Arnaud et al., 1979; (4) Genz et al., 1977; (5) Cox, 1977; (6) Kueppers and Christopherson, 1978.

North American populations, but is still only about half of that found in Spain and Portugal. Selective forces may have produced some of the population differences, particularly for the Z allele.

Subvariants of M (M1, M2, M3) are polymorphic in most populations studied, but M2 and M3 are relatively infrequent in blacks (Table III).

Linkage and Mapping for the Pi Locus

Linkage Studies

The only established linkage for the Pi locus is with Gm, the locus for the constant region of the heavy chain of IgG. This linkage, first recorded by Gedde-Dahl et al. (1972), was confirmed with additional data to produce, in the combined data, a lod score of 16.3 at 23% recombination (Gedde-Dahl et al., 1975). An allele-specific difference in recombination frequency for Pi^Z and other variant alleles was noted by Gedde-Dahl et al. (1972) and persisted in their later data. For the combined data for males, the estimates of recombination were 26% for MS, 25% for "non-Z" variants, and 11% for MZ. The likelihood difference, calculated as the sum of the peak lod scores for the MZ and MS variants separately minus the peak score for the two variants combined ($7.34 + 2.41 - 8.45$) is 1.30 (Weitkamp *et al.*, 1978). Using the logarithm of this likelihood difference as the homogeneity criterion (Morton, 1956), the data appear to be significantly heterogeneous. Further data of Weitkamp *et al.* (1978) have supported this heterogeneity. Combining data from all three reports, they reported the peak lod score for MZ males as 8.58 at 11% recombination and for MS males 1.46 at 32% recombination. The peak score for the combined data, assuming no heterogeneity, is 7.55 at 23% recombination. The likelihood difference is 2.49 and, using the criterion of Morton, the difference is significant.

The finding of an allele-specific effect on recombination frequency has not been reported for other gene loci in man. Further data will be required to ensure that this unusual finding is not due to chance.

Several possible explanations for heterogeneity were proposed by Gedde-Dahl *et al.* (1975). There is no evidence for more than one *Pi* locus. The specific amino acid substitutions in Pi S and Z types of $\alpha_1 AT$ have now been identified. It seems unlikely that the altered gene itself would affect recombination. There is insufficient data to determine if a generalized decrease in recombination frequency occurs at other loci in MZ males (Weitkamp *et al.*, 1978). If the $\alpha_1 AT$ gene is expressed in the spermatocyte, a decreased level of $\alpha_1 AT$ in MZ males might be related to a reduction in recombination. Recombination-controlling genes in linkage disequilibrium with the Pi locus have also been suggested. These possibilities are entirely speculative at present.

Chromosome Mapping

Several proposals have been made for the chromosome location of the Gm–Pi linkage pair. Using B-cell hybrids, Bennick *et al.* (1978) suggested that the *Gm* locus was on chromosome 8. Subsequent studies, using specific antibodies to the hinge region of IgG 3, shed doubt on their conclusions. Smith and Hirschhorn (1978) using human lymphoid × mouse RAG cell hybrids showed that chromosome 6 is necessary for immunoglobulin (Ig) heavy-chain production. However, the possibility that chromosome 6 codes for a regulator rather than a structural gene for IgG has not been excluded. Much of chromosome 6 has been excluded from having Gm–Pi by a number of studies using pedigree linkage analysis and deletion mapping (Noades and Cook, 1976).

A linkage of acid phosphatase (ACP_1) on chromosome 2 to Gm–Pi was not supported by further data (Noades and Cook, 1976).

A dominant form of spherocytosis has been localized to chromosome 8 or 12p (Kimberling et al., 1975). Gm and dominant spherocytosis are possibly linked, with 30% recombination (Noades and Cook, 1976).

In summary, the Gm–Pi linkage group may be on chromosome 6, 8, or 12. (See Addendum.)

Selective Mechanisms

The Z allele is polymorphic in most populations, in spite of its reduction of fitness for homozygotes. This is suggestive of a selective advantage for heterozygotes. Increased fertility of heterozygotes or of surviving homozygotes is one mechanism by which the Pi^Z allele might be maintained. Proteases are important in the fertilization process. Acrosin, a protease capable of dissolving the zona pellucida of the ovum, is apparently essential for penetration by the spermatozoa during the fertilization process. $\alpha_1 AT$ is one of the inhibitors which inhibits acrosin *in vitro* (Schumacher, 1971) and reduces the fertilization capacity of capacitated spermatozoa *in vivo* (Yang et al., 1976). $\alpha_1 AT$ has been identified in cervical mucus (Schumacher, 1970). These studies would tend to support the suggestion that reduced levels of $\alpha_1 AT$ in homozygous or heterozygous females might increase their reproductive fitness (Fagerhol, 1972a; Kueppers, 1972).

Preferential survival and increased good health of MZ individuals have been suggested in the data of Pierce et al. (1975), in which the frequency of MZ individuals is increased among older blood donors. An unexplained increase in frequency of MZ individuals has been found in adults when compared with children (Cox and Huber, 1980).

Another mechanism for increasing the Z allele frequency is that of preferential transmission of the Z allele from MZ heterozygotes (Chapuis-Cellier and Arnaud, 1979; Iammarino et al., 1979). Their data indicate that MZ fathers, but not mothers, and their M spouses, have produced a higher-than-expected proportion of MZ offspring. An analysis of our own data, based on a larger number of matings, does not support these conclusions (Cox, 1980a). Our data, combined with that from the literature, are shown in Table IV. The data of Iammarino et al. (1979) have been modified to exclude duplicated families and use data obtained di-

TABLE IV. Offspring of Heterozygous Matings

| | Pi type of offspring | | | | | |
| | MZ♂ × M♀ | | | MZ♀ × M♂ | | |
Reference	M	MZ	Total	M	MZ	Total
Chapuis-Cellier and Arnaud (1979)	10	31	41	16	11	27
Iammarino et al. (1979)[a]	9	20	29	21	14	35
Cox[b]	37	31	68	18	20	38
Total	56	82	138	55	45	100
Total known to be unbiased	46	51	97	39	34	73

[a] Modified: duplicated family removed; unbiased figures obtained from all pedigrees.
[b] Data from Hospital for Sick Children, Toronto.

rectly from the source material, excluding ascertainment bias. Some of the errors inherent in such analyses have been discussed in greater detail (Cox, 1980). Where an MZ individual (parent of a ZZ patient) has served as a proband to identify an MZ × MM mating (e.g., in grandparents), that MZ "proband" must then be excluded from the analysis of offspring to avoid ascertainment bias. This has not necessarily been done in the two reported studies.

The possibility of preferential transmission of the Z allele is of great interest, as it suggests that a serum protease inhibitor (α_1AT) might be active in or on spermatozoa. However, the data at present are inadequate to warrant acceptance of this possibility. Further direct studies of spermatozoa and the addition of pedigree data from other centers will be important for providing conclusions.

BIOCHEMICAL ASPECTS

A prerequisite for studies on the molecular structure of a protein is an efficient and nondegrading combination of methods for its purification. Unfortunately, α_1AT has not been easy to purify. Early attempts included use of both low pH and high temperatures which, we now know, will denature the protein. The studies by Jirgensons (1977) have shown that the tertiary structure of α_1AT is dependent upon proper hydration of parts of the molecule. For this reason, even salting-out methods may not be sufficiently mild.

The starting material for purification work is also very important. Use of plasma from outdated blood may not be appropriate since the protein may be affected by enzymes released from granulocytes, which have a life-span of only a few days *in vitro*. Pooled normal plasma or serum should also be avoided because of the high frequency of genetic variants. Preferably one should use reliably Pi-typed (by PIEF) plasma from single donors.

In recent years many groups have used ion exchange chromatography in combination with $(NH_4)_2SO_4$ or ethanol precipitation, gel filtration, and preparative electrophoresis (Crawford, 1973; Kress and Laskowski, 1973; Chan *et al.*, 1973; Pannell *et al.*, 1974; Cohen and Fallat, 1974). However, the introduction of different types of affinity chromatography procedures has considerably increased the efficiency and gentleness of the purification systems. Pannell *et al.* (1974) have removed contaminant albumin by use of blue dextran coupled to sepharose beads, while Laurell *et al.* (1975) removed albumin, prealbumin, and haptoglobin by use of specific antibodies against these proteins coupled to sepharose. The latter group also introduced a simple, gentle, and efficient purification principle based upon thiol–disulfide interchange. This method takes advantage of the fact that $\alpha_1 AT$ carries a reactive thiol group by which the protein can reversibly interact with other thiol compounds that have been immobilized on sepharose. In contrast to ion exchange chromatography, the thiol method is independent of the variation in charge of the genetic variants and the charge microheterogeneity which is a normal property of $\alpha_1 AT$. Buffers with a high pH are used, so acid denaturation can be avoided.

Data on the molecular structure of $\alpha_1 AT$ should be regarded with caution unless the purified material is shown, by methods such as acid starch gel electrophoresis or isofocusing, to have the normal microheterogeneity shown by the native protein in fresh plasma. If the protein has been exposed to a pH below 4.9, the microheterogeneity pattern is distorted (Jeppsson *et al.*, 1978a).

When the protein has been appropriately purified from suitable starting material, the molecular weight of $\alpha_1 AT$ has been estimated to be about 55,000 (Chan *et al.*, 1973; Horng and Gan, 1974; Kress and Laskowski, 1973; Pannell *et al.*, 1974; Jeppsson *et al.*, 1978a). These reports give data on the amino acid composition that are in quite good agreement, the mean composition being as follows (expressed as numbers of residues per molecule): asp 44, thr 29, ser 23, glu 51, pro 20, gly 22, ala 25, val 25, met 8, isoleu 19, leu 47, tyr 6, phe 26, lys 37, his 13, arg 8, ½ cys 2, try 2.

Jeppsson *et al.* (1978a) did not find a significant difference between the overall amino acid composition of the M and Z proteins, but as discussed below, amino acid substitutions have been identified for the S and Z variants. α_1AT apparently consists of one polypeptide chain. There is agreement that the molecule includes about 12% carbohydrate, but as shown in Table V, the data for the composition of the carbohydrate side chains are conflicting. Several Pi variants (M1, M2, M3, and S) have been reported to have similar carbohydrate compositions (Yoshida and Mega, 1979).

Chan *et al.* (1976) have suggested that two types of side chains occur, each with branches half way up from the asparagine attachment points. According to this model, one type has three and the other two branches, all of which have sialic acid terminals attached to galactose. While this hypothesis suggests a total of ten sialic acid residues, most reports are compatible with only six to eight of such residues.

The Z protein which has been isolated from hepatocytes has carbohydrate side chains which differ markedly from the normal, with an increase in mannose, and a lack of sialic acid and galactose (Hercz *et al.*, 1978). The liver protein has no trypsin inhibitory activity but is antigenically competent. In contrast, the Z protein in plasma has both inhibitory and antigenic activity and differs very little in carbohydrate content from

TABLE V. Carbohydrate Content of
α_1-Antitrypsin

	Number of residues per molecule[a]
Mannose	8–12
Galactose	6–9
Fucose	0–1
N-Acetyl glucosamine	9–19
Sialic acid	2–10

[a] The ranges given were compiled from the following reports: Chan *et al.* (1973); Chan *et al.* (1976); Hercz *et al.* (1978); Horng and Gan (1974); Jeppsson *et al.* (1978a); Kress and Laskowski (1973); Roll *et al.* (1978); Yoshida and Mega (1979).

the M protein (Jeppsson *et al.*, 1978*a*). This may suggest that the side chains contribute significantly to the molecular configuration necessary for protease inhibition.

Microheterogeneity

Since the introduction of acid starch gel electrophoresis, a characteristic microheterogeneity has been observed for each variant of $\alpha_1 AT$, although the protein appears homogenous at alkaline pH (Fagerhol and Laurell, 1967; Fagerhol, 1969). This phenomenon was later found to occur also in PIEF, and even in agarose gel electrophoresis at pH 5 (Laurell and Persson, 1973). One common factor for these methods is exposure of the protein to a pH of 5 or lower.

Figure 5 shows the typical Pi pattern by crossed immunoelectrophoresis. By this method, one can detect and measure eight bands containing $\alpha_1 AT$. These have been given serial numbers from the anode towards the cathode, which is convenient when discussing specific bands of various allele products (Fagerhol, 1969). A similar pattern can be obtained by densitometric scanning of stained PIEF gels, but by this method other proteins will also be recorded (Allen *et al.*, 1974).

For all variants, band numbers 4 and 6 predominate; they contain about 40% and 35%, respectively, of the total $\alpha_1 AT$, and their isoelectic points are 4.52 and 4.59 (Jeppsson *et al.*, 1978*a*). The Z protein in plasma also shows this microheterogeneity, but the low level of this variant is reflected by the Z_4 and Z_6 peaks being about as high as the M_7 and M_8 peaks when an MZ plasma sample is tested.

The distribution of $\alpha_1 AT$ among the eight zones of the Pi patterns

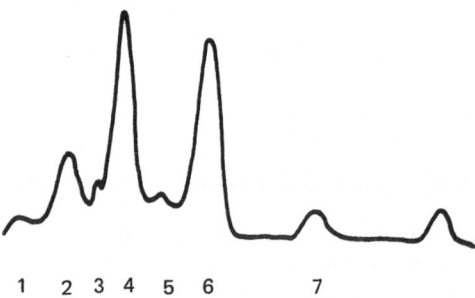

1   2   3   4   5   6   7

Fig. 5. The microheterogeneity of α_1-antitrypsin as revealed by antigen–antibody crossed electrophoresis with acid starch gel electrophoresis in the first dimension. The pattern shown is the outline of the immunoprecipitates in the second dimension. The heights of the peaks correspond to the α_1-antitrypsin content of the Pi bands. Anode to left.

is quite characteristic and constant with a few notable exceptions: (a) In the Pi^1 allele product, the I_4 and I_7 are considerably increased with a corresponding decrease of the I_6 and I_8 bands; (b) Estrogens and corticosteroids in high doses can increase bands 6 and 8 relative to bands 4 and 7. A similar pattern is seen in newborns (Headings and Bose, 1974; Evans et al., 1976, Arnaud, 1977); (c) In diffuse intravascular coagulation associated with terminal leukemia, high peaks in the anodal region (corresponding to, or anodal to, peaks 1 to 4) can occur (Inokuma, 1978). A similar pattern can occur in sera due to microbial growth or exposure to room temperature or higher for many days. Such patterns may be misinterpreted as due to the presence of genetic variants.

At an early stage, it was thought that interaction between $\alpha_1 AT$ and the starch gel and/or borate ions was responsible for the microheterogeneity because substitution of the starch gel with polyacrylamide gel of a similar pore size and buffer composition did not give the characteristic Pi pattern. This hypothesis was abandoned, since the typical Pi pattern could be obtained with other buffers and gels and by PIEF.

In recent years, several groups have tried to explain the microheterogeneity by variation in the content of sialic acid residues, but the results are equivocal. For instance, Arnaud (1977) found that after exhaustive desialylation of Pi M and Z sera the microheterogeneity remained although all bands had shifted towards the cathode. Hercz et al. (1978) found microheterogeneity in the sialic-acid-free liver $\alpha_1 AT$. On the other hand, Jeppsson et al. (1978a) found that the major bands M_4 and M_6, isolated by preparative isofocusing, contained 7 and 6 sialic acid residues, respectively, in agreement with earlier studies on $\alpha_1 AT$ in serum (Cox, 1975). After complete desialylation, Jeppsson et al. (1978a) found that 90% of the protein focused in one band. Band displacements after short periods of exposure to neuraminidase corresponded to stepwise displacement of the M_2 band to M_4, band M_4 to M_6, etc. But after complete desialylation, about 10% of the protein focused in separate bands, suggesting that other factors, in addition to variation in sialic acid content, contribute to the microheterogeneity. Since microheterogeneity becomes evident at a pH known to cause unfolding of the $\alpha_1 AT$ molecule, changes in the tertiary structure may be of importance.

With recently developed techniques like preparative IEF, each of the major bands in the Pi pattern can be isolated and the cause of the microheterogeneity analyzed directly.

Antigenicity

All the α_1 AT variants discovered up to now carry a common antigenic determinant, as judged from their reactions with antisera from commonly used experimental animals such as rabbit, goat, sheep, and pig. In no instance has isoimmunization against α_1 AT been reported in man.

Several laboratories have tried to prepare specific antibodies against the Z protein and other variants, but without success. One might suspect that even if such antibodies were made by animals they would probably have been overlooked, since the commonly used immune-precipitation methods in gels require that each antigenic molecule carries at least two antigenic determinants of the same type. If the Z protein differs from the M protein by substitution of a single amino acid, there will be only one Z-specific antigenic site on each molecule. As a consequence, in the search for antibodies against the Z typical site, one must use other methods, such as radioimmunoassay or passive hemagglutination.

Kueppers (1968a) has shown that there is a high degree of cross-reactivity between human and subhuman primate α_1 AT in reactions with rabbit antibodies against the human protein. The degree of cross-reactivity was estimated as almost 100% with sera from Hominoidea and about 70% with Cercopithecoidea.

Inhibition of Proteases

α_1 AT can inhibit a series of proteolytic enzymes by formation of $1:1$ molar complexes. Among the enzymes that can be inhibited are particularly the serine proteases which are characterized by a serine residue at their active site (Matthews *et al.*, 1967; Blow *et al.*, 1969; Hunkapiller *et al.*, 1973; Stroud, 1974). Although α_1 AT is responsible for about 90% of the serum trypsin inhibitory capacity, the inhibition of granulocyte and macrophage elastase, collagenase, and cathepsin G is probably its major physiological function (Janoff, 1971; Ohlsson, 1975; Travis *et al.*, 1978). These enzymes constitute a major portion of the total protein in leukocyte granules. In the absence of inhibition, the proteolytic enzymes can digest elastin, collagen, microfibrillar component, and proteoglycans.

Other enzymes that are inhibited by α_1 AT are pancreatic chymotrypsin and elastase (Schwick *et al.*, 1966; Kueppers and Bearn, 1966; Ohlsson, 1971b), collagenase from skin (Tokoro *et al.*, 1972) and synovia

(Harris *et al.*, 1969), plasmin and thrombin (Rimon *et al.*, 1966), acrosin (Fritz *et al.*, 1972*a*), kallikrein (Fritz *et al.*, 1972*b*), urokinase (Clemmensen and Christensen, 1976), rennin (Scharp *et al.*, 1976), coagulation factor XI (Heck and Kaplan, 1974), Hageman factor cofactor (Crawford and Ogston, 1974), and cathepsin G (Travis *et al.*, 1978). For most of these enzymes there are other probably more important plasma inhibitors than α_1AT.

A general model for inhibition of proteases, based upon studies of the interaction between trypsin and soybean trypsin inhibitor, was formulated by Laskowski *et al.* (1971). According to this model, the active site of the inhibitor contains either a single Arg-X or Lys-X bond. The inhibitory activity of α_1AT against trypsin, chymotrypsin, and elastase is lost after acylation of lysyl residues and can be regained by deacylation (Heimburger *et al.*, 1971; Johnson and Travis, 1975). Furthermore, the appearance of a new threonine NH_2 terminus after interaction of α_1AT with enzymes suggests that the inhibitor site of α_1AT might be a lysyl-threonyl bond (Johnson and Travis, 1976; Travis *et al.*, 1978; Cohen *et al.*, 1977, Johnson and Travis, 1977). However, results from these previous experiments might have alternative explanations: studies on the amino acid sequence of an undecapeptide with the active site indicate that the active site includes a methionine (Johnson and Travis, 1978). In fact, they find a remarkable homology in the reactive sites of α_1AT, the trypsin-binding site of the lima bean inhibitor, and the elastase-binding site of the garden bean inhibitor. They also found an ambiguity adjacent to the methionyl residue in that some peptides had methionyl-seryl and others methionyl-threonyl bonds. This ambiguity is not related to M subvariants and could be a technical artifact [Travis, 1979 (personal communication)]. Since James and Cohen (1978) concluded that elastase interacted with an X-serine bond in α_1AT, a methionyl-threonyl variant, if it exists, could be a relatively poor elastase inhibitor *in vivo*.

Janoff and Carp (1977) have reported that α_1AT can be inactivated by various types of oxidants including those present in tobacco smoke. A similar inactivation occurs when potent oxidants (probably $\cdot$OH) are released from neutrophils during phagocytosis of antigen–antibody complexes (Carp and Janoff, 1979). Myeloperoxidase from neutrophils inactivates α_1AT (Matheson *et al.*, 1979).

The question of the structure and number of active sites on the α_1AT molecule is not yet settled. James and Cohen (1978) have suggested that

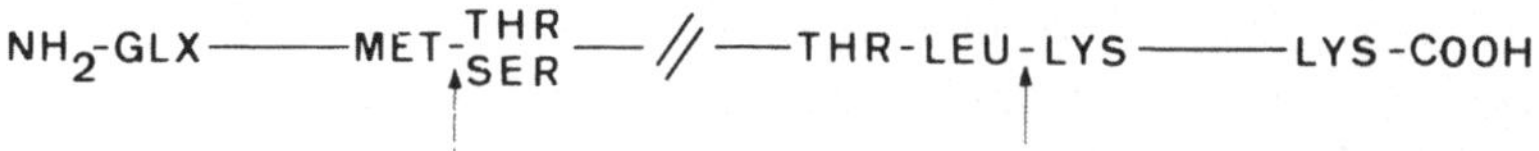

$$NH_2\text{-GLX}\text{——}MET\text{-}\begin{matrix}THR\\SER\end{matrix}\text{——}//\text{——}THR\text{-}LEU\text{-}LYS\text{——}LYS\text{-}COOH$$

Fig 6. The amino acid sequences at the two active sites of α_1-antitrypsin (arrows) where elastase attacks the inhibitor. The length of the lines do not reflect the length of the polypeptide segments.

elastase can attack two different sites, and that the order in which they are attacked will decide whether the enzyme will be inhibited or not. According to their model (Fig. 6), a stable complex will be formed only if the elastase first splits an X-serine bond close to the NH_2 terminus and subsequently binds to a leucyl after splitting a Y-leucine bond in the COOH terminal part of the polypeptide.

The question then arises whether all these data can be reconciled and included in a single consistent hypothesis. If the X-seryl bond is the methionyl-seryl bond suggested by Johnson and Travis (1978), and the leucyl-Y bond is a leucyl-lysine, most of the biochemical data have been accounted for. This means that the previous and the more recent hypotheses of lysyl and methionyl, respectively, at the active site are *both* correct by assuming *two* active sites as suggested by James and Cohen (1978).

The biological implications of having two binding sites on α_1AT and the necessity of a certain order of their interaction with enzyme for inhibition to occur are still obscure. One should take into consideration the reports on reactivation of elastase from α_1AT–elastase complexes in mixtures with enzyme excess (Johnson and Travis, 1976; Arnaud, 1977), the possible storage of inactive enzymes in complexes with α_1AT in neutrophils and mast cells (Benitez-Bribiesca and Freyre-Horta, 1978; Benitez-Bribiesca *et al.*, 1973), and the temporary inhibition of neutrophil lysosomal enzymes by a cytosol inhibitor from the same cells (Steven *et al.*, 1976). According to the latter authors, proteolytic activity is regained after 1.5–2 hr of incubation with excess of inhibitor. *In vivo*, that period of time would be sufficient for the complexes to be removed by the reticuloendothelial system.

Laurell (1975) has suggested that α_1AT, due to its small size and wide distribution in the extravascular fluid, can function as a shuttle for proteases. When such enzymes are released in the tissues, they are rapidly complexed with α_1AT, and, later, the proteases are transferred to a second inhibitor, α_2-macroglobulin, which acts as a scavenger (Ohlsson, 1971a). The elastase–α_2–macroglobulin complexes have a half-life in the circulation of about 12 min compared with 60 min for α_1AT–elastase complexes in man (Ohlsson and Laurell, 1976).

Unlike these *in vivo* findings, Aubry and Bieth (1977) found that the transfer of human and bovine trypsin and chymotrypsin from α_1AT to $\alpha_2$macroglobulin *in vitro* was very slow. The original complexes had half-lives of a least one week. Leukocyte elastase and collagenase may behave differently, but probably the *in vivo* situation includes additional enzymatic attack on the enzyme–inhibitor complexes as mentioned above.

As long as proteases are bound to α_1AT, they are unable to interact with substrates. This is in contrast to the situation when enzymes are bound to α_2-macroglobulin. The latter is more like an entrapment within crypts of the large molecule and does not involve binding of the active site of the enzyme. Therefore the bound enzyme can digest several substrates. In fact, such complexes have higher activities against some small synthetic substrates (Twumasi *et al.*, 1977). This phenomenon may well be of pathogenetic importance since Galdston *et al.* (1979) have shown that human granulocyte elastase bound to α_2-macroglobulin retained 6% of its activity against tropoelastin and solubilized elastin in the presence of normal serum and about 12% in the presence of α_1AT-deficient (Pi type Z) serum. They suggest that such complexes may contribute to the development of emphysema by digesting soluble precursors of elastin.

α_1-*Antitrypsin and the Clotting System*

Recent reports on the presence of α_1AT within platelets (Bagdasarian and Colman, 1975, 1978) and the inhibition of enzymes belonging to the clotting and fibrinolytic systems (Rimon *et al.*, 1966; Clemmensen and Christensen, 1976; Heck and Kaplan, 1974; Crawford and Ogston, 1974) represent at least theoretical links between α_1AT and these systems. Neither bleeding nor thrombosis have been typical findings in patients with α_1AT deficiency, but this may not be a definite argument against a place for α_1AT in clotting/fibrinolysis. After all, few will deny factor XII a place in the clotting system despite the fact that Mr. Hageman did not have a clinical bleeding tendency (Ratnoff, 1976).

An interesting possibility is that the slow degradation of pulmonary tissue in α_1AT deficiency, starting in the capillaries (Martin and Boatman, 1965) and leading to early emphysema, might be secondary to a slightly increased sequestration of activated platelets trapped in the pulmonary capillaries. The presence of α_1AT in the platelets may contribute to the inhibition of the clotting and fibrinolytic enzymes as well as inhibition of leukocyte and macrophage proteases that participate in the removal of aggregated platelets. A shortened life-span of platelets in α_1AT deficiency should be sought in support of this hypothesis.

α_1-Antitrypsin and the Immune Response

Mazzei *et al.* (1966) showed that trypsin and chymotrypsin can act as mitogens on lymphocytes. Later, a selective stimulation of B lymphocytes was demonstrated (Vischer, 1974; Kaplan and Bona, 1974). The *in vivo* and *in vitro* immune response of mouse spleen cells against sheep red cells was suppressed by α_1AT. This suppression was regulated by B-cell response, and not through an effect on adherent cells or T cells (Arora *et al.*, 1978).

Gisler *et al.* (1976) concluded that trypsin may substitute for helper T cells. Elastase and cathepsin G from human polymorphonuclear leukocytes had similar effects (Vischer *et al.*, 1976), and again no effect was seen on thymic cells. Stimulation apparently acted through direct proteolytic action on the leukocyte surface (Bretz *et al.*, 1976). On the other hand, Yoshinaga *et al.* (1975) found that thymocytes from mice were triggered to synthesize DNA by small numbers of peritoneal exudate neutrophils or macrophages or even by substances released from such cells.

Those results leave no doubt that proteases from leukocytes, and by consequence their potent inhibitor α_1AT, may modulate the function of immunologically competent cells. A more direct link to α_1AT has been presented by Lipsky *et al.* (1979) who have demonstrated that human lymphocytes have α_1AT bound to their plasma membrane after concanavalin-A-induced blastogenic transformation.

Complement activation via the alternate pathway has been shown to stimulate the release of lysosomal enzymes from granulocytes (Goldstein *et al.*, 1973). The secreted proteases can, at low concentrations, release from complement components factors which promote chemotaxis and secretion. High proteolytic activity will rapidly inactivate these components (Venge, 1978). Again, a modulation of the proteolytic activity by α_1AT may regulate the inflammatory response.

Folds *et al.* (1978) have shown that elastase from human neutrophils can split IgG molecules leaving a Fab-like fragment. This may be of importance in relation to the activation of complement by antigen–antibody complexes and for the formation and fate of such complexes.

Biochemical Characterization of Genetic Variants of α_1-Antitrypsin

From the behavior of many genetic variants of α_1AT on electrophoresis and by PIEF, and from analogy with other polymorphisms, it has been assumed that the variants have arisen by point mutations and single amino acid substitutions. Up to now there has been no evidence for gene duplications or deletions. The pi^{null} gene resulting in a complete lack of α_1AT in plasma, could be a deletion.

Two independent groups (Jeppsson, 1976; Yoshida et al., 1976) have reported that a glutamic acid in the normal α_1AT (Pi type M) had been substituted by a lysine in the Z protein. At the same time, Owen et al. (1976) reported a substitution of a glutamic acid by valine in the S protein. These substitutions occur in the same CNBr fragment, the "C fragment" (Jeppsson et al., 1978a). This 109-residue fragment has been sequenced. The S-variant substitution is in residue 22 (Glu $\rightarrow$ Val) and the Z-variant substitution in residue 100 (Glu $\rightarrow$ Lys) (Owen et al., 1978). The fragment in question also contains oligosaccharides and no evidence has been found for lack of any carbohydrate side chain in the S- or Z-specific fragments. The estimate of one sialic acid residue less in the isolated Z protein compared with the M protein might be explained by a slightly different acetylation or glycosylation of the sialic acid in the Z protein leading to a lower color yield with the thiobarbituric reaction used for sialic acid quantitation (Jeppsson et al., 1978a).

In contrast to other variants, the Z protein accumulates in large amounts in the hepatocytes where it is synthesized (Sharp et al., 1969). Storage does not seem to be responsible for cirrhosis.

The Z protein in the liver differs structurally from the Z protein in plasma. While the latter has a normal protease-inhibitory capacity, the former is inactive, although it is antigenically competent (Jeppsson et al., 1975; Hercz et al., 1978). The latter authors have also shown that the Z protein extracted from the liver contains more mannose than the normal protein and lacks sialic acid and galactose. They suggest that, after synthesis of the polypeptide backbone, some primordial carbohydrate side chains are attached. Only after remodeling of these, including removal of some mannose residues and addition of galactose and sialic acid, can the protein be secreted in its native plasma-protein configuration.

Several authors have stressed that the Z protein accumulates in the rough endoplasmic reticulum. After extraction, it has a strong tendency to aggregate (Jeppsson *et al.*, 1975). This raises the possibility that the accumulation may be the consequence of intracellular aggregation, which secondarily may block the remodeling of the side chains and further transport through the endoplasmic system.

There has been no report of any α_1AT variant in plasma with decreased inhibitory capacity relative to the amount of protein, although few variants have been adequately tested. The variants F, M, S, and Z have a comparable reactivity of their thiol groups (Jeppsson *et al.*, 1978a).

Normally less than 10% of the plasma α_1AT occurs as complexes with other proteins with reactive thiols, such as IgA, κ chains, fibrinogen, and globulin HC (Laurell and Thulin, 1975; Tejler and Grubb, 1976). In patients with IgA myelomas there is a linear relation between the concentration of the myeloma protein and the percentage of α_1AT bound to IgA. Furthermore, the inhibitory capacity of the bound α_1AT is retained (Musiani *et al.*, 1978). This phenomenon may represent a protective mechanism for circulating IgA against proteolytic attack.

METABOLISM OF α_1-ANTITRYPSIN

Synthesis and Distribution

Many observations (Schultze and Heremans, 1966; Sharp *et al.*, 1969; Aagenaes *et al.*, 1972; Bhan *et al.*, 1976; Gautier *et al.*, 1977) suggest that α_1AT is synthesized in the parenchymal cells of the liver and that it is rapidly released into plasma. In fact, the release is so rapid that α_1AT is not normally detectable in hepatocytes by standard immunofluorescent techniques. The Z variant represents an exception to this, being readily seen both in homozygotes and heterozygotes.

One direct proof of the origin of the plasma α_1AT is the observation (Sharp, 1971) that after liver transplants to α_1AT-deficient patients, their plasma α_1AT levels returned to normal and converted to the Pi M type of the donors.

The synthesis and release of M and of Z α_1 AT has been demonstrated in short- and long-term liver-cell cultures (Bhan *et al.*, 1976; Gautier *et al.*, 1977). Eriksson *et al.* (1978) have shown that α_1 AT synthesis *in vitro* by cultured fetal human liver cells is sensitive to α_1 AT in the medium. The release of α_1 AT from the cells was suppressed by increasing concentrations of α_1 AT in the medium.

α_1 AT is also present in platelets (Bagdasarian and Colman, 1975; Nachman and Harpel, 1976; Nalli *et al.*, 1977, Bagdasarian and Colman, 1978), alveolar macrophages (Cohen, 1973), mast cells (Benitez-Bribiesca *et al.*, 1973), neutrophils (Benitez-Bribiesca and Freyre-Horta, 1978) and lymphocytes (Lipsky *et al.*, 1979).

The studies by Bagdasarian and Colman (1978) included platelets from one patient with the Pi type Z and one with absolute α_1 AT deficiency (Pi type null) and suggest that the α_1 AT in the granular and soluble subcellular fractions may be synthesized by the platelets while membrane-bound α_1 AT may be adsorbed from plasma.

This seems to be a situation similar to that of antithrombin-III for which there is platelet antithrombin in addition to the plasma inhibitor. Recent data (Tullis and Watanabe, 1978) suggest that among those who have inherited antithrombin-III deficiency, those who also have a low platelet antithrombin are prone to recurrent thrombosis.

The intracellular occurrence of α_1 AT is of particular interest. The possible increased fertilizing capacity of sperm cells carrying the Z gene points to the possibility that such cells can synthesize α_1 AT and that due to the Z-gene acrosomal proteases are less efficiently inhibited.

The finding of α_1 AT in various other cell types also raises the question of whether the Z protein found in plasma can be synthesized outside the liver, but no data on this subject are available yet.

α_1 AT is found in many different body fluids such as cerebrospinal fluid (Schuller *et al.*, 1970), perilymph (Chevance *et al.*, 1976), tears (Rennert *et al.*, 1974), cervical mucus and seminal fluid (Schumacher, 1970), amniotic fluid (Sutcliffe and Brock, 1973), and duodenal fluid and bile (Kyaw-Myint *et al.*, 1975).

About 40% of the protein is found in plasma, and most of the remaining 60% is in the extravascular compartment (Laurell, 1975).

Because of its small molecular size, α_1 AT can easily leave the blood vessels and enter the tissues and body fluids. However, α_1 AT does not cross the placenta in significant amounts since the Pi type of the fetal or cord serum is that of the fetus (Talamo *et al.*, 1975; Arnaud, 1977). Prenatal detection of α_1AT deficiency has therefore been shown to be possible (Jeppsson *et al.*, 1979*a*).

The turnover of isolated M and Z proteins has been studied simultaneously in healthy Pi type M individuals by Laurell *et al.* (1977). The half-lives and fractional catabolic rates were 7 days and 0.26 for the M protein and 5 days and 0.40 for the Z protein. The Z protein has a shorter life-span, but decreased release from the hepatocytes is probably more important for the low plasma level in Pi type Z individuals. Likewise, the finding of fractional catabolic rates of 0.36 and 0.34 for the S and Mmalton proteins (Jeppsson *et al.*, 1978*b*) is not sufficient to account for the reduced concentrations of these variants in plasma. These authors also showed that removal of sialic acid from the M protein resulted in an extremely rapid removal of α_1AT from the circulation. Even a 20% desialylation doubled the fractional catabolic rate. This is in keeping with the demonstration of receptors on the hepatocyte membrane for glycoproteins having exposed galactose residues (Hudgin *et al.*, 1974).

Concentration of α_1-Antitrypsin in Plasma

In healthy unmedicated individuals, the plasma α_1AT concentrations are, to a great extent, determined by the individual's alleles at the Pi locus. Decreased concentrations of α_1AT are associated with the following Pi alleles, where concentrations are expressed as % of the normal M:

null	0%	(Talamo *et al.*, 1973)
Mmalton	12%	(Cox, 1976)
Z	15%	(Laurell and Eriksson, 1963)
P	30%	(Fagerhol and Hauge, 1968)
S	60%	(Fagerhol, 1969)
I	68%	(Arnaud *et al.*, 1978)

α_1AT can be measured either by functional assays of its trypsin- or elastase-inhibitory capacity using synthetic or protein substrates (Erlanger *et al.*, 1961; Senior *et al.*, 1972), or by immunochemical methods such as radial immunodiffusion (Mancini *et al.*, 1965) or electroimmunoassay (Laurell, 1966). These methods have been assessed by Talamo *et al.* (1978). At least for the most common Pi types, there is a good correlation between functional and immunological assays (Talamo *et al.*, 1972; Billingsley and Cox, 1980; Vercaigne *et al.*, 1979). All the described variants apparently carry the same antigenic determinant. No functionally deficient variants have been identified.

Attempts at determination of absolute $\alpha_1 AT$ concentrations have been hampered by difficulty in obtaining pure, native $\alpha_1 AT$ and trypsin preparations with 100% of active molecules. For this reason, many laboratories have expressed $\alpha_1 AT$ concentrations as % of that in large standard pools of normal sera. Comparison of such pools prepared in the United States and Europe has given nearly identical concentrations.

Recent studies (Jeppsson *et al.*, 1978*a*) suggest that the normal concentration of $\alpha_1 AT$ in plasma is about 1.3 g/liter.

Besides the presence of certain genetic variants, low levels of $\alpha_1 AT$ have been detected secondary to the respiratory distress syndrome in newborns (Evans *et al.*, 1970), renal homograft rejection (Tyler *et al.*, 1962), severe protein-losing conditions (Eriksson, 1965), and terminal liver failure (Talamo, 1975). In these conditions an increased consumption in association with decreased synthesis or release of the protein probably leads to decreased plasma levels. Talamo (1975) has also seen severe lowering of $\alpha_1 AT$ at some point during the course of cystic fibrosis in twelve out of 105 patients, but the mechanism remains obscure.

$\alpha_1 AT$ can be inactivated by the bacteria *Pseudomonas aeruginosa* and *Proteus mirabilis* (Moskowitz and Heinrich, 1972), snake venom enzymes (Kurecki *et al.*, 1978), thiol proteinases (Johnson and Travis, 1977), and cigarette-smoke condensate (Janoff and Carp, 1977).

Increased levels of $\alpha_1 AT$ are much more frequently seen than low levels, because this protein is one of the acute-phase reactants which are elevated in response to inflammation, especially when associated with tissue damage. A two- to threefold increase is also found during pregnancy (Ganrot and Bjerre, 1967), malignant disease (Harris *et al.*, 1974), following severe burns (Busch and Wilms, 1974), or in response to intravenous injection of typhoid vaccine (Kueppers, 1968*b*). Increases of 20% and 50% are seen in response to smoking and estrogen contraceptive pills, respectively (Elson *et al.*, 1973; Laurell *et al.*, 1968).

The plasma level does not vary significantly during the menstrual cycle (Kueppers *et al.*, 1972), but a drop to about 10% of the initial level has been found in cervical mucus at the time of ovulation (Schumacher and Pearl, 1968). This may be part of the physiological changes to facilitate fertilization.

Local increase in the α_1AT concentration has been reported in synovial fluids from some patients with rheumatoid arthritis (Swedlund *et al.*, 1969) and in cerebrospinal fluid from patients with intracranial tumors (Galvez et al., 1979).

There are numerous observations to suggest that the levels of α_1AT in plasma and other body fluids vary according to need, which seems to be primarily the appropriate modulation of proteolytic activity of leukocyte proteases.

ASSOCIATION BETWEEN Pi TYPES AND DISEASE

Pulmonary Disease

Pulmonary lesions were noted in the first patients found, by agarose electrophoresis, to have α_1AT deficiency (Laurell and Eriksson, 1963). Of 33 individuals with α_1AT deficiency, 23 were found to have chronic obstructive lung disease (COPD), often with an early age of onset: more than 50% were less than 40 years of age (Eriksson, 1965). The association of α_1AT deficiency and COPD has now been confirmed in studies of hundreds of patients. Characteristically, the emphysema is of a diffuse panacinar type, beginning at the lung bases and progressing towards the apices. Microscopically, the peribronchial elastica is disrupted.

There is a wide range of clinical variation in lung disease among individuals with α_1AT deficiency. Even among those Pi type Z individuals who are asymptomatic, abnormalities in lung function are usually observed after 26 years of age, particularly in lung mechanics and flow rates (Larsson et al., 1976; Rawlings et al., 1976). Smoking has a pronounced effect on the development of lung disease. The onset of emphysema is from 13–15 years earlier in Pi Z individuals who smoke (Black and Kueppers, 1978; Larsson, 1978). Nonsmokers can live into their sixth and seventh decades. The extent of clinical variation is not entirely accounted for by smoking history, and other factors, possibly genetic, must be involved (Black and Kueppers, 1978).

The relative risk for Pi Z individuals to develop COPD has not been determined. However, in a Swedish study of individuals of Pi type Z, 39% of patients under 40, and 85% of patients over 40, developed COPD (Larsson, 1978). However, this study is biased because of ascertainment through hospital admission, so the risk figures are too high. The risk for developing symptomatic lung disease during infancy and childhood is low and there are only a few reports in the literature, which have been recently summarized (Cox and Talamo, 1979).

The extent of the risk for Pi MZ heterozygotes to develop COPD has been the subject of considerable controversy. Assessment of such risk is important since MZ individuals constitute about 3–5% of most populations. The proportion of MZ individuals among all patients with COPD has been studied in several countries. In a review and evaluation of these studies, plus our own study, we have concluded that adults of Pi type MZ have about three times the risk of the normal population for developing obstructive lung disease (Cox *et al.*, 1976). In general, these studies have not evaluated the effects of smoking, and the risks will vary with smoking history.

Another approach is to study the lung function of Pi MZ adults compared with those of Pi type M. No differences are found by routine spirometry (Webb *et al.*, 1973). Pi MZ adults show a greater deterioration of lung elasticity with age than do adults of Pi type M, further accelerated by the effects of smoking (Cooper *et al.*, 1974). A study of MZ Swedish males 50 years of age showed impaired lung function in smoking heterozygotes but not in nonsmoking heterozygotes (Larsson *et al.*, 1977). While differences in the mechanical properties of the lungs of MZ individuals, particularly in those who smoke, have been identified, there does not seem to be a strikingly great increase in clinical disease among these individuals.

Variation in the concentration of leukocyte proteases has been proposed as a factor to influence the extent of lung damage in both Pi MZ and Z individuals. Data are discussed in the subsequent section.

Studies on the distribution of Pi types in asthma have been conflicting. A high proportion of Pi types MS and S in patients with bronchial asthma was reported by Fagerhol and Hauge (1969). A normal distribution of Pi types was found in adults with asthma in Poland (Szczeklik *et al.*, 1974) and in the United States (Webb *et al.*, 1973; Ihrig *et al.*, 1975). A normal distribution of Pi types was also found in 109 asthmatic children at The Hospital for Sick Children, Toronto, (Cox and Cooper, 1976, unpublished data) and in 151 child patients in the United States (Katz *et al.*, 1976). Arnaud *et al.* (1976b) found an increased frequency of Pi type MS among 298 asthmatic children when compared with controls. This increase was found to be entirely in the group of 65 patients with nonatopic asthma. The association of Pi heterozygosity with asthma, if it exists, appears to be with the S rather than the Z allele, suggesting that the association is not related to a reduced concentration of $\alpha_1 AT$. Because the S is variable between populations, it is possible that the asthmatic patients could be drawn from some subgroup of the population which is more susceptible to asthma.

Deficiency types of $\alpha_1 AT$ may lead to more severe asthmatic disease and particularly to the development of emphysema (Szczeklik et al., 1974; Chapuis-Cellier, 1975). Z heterozygotes were more frequent among asthmatics requiring steroid treatment than in those not requiring such treatment (Katz et al., 1976). Earlier onset, greater severity of disease, and more frequent pulmonary overinflation were observed in patients with deficient phenotypes (P, Z, S, and I) than among nondeficient phenotypes (Arnaud et al., 1976b).

Several review articles have discussed lung disease and $\alpha_1 AT$ deficiency (Kueppers and Black, 1974; Morse, 1978; Kueppers, 1978).

Liver Disease

An association of $\alpha_1 AT$ deficiency and liver disease in children was reported by Sharp et al. (1969). The liver disease in infancy frequently presents as "neonatal hepatitis syndrome," with prolonged jaundice, failure to thrive, hepatomegaly, and, sometimes, splenomegaly (Moroz et al., 1976b; Cutz and Cox, 1979). The abnormalities of liver function tests, particularly conjugated hyperbilirubinemia, suggest intrahepatic cholestasis. From 13 to 30% of infants with neonatal hepatitis syndrome have been found to be of Pi type Z (Cottrall et al., 1974; Moroz et al., 1976b).

A follow-up study of 122 children of Pi type Z, detected through a screening program, has been reported (Sveger, 1976). Approximately 17% of such children developed clinically recognizable liver abnormalities: 7.4% had prolonged obstructive jaundice with severe liver disease, 4.1% had prolonged jaundice with mild liver disease, and 6.5% had abnormalities suggestive of liver disease. The prognosis for this group of children was originally thought to be poor, with progressive cirrhosis leading to early death. However, perhaps as many as two-thirds of these children show considerable recovery from their liver damage (Moroz et al., 1976b; Odièvre et al., 1976). It should be kept in mind, for purposes of genetic counseling, that the risk for two Pi MZ parents to have a child of Pi type

ZZ with severe liver disease is relatively small, in the order of 1–2%. The assumption is made, as yet with inadequate data, that the risk for developing liver disease does not vary widely between families.

The inclusions of $\alpha_1 AT$ in the liver are present in every individual of Pi type Z examined to date and are not believed to contribute to the liver disease. Children of Pi types MZ and SZ with various types of liver disease have been described. Much less $\alpha_1 AT$ was present in their hepatocytes (Cutz and Cox, 1979). The pathological and clinical features of liver disease in children has been reviewed (Cutz and Cox, 1979).

Cirrhosis and fibrosis of the liver were reported in Swedish adults with $\alpha_1 AT$ deficiency (Berg and Eriksson, 1972). In 11 randomly selected individuals with $\alpha_1 AT$ deficiency and with no clinical signs of liver disease, no evidence of liver disease was found by intensive tests of liver function (Larsson and Eriksson, 1977). Only three of these 11 patients were over 50 years of age. Among a series of 246 Pi Z patients ascertained through hospital admission, 12% had liver cirrhosis (Larsson, 1978). Of the Pi Z individuals over 50 years of age in that series, 19% had a diagnosis of liver cirrhosis. Neonatal hepatitis was not recalled as being present in any of the patients who later developed cirrhosis. These figures may be biased upward because of the use of hospitalized patients, but they demonstrate an increased risk for liver disease in $\alpha_1 AT$-deficient patients in later years.

An increased risk for Pi MZ individuals to develop cirrhosis was suggested by isolated reports of this association. In a study of 159 adults with liver cirrhosis, 132 of which had chronic alcoholism, the frequency of Pi MZ individuals was numerically increased in both the alcoholic and nonalcoholic group. However, these differences were relatively small and were not significant (Morin et al., 1975). The number of individuals with nonalcoholic cirrhosis is too small for reaching firm conclusions. One report of patients with hepatitis and cirrhosis found no patients of Pi type MZ or Z and a high frequency of FM (Theodoropoulos et al., 1976). However, technical problems make this study unreliable. Sera were Pi-typed by starch gel electrophoresis only, without crossed immunoelectrophoresis. Because the concentration of $\alpha_1 AT$ is frequently elevated in liver disease, many of the MZ individuals would be missed. The high frequency of FM is probably due to the pattern frequently produced in liver disease, where the M2 band is increased and can be mistaken for FM.

Primary Liver Carcinoma

In a study of hepatic tissue from 14 adults of Pi type Z, cirrhosis of the liver was found in five and fibrosis in three. Three of the eight also had primary liver carcinoma (hepatoma) (Berg and Eriksson, 1972). In a later study from Sweden, which may have included some of the previous patients, six of nine patients of Pi type Z with cirrhosis also had hepatoma. All were over 50 years of age (Eriksson and Hagerstrand, 1974). α-Fetoprotein, assayed in four of the patients with hepatoma, was detected in only one. In a study of 19 European and 13 African patients with primary hepatic carcinoma, no patients were of Pi type MZ or Z (Charlionet et al., 1976). However, hepatoma was detected in these patients by the presence of a positive test for α-fetoprotein. Nine of 12 black Africans with primary liver carcinoma were of Pi type MZ (Clerc et al., 1977). Six of their eight MZ patients in whom α-fetoprotein was assayed by radioimmunoassay had abnormally high concentrations. Studies of pathological tissues have yielded conflicting results. Among 56 Danish patients with primary liver carcinoma, 23% were found to have PAS-positive globules indicative of the presence of Z type $\alpha_1 AT$ (Reintoft and Hagerstrand, 1979). Livers of two of 42 patients in England with hepatocellular carcinoma had PAS-positive granules, not more than expected in the normal population (Kelly et al., 1979). Such pathological studies are difficult to interpret because Pi MZ individuals do not always show PAS granules in the liver, and tumor tissues can show PAS granules where no such granules are present in the nontumor tissue (Palmer and Wolfe, 1976). Hepatoma was present in 3% of 246 Pi Z patients over 20 years of age admitted to Swedish hospitals (Larsson, 1978). This figure would give an underestimate of the risk, as 40% of the patients were less than 50 years of age. In general, the data suggest an increased frequency of hepatocellular carcinoma for individuals of both Pi types MZ and Z.

Since α-fetoprotein is usually elevated in primary liver carcinoma, it is unusual that patients with $\alpha_1 AT$ deficiency do not show this elevation.

Among a series of adult patients of Pi type Z compiled by one of us (D.C.), two patients had primary liver carcinoma; one of these had cirrhosis without COPD, the other had COPD without evidence of liver disease. Another patient with emphysema and no evidence of cirrhosis has also been reported to have primary liver carcinoma (Schleissner and Cohen, 1975). Hepatoma can apparently occur in $\alpha_1 AT$-deficient individuals without preexisting cirrhosis.

Other Malignancies

In addition to the risk of developing hepatoma, MZ heterozygotes may be more susceptible to some other malignancies.

Arnaud (1977) reported a significant increase of Z heterozygotes (Pi types MZ and SZ) in 123 patients with squamous-cell carcinoma: 7.3% in patients compared with 3.1% in controls. In another study, however, two of 23 patients with squamous-cell carcinoma were of Pi type MZ, but the total group of 72 patients with all types of lung cancer did not show an increase in Z heterozygotes (Harris *et al.*, 1976).

In a series of 33 patients with paraproteinemias (myeloma and lymphoma), three of 33 patients were of Pi type MZ compared with nine of 515 controls. Further studies are needed to confirm this association (Ananthakrishnan *et al.*, 1979).

These associations with a possible increased risk for malignancy could be explained by linkage disequilibrium between the Pi locus and a locus for increased susceptibility to malignancy. Alternatively, α_1AT could be important in the control of cell proliferation or in the control of the immune response which would eliminate aberrant cells. The absence of Pi Z individuals among the patients with malignancy could be due to death from other causes before malignancy develops (Ananthakrishnan *et al.*, 1979).

Kidney Disease

Kidney disease was first reported in association with α_1AT deficiency in an adult with severe emphysema, necrotizing angiitis, and mild liver disease (Miller and Kuschner, 1969). Subacute glomerular nephritis was diagnosed at autopsy. Membranoproliferative glomerulonephritis has been identified in three Pi Z children who died with cirrhosis (Moroz *et al.*, 1976a). Similar renal changes have been reported by Milford Ward *et al.* (1975) and Sharp (1976). Glomerular renal damage, manifested by constant or recurrent proteinuria or hematuria, was found in 15% of the 246 Swedish Pi Z patients ascertained through hospital admission (Larsson, 1978).

One unique feature of the membranoproliferative glomerulonephritis is that deposits of α_1AT have been detected along the glomerular basement membranes, using immunofluorescent techniques (Moroz *et al.*,

1976*a*). these deposits were found in addition to those of immunoglobulins IgG, M, and A and complement (C3) which are frequently found in cases of membranoproliferative glomerulonephritis.

Rheumatoid Arthritis

Patients of Pi types MZ and SZ were reported to be more frequent among adults with rheumatoid arthritis than among controls (Cox and Huber, 1976). Other studies have presented conflicting results. In a United States study, no increase in Z heterozygotes was found (Collins *et al.*, 1976). These patients were off almost all medication and so may not have been severely affected. No increase in Z heterozygotes was found in Swedish (Sjoblom and Wollheim, 1977) or Swiss (Brackertz and Kueppers, 1977) patients not selected for disease severity. However, a significant increase in MZ patients has been shown in three British studies (Arnaud *et al.*, 1979; Buisseret *et al.*, 1977; Geddes *et al.*, 1977), in one associated with ankylosing spondylitis (Buisseret *et al.*, 1977) and in one with fibrosing alveolitis (Geddes *et al.*, 1977). An association of Pi MZ with acute anterior uveitis, which is associated with fibrosing alveolitis and ankylosing spondylitis, may account for the association of MZ with the latter disease (Brewerton *et al.*, 1978).

The increase in Z heterozygotes has been supported by data from 108 patients with classical rheumatoid arthritis, of whom 9.2% were Z heterozygotes compared with 3.5% of controls (Cox and Huber, 1980). This association apparently occurs in individuals with seropositive erosive arthritis. Disagreement in the literature could be due to geographical differences or to differences in patient selection.

In a study of 98 children with juvenile rheumatoid arthritis, 4.1% of patients and 1.3% of controls were Z heterozygotes, a nonsignificant difference (Cox and Huber, 1980). A series of 98 British patients, who were generally more severely affected than the former patients, did show a significantly increased MZ frequency in the patients with arthritis: 10.4% in patients vs. 1.6% in controls (Arnaud *et al.*, 1977*b*).

The hypothesis that $\alpha_1 AT$ is important in controlling tissue destruction after the initiation of the rheumatoid process is plausible. Phagocytosis of polymorphonuclear leukocytes in the presence of rheumatoid factor and complement components causes a release of elastase and other proteolytic enzymes which attack joint cartilage (Janoff and Blondin, 1970). Leukocyte elastase and cathepsin G degrade proteoglycans of cartilage, then attack the terminal peptides of collagen (Keiser *et al.*, 1976; Starkey *et al.*, 1977). α_1AT, present in synovial fluid at about 75% of the concentration in serum (Brackertz *et al.*, 1975) could be too limited in MZ heterozygotes to prevent joint destruction.

Miscellaneous Diseases

α_1AT deficiency, Pi type Z, has been reported in association with several disorders of an autoimmune type or associated with inflammation. Persistent cutaneous vasculitis in a child (Brandrup and Østergaard, 1978), panarteritis in association with emphysema and glomerulonephritis (Miller and Kuschner, 1969), and pancreatic fibrosis (Freeman *et al.*, 1976) have been reported. Among five patients with panniculitis (Weber–Christian disease), two, severely affected, had α_1AT deficiency (Rubinstein *et al.*, 1977). Hashimoto's thyroiditis (Nicholls and Janus, 1973) and severe combined immunodeficiency (Gelfand *et al.*, 1979) have each been reported in one Pi Z individual.

Multisystem fibrosis has been reported in a patient of Pi type SZ (Palmer *et al.*, 1978). Twenty-five percent of 80 British patients with acute anterior uveitis were of Pi type MZ (Brewerton *et al.*, 1978). However, none of 57 white American patients with anterior uveitis were of Pi type MZ (Brown *et al.*, 1979). This condition is also associated with HLA-B27, ankylosing spondylitis, and fibrosing alveolitis. Further studies will be required to clarify these associations.

An increased frequency of Pi MZ individuals was found in one series of patients with pancreatitis (Novis *et al.*, 1975). However, in this study, Pi typing was carried out only by starch gel electrophoresis with no crossed immunoelectrophoresis, a system now known to be unreliable for detecting all individuals of Pi type MZ.

In some cases, these reported associations may be coincidental. Further studies will be required. However, individuals with α_1AT deficiency, either because of their low concentrations of α_1AT or the presence of Z type α_1AT may be more prone to the formation of immune complexes or may be unable to counteract inflammation.

Chromosome Aberrations

An increased frequency of Pi phenotypes other than M associated with chromosome aberrations was reported by Aarskog and Fagerhol

(1970). Of their seven families with a proband with sex chromosome mosaicism, five had variant Pi types in one parent. These variants were all F or S, Pi variants associated with only a slight decrease in concentration of α_1AT. Parents of patients with Turner syndrome and trisomy 21 were not unusual. Among 21 patients with X-chromosome mosaicism, an increase of Z heterozygotes was found among parents and probands (Kueppers et al., 1975). Pi types among patients with Down syndrome have been reported (Arnaud, 1976a; Fineman et al., 1976). The frequency of Z heterozygotes among all patients with Down syndrome in all the above studies is 2.6%, similar to figures reported for normal populations. The combined frequencies for patients with sex chromosome mosaicism is 4.2%, representing two in 47 patients. While there does not appear to be any evidence for an increased frequency of the Z allele in Down syndrome, the possibility of an association with X mosaicism has not yet been entirely eliminated and requires further data. A possible increase in the S and Z alleles among 57 couples with a history of recurrent miscarriages has been reported (Aarskog et al., 1978).

The rationale for involvement of a protease inhibitor in the control of chromosome disjunction and the fertilization processes has been discussed (Aarskog and Fagerhol, 1970). Proteolytic enzymes are involved with the fertilization and cell-division mechanisms and, therefore, perhaps can be disturbed in the presence of inadequate protease inhibition.

PROTEASE INHIBITOR SYSTEMS AND DISEASE MECHANISMS

Interrelated Systems

Homeostasis in the body is dependent upon a delicate balance between a number of proteases and inhibitors which, for functional reasons, have been collected into separate but interrelated systems, including the clotting, fibrinolytic, kinin, and complement systems. Figure 7 shows some the possible pathways for activation of these systems, leading to inflammation and tissue damage.

The need for neutralization of proteases is dramatically exemplified in the clotting system. Here the clotting enzyme thrombin, after its contribution to hemostasis, must be inactivated by antithrombin-III. If not inactivated, the clot may continue to grow, resulting in thrombosis and even embolism (Egeberg, 1965).

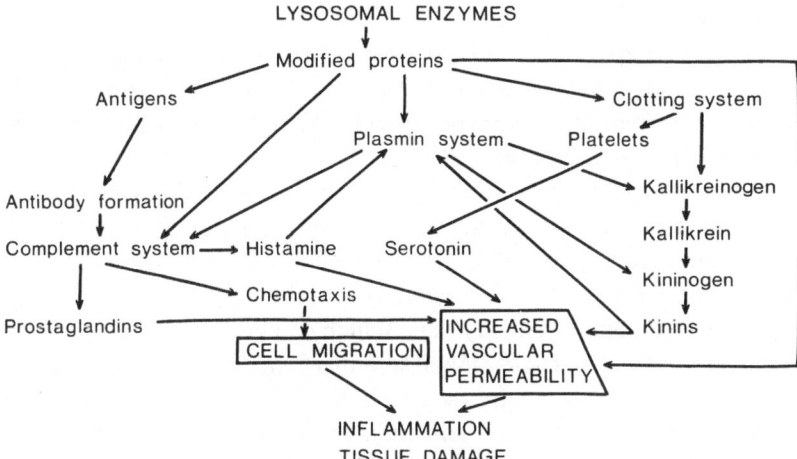

Fig. 7. Enzymatic pathways by which activation of leukocyte lysosomal proteases may lead to inflammation and tissue damage.

The clotting system includes a very efficient "cascade" or chain reaction for amplification, and the complement system may be regarded as an effector and amplification system for the immune response. In both cases inhibition at an early step will be particularly effective. Many of the enzymes involved may, if not properly inhibited, cause such severe tissue damage that one can assume that the evolution of alternative pathways of inhibition have been favored.

Detailed knowledge of the interaction between various proteases and inhibitors in health and disease in various organs is still lacking, but the type of lesions associated with the common deficiency states probably point to the major functions of the different factors. This does not exclude the possibility that in some patients who have additional predisposing disturbances inherited or acquired, the factor in question may participate in secondary reactions.

The neutral proteases from leukocytes and macrophages are apparently the main targets of $\alpha_1 AT$. Secondary targets may be Hageman factor cofactor, factor XI, urokinase, kallikrein, thrombin, and renin.

Leukocyte Proteases and $\alpha_1 AT$

Elastase, one of the several types of neutral proteases found in leukocytes, is generally believed to be the enzyme mainly responsible for the lung tissue destruction characteristic of emphysema. Porcine pancreatic elastase has produced a lesion in hamsters, resembling human panlobular emphysema (Kaplan *et al.*, 1973). Destruction from such elastase was abolished by administering elastase mixed with normal human serum, but there was no alteration of elastase damage with serum deficient in $\alpha_1 AT$. Homogenates of human leukocytes produced emphysema-like lesions when infused by aerosol into the dog lung (Janoff *et al.*, 1977).

Enzymes from neutrophils are released during phagocytosis, inflammation, or at the end of their life-span. Approximately 25% of the enzymes can be released during phagocytosis, which can be stimulated by immune complexes or inert particles (Weissmann *et al.*, 1971). Such enzymes, after release, would normally be controlled by protease inhibitors. Any imbalance between proteases and their inhibitors could be expected to lead to tissue destruction. Such an imbalance could be produced by a deficiency of the major inhibitor, α_1-antitrypsin, or by an excessively high concentration of neutrophil proteases, particularly elastase. Studies have therefore been carried out to quantitate leukocyte proteases in individuals.

On the basis of data from two families, Galdston *et al.* (1973) suggested that low levels of leukocyte protease activity might be associated with favorable clinical course in patients with reduced $\alpha_1 AT$ and that a low level of such activity is inherited, a conclusion not confirmed in later studies. Nine patients of Pi type Z and with COPD were studied by Kidokoro *et al.* (1977). They reported a significant correlation between increasing esterase activity of the elastase and the severity of lung disease as measured by tests of pulmonary function. In a study using both immunological and functional assays for elastase, Cox and Cornies (1980) examined leukocyte concentrations in ten normal Pi type M or MS individuals and ten Pi type Z individuals, classified according to the presence or severity of lung disease. There were no differences suggesting that those who had no or mild lung disease differed from those with severe lung disease. Klayton *et al.* (1975) studied 16 individuals of Pi type MZ, seven with normal pulmonary function tests and nine with COPD. No significant differences in elastase as measured by esterase activity were found. The individuals with COPD were mostly individuals who did not have clinically obvious lung problems and showed abnormalities only by tests of lung function. In another study of two groups of MZ and Z individuals, there was no significant difference between those with COPD and those with normal lung function, when elastase was measured as esterase activity (Galdston *et al.*, 1977).

In several studies, a comparison has been made between individuals of Pi type M with normal lung function and those who have COPD. A wide variation in concentration of elastase or esterase activity has been found in both control and COPD groups. The patients included in these studies are as follows: Galdston *et al.* (1977): 21 control and 13 COPD; Taylor and Kueppers (1977): 27 control and 17 COPD, studied by both esterase activity and immunological assay of elastase; Rodriguez *et al.* (1979): 31 control and 35 COPD, using elastin substrates: Abboud *et al.* (1979): 26 control and 42 COPD, using tritiated elastin substrate. The conclusions from these studies of a large number of individuals with both COPD and normal lung function is that there is no striking difference in the concentration of leukocyte elastase in individuals with COPD compared to normal controls. However, the values are generally somewhat increased in patients with COPD, although not significantly so, in these studies and may indicate that certain individuals among those with COPD do have an exceptionally high concentration of leukocyte elastase.

When leukocytes are destroyed, the release of their myeloperoxidase could inhibit α_1AT and further accentuate possible lung damage (Matheson *et al.*, 1979).

While there is no evidence that smokers have an increased concentration of leukocyte proteases, an increase has been found in the elastase-like esterase activity of alveolar macrophages in smokers (Harris *et al.*, 1976). Furthermore, alveolar macrophages from smokers, but not from nonsmokers, release elastase into serum-free culture medium and therefore may behave differently *in vivo* (Rodriguez *et al.*, 1977). A further mechanism for the harmful effects of smoking could be the suppression of the elastase inhibitory activity of α_1AT through the oxidative effects of cigarette smoke (Carp and Janoff, 1978; Cohen, 1979).

While no consistent differences have been noted in leukocyte elastase concentration that might predispose to disease, other leukocyte-related factors, such as increased leukocyte turnover, could be involved in the acceleration of the lung-tissue damage.

UNRESOLVED PROBLEMS

Studies on α_1AT have now expanded into many different fields, from pure biochemistry via biochemical genetics, clinical genetics, immunology, clinical physiology, pathology, and clinical medicine to preventive medicine, so that it is impossible to adequately cover all aspects of the subject. Since good research is characterized by posing ten new questions for each question answered, a review of important unresolved problems is deemed to be selective and far from comprehensive. We will therefore limit ourselves to a discussion of some areas of particular interest and active research.

With new and gentle methods for isolation of $\alpha_1 AT$ available, the amino acid substitutions in the described variants should be identified. Further studies of Z protein synthesized by liver tissue cultures may help solve the problem of the composition and the number of carbohydrate side chains, and the mechanism for retention of Z $\alpha_1 AT$ in the hepatocytes. Such studies may also lead to a better understanding of the regulation of synthesis and secretion mechanisms, and eventually to ways to alter them to the benefit of those with $\alpha_1 AT$ deficiency.

Variants with amino acid substitutions near or at the active site will surely be found. Further studies on the number and structure of active sites, including sequencing of isolated peptides will be of prime importance. Since a complete lack of $\alpha_1 AT$ is compatible with life, one would expect that so too will be the presence of functionally defective molecules.

The possible synthesis of $\alpha_1 AT$ by platelets, leukocytes, mast cells, sperm cells, and other cell types should soon be an active field of research touching upon vital functions of the protein as well as the selective forces behind the Pi polymorphism.

The Pi locus with the extended polymorphism due to the discovery of the subtypes of M will undoubtedly be useful for further attempts at chromosome mapping. Also, the data suggesting an effect on the crossing-over processes by the presence of the Pi^Z allele in males should give a clue for further studies on the possible roles of proteases and inhibitors in meiosis.

The suggested increase in fertilization potential of sperm cells carrying the Z allele requires further investigation. Immunofluorescence or immunoelectron-microscopy might be used to determine if $\alpha_1 AT$ is synthesized by sperm cells and if this inhibitor is removed or inactivated during the capacitation process.

The possible associations between $\alpha_1 AT$ variants, chromosome aberrations, recurrent abortions, and malignant diseases may have a common denominator in the modulation of intracellular proteases related to processes such as double fertilization, anaphase lag, and DNA repair and perhaps to immunological mechanisms. The findings of various effects of proteases on immunoglobulins and lymphocytes should be studied further as this may indicate links between α_1AT and diseases, such as rheumatoid arthritis and asthma.

The Kupffer cells in the liver are known to be vital. However, little is known about possible variability in their functional capacity in relation to diseases associated with α_1AT deficiency. If increased proteolytic activity can both stimulate the immune system and alter the handling of immune complexes, overloading of the Kupffer cells might contribute to the development of cirrhosis in Pi Z individuals. α_1AT might even be synthesized by Kupffer cells to modulate proteolysis in such cells. If so, insufficient inhibition due to the presence of the Pi^Z gene might affect their normal function.

Of utmost concern is the fact that despite considerable scientific advance in this field we still can offer those who suffer from "α_1AT-deficiency diseases" little in the way of treatment, although prevention of disease is possible to some extent. Avoidance of tobacco smoke is the only prophylaxis we can suggest, but if we are able to change the smoking habits of a Pi Z individual, this may add 15 years to his life.

Substitution therapy, either by giving α_1AT isolated from donor plasma or synthetic inhibitors, has been proposed for a long time. Thus use of microspheres, taken up by the lung and containing synthetic protease inhibitor is an interesting approach (Martodam et al., 1979). Possible side effects of new synthetic compounds, which must be taken for many years, would require evaluation.

Large amounts of α_1AT would have to be given intravenously at frequent intervals for many years before one might be able to tell whether the organ degeneration has been slowed down or even halted. The efficacy of such a treatment might be increased by chemical modification of the α_1AT molecule. If a significant amount of proteases bound to inhibitors can be reactivated by proteolytic attack on the complexes, the site for this attack might be modified and protected. Alternatively other parts of the molecule might be stabilized so that the conformational change needed for the release of the protease might be prevented. As mentioned above, recent data may suggest that α_1AT has two active sites for enzyme inhibition, and that attack on one must precede attack on the other for inhibition to occur. Interaction in the opposite sequence will inactivate the inhibitor instead. If this hypothesis is proven to be correct there may be another possibility of making each inhibitor molecule more efficient by chemical modification.

These problems represent only a few of the challenges that lie ahead for adding to our knowledge of the protease inhibitor systems.

ACKNOWLEDGMENTS. Studies reported in this chapter have been partly supported by grants to D.C. from the Department of National Health and Welfare Canada (#605-7-606) and the Medical Research Council of Canada (MA5426). The authors express thanks to Gail Billingsley for preparing the IEF gels and for constructive review of the manuscript, and Debra Edwards for preparation of the manuscript. D.C. is especially grateful to Dr. Andrew Sass-Kortsak, who initiated interest in this area.

REFERENCES

Aagenaes, O., Matlary, A., Elgjo, K., Munthe, E., and Fagerhol, M. K., 1972, Neonatal cholestasis in alpha-1-antitrypsin deficiency children, *Acta Paediatr. Scand.* **61:**632–642.

Aarskog, D., Aarseth, P., and Fagerhol, M. K., 1978, Alpha-1-antitrypsin (Pi)-types in recurrent miscarriages, *Clin. Genet.* **13:**81–84.

Aarskog, D., and Fagerhol, M. K., 1970, Protease inhibitor (Pi) phenotypes in chromosome aberrations, *J. Med. Genet.* **7:**367–370.

Abboud, R. T., Rushton, J.-M., and Grzybowski, S., 1979, Interrelationships between neutrophil elastase, serum alpha₁-antitrypsin, lung function and chest radiography in patients with chronic airflow obstruction, *Am. Rev. Resp. Dis.* **120:**31–40.

Allen, R. C., Harley, R. A., and Talamo, R. C., 1974, A new method for determination of alpha₁-antitrypsin phenotypes using isoelectric focusing on polyacrylamide gel slabs, *Am. J. Clin. Path.* **62:**732–739.

Ananthakrishnan, R., Biegler, B., and Dennis, P. M., 1979, Alpha₁-antitrypsin phenotypes in paraproteinaemias, *Lancet* **1:**561–562.

Arnaud, P., 1977, The Pi system of alpha-1-protease inhibitor (alpha-1-antitrypsin). Biochemical, immunological and genetic studies in health and disease, Ph.D. Dissertation, Medical University of South Carolina.

Arnaud, P., Chapuis-Cellier, C., and Creyssel, R., 1974, Polymorphisme de l'alpha-1-antitrypsin plasmatique (système Pi). Mise en évidence par électrofocalisation sur gel de polyacrylamide, *C. R. Soc. Biol., Paris* **168:**58–64.

Arnaud, P., Burdash, N. M., Wilson, G. B., and Fudenberg, H. H., 1976a, Alpha-1-antitrypsin (Pi) types in Down's syndrome, *Clin. Genet.* **10:**239–243.

Arnaud, P., Chapuis-Cellier, C., Souillet, G., Carron, R., Wilson, G. B., Creyssel, R., and Fudenberg, H. H., 1976b, High frequency of deficient Pi phenotypes of alpha-1-antitrypsin in nonatopic infantile asthma, *Trans. Assoc. Am. Physiol.* **89:**205–214.

Arnaud, P., Chapuis-Cellier, C., Vittoz, P., and Creyssel, C., 1977a, Alpha-1-antitrypsin phenotypes in Lyon, France, *Hum. Genet.* **39:**63–68.

Arnaud, P., Galbraith, R., Faulk, W. P., and Ansell, B. M., 1977b, Increased frequency of the MZ phenotype of alpha-1-protease inhibitor in juvenile chronic polyarthritis, *J. Clin. Invest.* **60:**1442–1444.

Arnaud, P., Koistinen, J., Wilson, G. B., and Fudenburg, H. H., 1977c, Alpha-1-antitrypsin (Pi) phenotypes in a Finnish population, *Scand. J. Clin Lab. Invest.* **37**:339–343.

Arnaud, P., Wilson, G. B., Koistinen, J., and Fudenberg, H. H., 1977d, Immunofixation after electrofocusing: Improved method for specific detection of serum proteins with determination of isoelectric points. I. Immunofixation print technique for detection of Alpha-1-Protease Inhibitor, *J. Immunol. Methods* **16**:221–231.

Arnaud, P., Chapuis-Cellier, C., Vittoz, P., and Fudenberg, H. H., 1978, Genetic polymorphism of serum α_1-protease inhibitor (α_1-antitrypsin): Pi^1, a deficient allele of the Pi system, *J. Lab. Clin. Med.* **92**:177–184.

Arnaud, P., Galbraith, R. M., Faulk, W. P., and Black, C., 1979, Pi phenotypes of $alpha_1$-antitrypsin in Southern England: Identification of M subtypes and implications for genetic studies, *Clin. Genet.* **15**:406–410.

Arora, P. K., Miller, H. C., and Aronson, L. D., 1978, α_1-antitrypsin is an effector of immunological stasis, *Nature* **274**:589–590.

Aubry, M., and Bieth, J., 1977, Kinetics of the inactivation of human and bovine trypsins and chymotrypsins by α_1-proteinase inhibitor and of their reactivation of α_2-macroglobulin, *Clin. Chim. Acta* **78**:371–380.

Axelsson, U., and Laurell, C.-B., 1965, Hereditary variants of serum α_1-antitrypsin, *Am. J. Hum. Genet.* **17**:466–472.

Bagdasarian, A., and Colman, R. W., 1975, Proteolytic inhibitors of human platelets, *Fed. Proc.* **34**:242.

Bagdasarian, A., and Colman, R. W., 1978, Subcellular localization and purification of platelet α_1-antitrypsin, *Blood* **51**:139–156.

Benitez-Bribiesca, L., and Freyre-Horta, R., 1978, Immunofluorescent localization of alpha-1-antitrypsin in human polymorphonuclear leukocytes, *Life Sci.* **22**:99–104.

Benitez-Bribiesca, L., Freyre, R., and de la Vega, G., 1973, Alpha-1-antitrypsin in human mast cells. Immunofluorescent localization, *Life Sci.* **13**:631–638.

Bennick, A., Gedde-Dahl, Jr., T., and Brogger, A., 1978, Mapping with B-cell hybrids, *Cytogenet. Cell Genet.* **22**:661–665.

Berg, N. O., and Eriksson, S., 1972, Liver disease in adults with $alpha_1$-antitrypsin deficiency, *N. Eng. J. Med.* **287**:1264–1267.

Bhan, A. K., Grand, R. J., Colten, H. R., and Alper, C. A., 1976, Liver in α_1-antitrypsin deficiency: Morphologic observations and *in vitro* synthesis of α_1-antitrypsin, *Pediat. Res.* **10**:35–40.

Billingsley, G. D., and Cox, D. W., 1980, Functional assay of α_1-antitrypsin in obstructive lung disease, *Am. Rev. Resp. Dis.* **121**:161–164.

Black, L. F., and Kueppers, F., 1978, $Alpha_1$-antitrypsin deficiency in nonsmokers, *Am. Rev. Resp. Dis.* **117**:421–428.

Blow, D. M., Birktoft, J. J., and Hartley, B. S., 1969, Role of a buried acid group in the mechanism of action of chymotrypsin, *Nature* **221**:337–340.

Blundell, G., Frazer, A., Cole, R. B., and Nevin, N. C., 1975, $Alpha_1$-antitrypsin phenotypes in Northern Ireland, *Ann. Hum. Genet., Lond.* **38**:289–294.

Brackertz, D., and Kueppers, F., 1977, Alpha-1-antitrypsin phenotypes in rheumatoid arthritis, *Lancet* **2**:934–935.

Brackertz, D., Hagmann, J., and Kueppers, F., 1975, Proteinase inhibitors in rheumatoid arthritis, *Ann. Rheum. Dis.* **34**:225–230.

Brandrup, F., and Østergaard, P. A., 1978, α_1-antitrypsin deficiency associated with persistent cutaneous vasculitis, *Arch. Dermatol.* **114**:921–924.

Bretz, U., Dewald, B., and Baggiolini, M., 1976, *In vitro* stimulation of lymphocytes by neutral proteinases from human polymorphonuclear leukocyte granules, *Schweiz. med. Wschr.* **106**:1373.

Brewerton, D. A., Webley, M., Murphy, A. H., and Milford Ward, A. M., 1978, The α_1-antitrypsin phenotype MZ in acute anterior uveitis, *Lancet* **i**:1103.

Brown, W. T., Mamelok, A. E., and Bearn, A. G., 1979, Anterior uveitis and alpha-1-antitrypsin, *Lancet* **ii**:646.

Buisseret, P. D., Pembrey, M. E., and Lessof, M. H., 1977, α_1-antitrypsin phenotypes in rheumatoid arthritis and ankylosing spondylitis, *Lancet* **ii**:1358–1359.

Bundy, H. F., and Mehl, J. W., 1959, Trypsin inhibitors of human serum. II. Isolation of the α-1-inhibitor, *J. Biol. Chem.* **234**:1124–1128.

Busch, R., and Wilms, K., 1974, Das verhalten des α_1-antitrypsin bei Verbrennungen und Verbrühungen, *Kinderaerztl. Prax.* **42**:352–356.

Carp, H., and Janoff, A., 1978, *In vitro* suppression of serum elastase-inhibitory capacity by fresh cigarette smoke and its prevention by antioxidants, *Am. Rev. Resp. Dis.* **118**:617–621.

Carp, H., and Janoff, A., 1979, *In vitro* suppression of serum elastase-inhibitory capacity by reactive oxygen species generated by phagocytosing polymorphonuclear leukocytes, *J. Clin. Invest.* **63**:793–797.

Chan, S. K., Luby, J., and Wu, Y. C., 1973, Purification and chemical composition of human α_1-antitrypsin of the MM type, *FEBS Lett.* **35**:79–82.

Chan, S. K., Rees, D. C., Li, S.-C., and Li, Y.-T., 1976, Linear structure of oligosaccharide chains in α_1-protease inhibitor isolated from human plasma, *J. Biol. Chem.* **251**:471–476.

Chapuis-Cellier, C., 1975, Etude biochimique et génétique de l'alpha-1-antitrypsine humaine, M.D. thesis, University of Lyons, Lyons, France.

Chapuis-Cellier, C., and Arnaud, P., 1979, Preferential transmission of the Z deficient allele of alpha$_1$-antitrypsin, *Science* **205**:407–408.

Charlionet, R., Martin, J.-P., Sesboué, R., and Ropartz, C., 1976, Is there a relationship between alpha-1-antitrypsin Pi MZ phenotype and hepatoma?, *Biomedicine* **25**:125–126.

Charlionet, R., Martin, J.-P., Sesboue, R., Madec, P. J., and Lefebvre, F., 1979, Synthesis of highly diversified carrier ampholytes. Evaluation of the resolving power of isoelectric focusing in the Pi system (alpha-1-antitrypsin genetic polymorphism), *J. Chromat.* **176**:89–101.

Chevance, L.-G., Causse, J. R., and Berges, J., 1976, α_1-antitrypsin activity of perilymph: occurrence during progression of otospongiosis, *Arch. Otolaryngol.* **102**:363–364.

Clemmensen, J., and Christensen, F., 1976, Inhibition of urokinase by complex formation with human α_1-antitrypsin, *Biochim. Biophys. Acta* **429**:591–599.

Clerc, M., Le Bras, M., Loubière, R., and Houvet, D., 1977, Cancer primitif du foie: Incidence de déficit en alpha-1-antitrypsin, *Nouv. Presse Med.* **6**:3061–3064.

Cohen, A. B., 1973, Interrelationships between the human alveolar macrophage and alpha-1-antitrypsin, *J. Clin. Invest.* **52**:2793–2799.

Cohen, A. B., 1979, The effects *in vivo* and *in vitro* of oxidative damage to purified α_1-antitrypsin and the enzyme-inhibiting activity of plasma, *Am. Rev. Resp. Dis.* **119**:953–960.

Cohen, A. B., and Fallat, R., 1974, Purification of phenotypically unaltered α-1-antitrypsin, *Biochim. Biophys. Acta* **336**:399–402.

Cohen, A. B., Gruenke, L. D., Craig, J. C., and Geczy, D., 1977, Specific lysin labelling by $^{18}OH^-$ during alkaline cleavage of the α_1-antitrypsin-trypsin complex, *Proc. Natl. Acad. Sci. USA* **74**:4311–4314.

Collins, R. L., Turner, R. A., and Johnson, A. M., 1976, Obstructive pulmonary disease in rheumatoid arthritis, *Arth. Rheum.* **19**:623–628.

Constans, J., and Viau, M., 1975, Une nouvelle mutation Pi^N au locus Pi dans les populations humaines, *C.R. Acad. Sci. (Paris)* **281**:1361–1364.

Cook, P. J. L., 1975, The genetics of α_1-antitrypsin: a family study in England and Scotland, *Ann. Hum. Genet., Lond.* **38**:275–287.

Cooper, D. M., Hoeppner, V. H., Cox, D. W., Zamel, N., Bryan, A. C., and Levison, H., 1974, Lung function in alpha$_1$-antitrypsin heterozygotes (Pi type MZ), *Am. Rev. Resp. Dis.* **110**:708–715.

Cottrall, K., Cook, P. J. L., and Mowat, A. P., 1974, Neonatal hepatitis syndrome and alpha-1-antitrypsin deficiency: an epidemiological study in south-east England, *Postgrad. Med. J.* **50**:376–380.

Cox, D. W., 1975, The effect of neuraminidase on genetic variants of α_1-antitrypsin, *Am. J. Hum. Genet.* **27**:165–177.

Cox, D. W., 1976, A new deficiency allele of alpha-1-antitrypsin: Pi^{Mmalton}, in: *Protides of the Biological Fluids* (H. Peeters, ed.), pp. 375–378, Pergamon Press, Oxford.

Cox, D. W., 1977, Genetic variants of α_1-antitrypsin by isoelectric focusing, *Am. J. Hum. Genet.* **29**:34A.

Cox, D. W., 1980, Transmission of Z allele from MZ heterozygotes for α_1-antitrypsin deficiency, *Am. J. Hum. Genet.* **32**:455–457.

Cox, D. W., 1981, New variants of α_1-antitrypsin: Comparison of Pi typing techniques, *Am. J. Hum. Genet.* **33** (in press).

Cox, D. W., and Celhoffer, L., 1974, Inherited variants of α_1-antitrypsin: A new allele Pi^N, *Can. J. Genet. Cytol.* **16**:297–303.

Cox, D. W., and Cornies, L., 1980, Leukocyte proteases in individuals with α_1-antitrypsin deficiency, Pi type Z, *Am. J. Hum. Genet.* **31**:43A.

Cox, D. W., and Huber, O., 1976, Rheumatoid arthritis and alpha-1-antitrypsin, *Lancet* **1**:1216–1217.

Cox, D. W., and Huber, O., 1980, Association of severe rheumatoid arthritis with heterozygosity for α_1-antitrypsin deficiency, *Clin. Genet.* **17**:153–160.

Cox, D. W., and Talamo, R. C., 1979, Genetic aspects of pediatric lung disease, *Ped. Clin. of N.A.* **26**:467–480.

Cox, D. W., Hoeppner, V. H., and Levison, H., 1976, Protease inhibitors in patients with chronic obstructive pulmonary disease: the alpha$_1$-antitrypsin heterozygote controversy, *Am. Rev. Resp. Dis.* **113**:601–606.

Cox, D. W., Simpson, N. E., and Jantti, R., 1978, Group-specific component, alpha$_1$-antitrypsin and esterase D in Canadian Eskimos, *Hum. Hered.* **28**:341–350.

Cox, D. W., Johnson, A. M., and Fagerhol, M. K., 1980, Report of nomenclature meeting for α_1-antitrypsin, INSERM, Rouen/Bois-Guillaume—1978, *Hum. Genet.* **53**:429–433.

Crawford, J. P., 1973, Purification and properties of normal human α_1-antitrypsin, *Arch. Biochem. Biophys.* **156**:215–222.

Crawford, G. P. M., and Ogston, D., 1974, Influence of α_1-antitrypsin on plasmin, urokinase and Hageman factor cofactor, *Biochim. Biophys. Acta* **354**:107–113.

Cutz, E., Cox, D. W., 1979, α_1-antitrypsin deficiency: The spectrum of pathology and pathophysiology, in: *Perspectives in Pediatric Pathology* (H. S. Rosenberg and R. P. Bolande, eds.), pp. 1–39, Masson, New York.

Egeberg, O., 1965, Inherited antithrombin deficiency causing thrombophilia, *Thromb. Diath. Haemorrh. (Stuttgart)* **13**:516–529.

Elson, L. A., Betts, T. E., and Darcy, D. A., 1973, α_1-antitrypsin in cigarette smokers, in: *Proceedings of the 5th International Symposium on the Biological Characterization of Human Tumors* (W. Davis and C. Maltoni, eds.), pp. 151–153, Excerpta Medica, Amsterdam.

Eriksson, S., 1965, Studies in α_1-antitrypsin deficiency, *Acta Med. Scand.* 177 suppl. 432:5–85.

Eriksson, S., and Hagerstrand, I., 1974, Cirrhosis and malignant hepatoma in α_1-antitrypsin deficiency, *Acta Med. Scand.* 195:451–458.

Eriksson, S., Alm, R., and Astedt, B., 1978, Organ cultures of human fetal hepatocytes in the study of extra-and intracellular α_1-antitrypsin, *Biochim. Biophys. Acta* 542:496–505.

Erlanger, B. F., Kokowsky, N., and Cohen, W., 1961, The preparation and properties of two new chromogenic substrates of trypsin, *Arch. Biochem.* 95:271–278.

Evans, H. E., Levi, M., and Mandl, I., 1970, Serum enzyme inhibitor concentrations in the respiratory distress syndrome, *Am. Rev. Resp. Dis.* 101:359–363.

Evans, H. E., Formaini, M., and Mandl, I., 1976, Prevalence of Pi types among newborns of different ethnic backgrounds, in: *Protides of the Biological Fluids* (H. Peeters, ed.), pp. 363–365, Pergamon Press, Oxford.

Evans, H. E., Bognacki, N. S., Perrott, L. M., and Glass, L., 1977, Prevalence of alpha[1]-antitrypsin Pi types among newborn infants of different ethnic backgrounds, *J. Ped.* 90:621–624.

Fagerhol, M. K., 1967, Serum Pi types in Norwegians, *Acta Path. Microbiol. Scand.* 70:421–428.

Fagerhol, M., 1968, The Pi-System. Genetic variants of serum α_1-antitrypsin, *Ser. Haematol.* 1:153–161.

Fagerhol, M. K., 1969, Quantitative studies on the inherited variants of serum α_1-antitrypsin, *Scand. J. Clin. Lab. Invest.* 23:97–103.

Fagerhol, M. K., 1972a, The serum alpha-1-antitrypsin polymorphism, in: *Proceedings of the IVth International Congress of Human Genetics, Paris, 1971* (J. de Grouchy, F. J. G. Ebling, and I. W. Henderson, eds.), pp. 279–285, Excerpta Medica, Amsterdam.

Fagerhol, M. K., 1972b, Genetics of the Pi system, in: *Pulmonary Emphysema and Proteolysis* (C. Mittman, ed.) pp. 123–131, Academic Press, New York.

Fagerhol, M. K., and Braend, M., 1965, Serum prealbumin: Polymorphism in man, *Science* 149:986–987.

Fagerhol, M. K., and Gedde-Dahl, Jr., T., 1969, Genetics of the Pi serum types. Family studies of the inherited variants of serum alpha[1]-antitrypsin, *Hum. Hered.* 19:3–8.

Fagerhol, M. K., and Hauge, H. E., 1968, The Pi phenotype MP. Discovery of a ninth allele belonging to the system of inherited variants of serum α_1-antitrypsin, *Vox Sang.* 15:396–400.

Fagerhol, M. K., and Hauge, H. E., 1969, Serum Pi types in patients with pulmonary diseases, *Acta Allergol.* 24:107–114.

Fagerhol, M. K., and Laurell, C.-B., 1967, The polymorphism of "prealbumins" and α_1-antitrypsin in human sera, *Clin. Chim. Acta* 16:199–203.

Fagerhol, M. K., and Laurell, C.-B., 1970, The Pi system—inherited variants of serum α_1-antitrypsin, in: *Progress in Medical Genetics* (A. Steinberg and A. Bearn, eds.), Vol. VII, pp. 96–111, Grune and Stratton, New York.

Fagerhol, M. K., and Tenfjord, O. W., 1968, Serum Pi types in some European, American, Asian and African populations, *Acta Path. Microbiol. Scand.* 72:601–608.

Fagerhol, M. K., Eriksson, A. W., and Monn, E., 1969, Serum Pi types in some Lappish and Finnish populations, *Hum. Hered.* 19:3–7.

Feldmann, G., Martin, J.-P., Sesboue, R., Ropartz, C., Perelman, R., Nathanson, M., Seringe, P., and Benhamou, J.-P., 1975, The ultrastructure of hepatocytes in alpha-1-antitrypsin deficiency with the genotype Pi−−. *Gut* 16:796–799.

Fineman, R. M., Kidd, K. K., Johnson, A. M., and Breg, W. R., 1976, Increased frequency of heterozygotes for α_1-antitrypsin variants in individuals with either sex chromosome mosaicism or trisomy 21, *Nature* 260:320–321.

Folds, J. D., Prince, H., and Spitznagel, J. K., 1978, Limited cleavage of human immunoglobulins by elastase of human neutrophil polymorphonuclear granulocytes. Possible modulator of immune complex disease, *Lab. Invest.* **39**:313–321.

Frants, R. R., and Eriksson, A. W., 1976, α₁-antitrypsin: Common subtypes of Pi M, *Hum. Hered.* **26**:435–440.

Frants, R. R., and Eriksson, A. W., 1978, Reliable classification of six Pi M subtypes by separator isoelectric focusing, *Hum. Hered.* **28**:201–209.

Frants, R. R., Noordhoek, G. T., and Eriksson, A. W., 1978, Separator isoelectric focusing for identification of α-1-antitrypsin (Pi M) subtypes, *Scand. J. Clin. Invest.* **38**:457–462.

Freeman, H. J., Weinstein, W. M., Shnitka, T. K., Crockford, P. M., and Herbert, F. A., 1976, Alpha₁-antitrypsin deficiency and pancreatic fibrosis, *Ann. Int. Med.* **85**:73–76.

Fritz, H., Heimburger, N., Meier, M., Arnold, M., Zaneveld, L. J. D., and Schumacher, G. F. B., 1972a, Human acrosin. kinetics of inhibition by inhibitors from human sera, *Hoppe-Seylers Z. Physiol. Chem.* **353**:1953–1956.

Fritz, H., Wunderer, G., Kummer, K., Heimburger, N., and Werle, E., 1972b, Antitrypsin und C1-Inaktivator. Progressiv-inhibitoren für Serumkallikreine von Mensch und Schwein, *Hoppe-Seylers Z. Physiol. Chem.* **353**:906–910.

Galdston, M., Janoff, A., and Davis, A. L., 1973, Familial variation of leukocyte lysosomal protease and serum α₁-antitrypsin as determinants in chronic obstructive pulmonary disease, *Am. Rev. Resp. Dis.* **107**:718–727.

Galdston, M., Melnick, E. L., Goldring, R. M., Levytska, V., Curasi, C. A., and Davis, A. L., 1977, Interactions of neutrophil elastase, serum trypsin inhibitory activity, and smoking history as risk factors for chronic obstructive pulmonary disease in patients with MM, MZ, and ZZ phenotypes for alpha₁-antitrypsin, *Am. Rev. Resp. Dis.* **116**:837–846.

Galdston, M., Levytska, V., Liener, I. E., and Twumasi, D. Y., 1979, Degradation of tropoelastin and elastin substrates by human neutrophil elastase, free and bound to alpha-2-macroglobulin in serum of the M and Z (Pi) phenotypes for alpha-1-antitrypsin, *Am. Rev. Resp. Dis.* **119**:435–441.

Galvez, S., Farcas, A., and Monari, M., 1979, The concentration of alpha-1-antitrypsin in cerebrospinal fluid and serum in a series of 40 intracranial tumors, *Clin. Chim. Acta.* **91**:191–196.

Ganrot, P. O., and Bjerre, B., 1967, α₁-antitrypsin and α₂-macroglobulin concentration in serum during pregnancy, *Acta Obstet. Gynecol. Scand.* **46**:126–137.

Gautier, M., Martin, J.-P., and Polini, G., 1977, *In vitro* synthesis of alpha-1-antitrypsin in long term monolayer human liver cultures, *Biomedecine* **27**:116–119.

Gedde-Dahl, Jr., T., Fagerhol, M. K., Cook, P. J. L., and Noades, J., 1972, Autosomal linkage between the Gm and Pi loci in man, *Ann. Hum. Genet., Lond.* **35**:393–399.

Gedde-Dahl, Jr., T., Cook, P. J. L., Fagerhol, M. K., and Pierce, J. A., 1975, Improved estimate of the Gm-Pi linkage, *Ann. Hum. Genet., Lond.* **39**:43–50.

Geddes, D. M., Webley, M., Brewerton, D. A., Turton, D. W., Turner-Warwick, M., Murphy, A. H., and Milford Ward, A., 1977, α₁-antitrypsin phenotypes in fibrosing alveolitis and rheumatoid arthritis, *Lancet* **2**:1049–1050.

Gelfand, E. W., Cox, D. W., Lin, M. T., and Dosch, H.-M., 1979, Severe combined immunedeficiency disease in patient with α₁-antitrypsin deficiency, *Lancet* **2**:202.

Genz, T., Martin, J.-P., and Cleve, H., 1977, Classification of α₁-antitrypsin (Pi) phenotypes of isoelectrofocusing: Distinction of six subtypes of the PiM phenotype, *Hum. Genet.* **38**:325–332.

Gisler, R. H., Vischer, T. L., and Dukor, P., 1976, Trypsin increases *in vitro* antibody synthesis and substitutes for help T cells, *J. Immunol.* **116**:1354–1357.

Goedde, H. W., Hirth, L., Benkmann, H.-G., Pellicer, A., Pellicer, T., Stahn, M., and

Singh, S., 1973, Population genetic studies of serum protein poly-morphisms in four Spanish populations. Part II, *Hum. Hered.* **23**:135–146.

Goldstein, I. M., Brai, M., Osler, A. G., and Weissman, G.,1973, Lysosomal enzyme release from human leukocytes: Mediation by the alternate pathway of complement activation, *J. Immunol.* **111**:33–37.

Harris, C. C., Primack, A., and Cohen, M. H., 1974, Elevated α_1-antitrypsin serum levels in lung cancer patients, *Cancer* **34**:280–281.

Harris, E. D., DiBona, D. R., and Krane, S. M., 1969, Collagenase in human synovial fluid, *J. Clin. Invest.* **48**:2104–2113.

Harris, J. O., Olsen, G. N., Castle, J. R., and Maloney, A. S., 1976, Comparison of proteolytic enzyme activity in pulmonary alveolar macrophages and blood leukocytes in smokers and nonsmokers, *Am. Rev. Resp. Dis.* **111**:579–586.

Headings, V. E., and Bose, S., 1974, Multiple molecular forms of serum α_1-antitrypsin, *Biochem. Genet.* **11**:629–636.

Heck, L. W., and Kaplan, A. P., 1974, Substrates of Hageman factor. I. Isolation and characterization of human factor XI (PTA) and inhibition of the activated enzyme of α_1-antitrypsin, *J. Exp. Med.* **140**:1615–1630.

Heimburger, N., Haupt, H., and Schwick, H. G., 1971, Proteinase inhibitors in human plasma, in: *Proteinase Inhibitors* (H. Fritz and H. Tschesche, eds.), pp. 1–21, Walter de Gruyter, New York.

Hercz, A., Katona, E., Cutz, E., Wilson, J. R., and Barton, M., 1978, α_1-Antitrypsin: The presence of excess mannose in the Z variant isolated from liver, *Science* **201**:1229–1232.

Hoffman, J. J. M. L., and van den Broek, W. G. M., 1976, Distribution of alpha-1-antitrypsin phenotypes in two Dutch population groups, *Hum. Genet.* **32**:43–48.

Horng, W. J., and Gan, J. D., 1974, Purification and characterization of human plasma α_1-antitrypsin, *Tex. Rep. Biol. Med.* **32**:489–504.

Hudgin, R. L., Pricer, W. E., Jr., and Ashwell, G., 1974, The isolation and properties of a rabbit liver binding protein specific for asialoglycoproteins, *J. Biol. Chem.* **249**:5536–5543.

Hug, G., Chuck, G., Bowles, B., and Fagerhol, M. K., 1979, Y_{Pratt}: New alpha$_1$-antitrypsin variant migrates slower than Z phenotype on polyacrylamide but faster on starch gel electrophoresis, *Ped. Res.* **13**:420.

Hunkapiller, M. W., Smallcombe, S. H., Whitaker, D. R., and Richards, J. H., 1973, Carbon nuclear magnetic resonance studies of the histidine residue in α-lytic protease. Implications for the catalytic mechanism of serine proteases, *Biochemistry* **12**:4732–4743.

Iammarino, R. M. Wagener, D. K. and Allen, R. C., 1979, Segregation distortion of the alpha-1-antitrypsin Pi Z allele, *Am. J. Hum. Genet.* **31**:508–517.

Ihrig, J., Schwartz, H. J., Rynbrandt, D. J., and Kleinerman, J., 1975, Serum trypsin inhibitory capacity and Pi phenotypes. Prevalence of α_1-antitrypsin deficiency in an allergy population, *Am. J. Clin. Path.* **64**:297–303.

Inokuma, S., 1978, Acquired aberrant expression of serum α_1-antitrypsin in disseminated intravascular coagulation, *Thromb. Res.* **13**:123–128.

James, H. L., and Cohen, A. B., 1978, Mechanism of inhibition of porcine elastase by human α_1-antitrypsin, *J. Clin. Invest.* **62**:1344–1353.

Janoff, A., 1971, Elastase-like proteases of human granulocytes and alveolar macrophages, in: *Pulmonary Emphysema and Proteolysis* (C. Mittman, ed.), pp. 205–224, Academic Press, New York.

Janoff, A., 1972, Inhibition of human granulocyte elastase by serum α_1-antitrypsin, *Am. Rev. Resp. Dis.* **105**:121–122.

Janoff, A., and Carp, H., 1977, Possible mechanism of emphysema in smokers. Cigarette smoke condensate suppresses protease inhibition *in vitro*, *Am. Rev. Resp. Dis.* **116**:65–72.

Janoff, A., Sloan, B., Weinbaum, G., Damiano, V., Sandhaus, R. A., Elias, J., and Kimbel, P., 1977, Experimental emphysema induced with purified human neutrophil elastase: Tissue localization of the instilled protease, *Am. Rev. Resp. Dis.* **115**:461.

Janus, E. E., Sheat, J. M., and Carrell, R. W., 1975, Alpha-1-antitrypsin variants in New Zealand, *N. Z. Med. J.* **82**:289–291.

Jeppson, J.-O., 1976, Amino acid substitution Gly-Lys in α_1-antitrypsin Pi Z, *FEBS Lett.* **65**:195–197.

Jeppsson, J. O., Larsson, C., and Eriksson, S., 1975, Characterization of α_1-antitrypsin in the inclusion bodies from the liver in α_1-antitrypsin deficiency, *N. Eng. J. Med.* **293**:576–579.

Jeppsson, J.-O., Laurell, C.-B., and Fagerhol, M. K., 1978a, Properties of isolated α_1-antitrypsin of Pi types M, S and Z, *Eur. J. Biochem.* **83**:143–153.

Jeppsson, J.-O., Laurell, C.-B., Nosslin, B., and Cox, D. W., 1978b, Catabolic rate of α_1-antitrypsin of Pi types S and Mmalton and of asialylated M protein in man, *Clin. Sci. Mol. Med.* **55**:103–107.

Jeppsson, J.-O., Franzén, B., Sveger, T., Cordesius, E., Strömberg, P., and Gustavii, B., 1979a, Prenatal exclusion of α_1-antitrypsin deficiency in a high risk fetus, *N. Eng. J. Med.* **300**:1441–1442.

Jeppsson, J.-O., Laurell, C. B., and Franzen, B., 1979b, Agarose gel electrophoresis, *Clin. Chem.* **25**:629–638.

Jirgensons, B., 1977, Circular dichroism and conformation of human α_1-antitrypsin, *Biochim. Biophys, Acta.* **493**:352–358.

Johansson, B. G., 1972, Agarose gel electrophoresis, *Scand. J. Clin. Lab. Invest.* 29, suppl. **124**:7–19.

Johnson, A. M., 1976, Genetic typing of α_1-antitrypsin by immunofixation electrophoresis. Indentification of subtypes of Pi M, *J. Lab. Clin. Med.* **87**:152–163.

Johnson, D. A., and Travis, J., 1975, Mechanism and structure of human α_1-proteinase inhibitor, in: *Protides of the Biological Fluids* (H. Peeters, ed.), pp. 35–38, Pergamon Press, Oxford.

Johnson, D., and Travis, J., 1976, Human α_1-proteinase inhibitor. Mechanism of action. Evidence for activation by limited proteolysis, *Biochem. Biophys. Res. Commun.* **72**:33–39.

Johnson, D., and Travis, J., 1977, Inactivation of human α_1-proteinase inhibitor by thiol proteinases, *Biochem. J.* **163**:639–641.

Johnson, D., and Travis, J., 1978, Structural evidence for methionine at the reactive site of human α_1-proteinase inhibitor, *J. Biol. Chem.* **253**:7142–7144.

Kaplan, J. G., and Bona, C., 1974, Proteases as mitogens, *Exp. Cell Res.* **88**:388–394.

Kaplan, P. D., Kuhn, C., and Pierce, J. A., 1973, The induction of emphysema with elastase. 1. The evolution of the lesion and the influence of serum, *J. Lab. Clin. Med.* **82**:349–356.

Katz, R. M., Lieberman, J., and Siegel, S. C., 1976, Alpha-1-antitrypsin levels and prevalence of Pi variant phenotypes in asthmatic children, *J. Allergy Clin. Immun.* **57**:41–45.

Keiser, H., Greenwald, R. A., Feinstein, G., and Janoff, A., 1976, Degradation of cartilage proteoglycan by human leukocyte granule neutral proteases—a model of joint injury, *J. Clin. Invest.* **57**:625–632.

Kellermann, G., and Walter, H., 1970, Investigations on the population genetics of the α_1-antitrypsin polymorphism, *Humangenetik* **10**:145–150.

Kelly, J. K., Davies, J. S., and Jones, A. W., 1979, Alpha-1-antitrypsin deficiency and hepatocellular carcinoma, *J. Clin. Path.* **32**:373–376.

Khitri, A., Benlatrache, K., and Martin, J.-P., 1977, Pi W3 constantine, nouvel allèle du système Pi, *Sem. Hop. Paris* **53**:909–910.

Kidokoro, Y., Kravis, T. C., Moser, K. M., Taylor, J. D., and Crawford, I. P., 1977, Relationship of leukocyte elastase concentration to severity of emphysema in homozygous α_1-antitrypsin-deficient persons, *Am. Rev. Resp. Dis.* **115**:793–803.

Kimberling, W. J., Fulbeck, T., Dixon, L., and Lubs, H. A., 1975, Localization of spherocytosis to chromosome 8 or 12 and report of a family with spherocytosis and a reciprocal translocation, *Am. J. Hum. Genet.* **27**:586–594.

Klasen, E. C., and de Brij, R.-J., 1979, Clearance of α_1-antitrypsin isoelectric focusing patterns in blood samples treated with heparin, *Clin. Chim.* **95**:391–394.

Klasen, E. C., Franken, C., Volkers, W. S., and Bernini, L. F., 1977, Population genetics of α_1-antitrypsin in the Netherlands: Description of a new electrophoretic variant, *Hum. Genet.* **37**:303–313.

Klasen, E. C., D'Andrea, F., and Bernini, L. F., 1978, Phenotype and gene distribution of alpha-1-antitrypsin in a North Italian population, *Hum. Hered.* **28**:474–478.

Klayton, R., Fallat, R., and Cohen, A. B., 1975, Determinants of chronic obstructive pulmonary disease in patients with intermediate levels of alpha$_1$-antitrypsin, *Am. Rev. Resp. Dis.* **112**:71–75.

Kress, L. F., and Laskowski, M., Sr., 1973, Large scale purification of α_1-trypsin inhibitor from human plasma, *Prep. Biochem.* **3**:541–552.

Kueppers, F., 1968a, Antigenic similarity of human α_1-antitrypsin to a corresponding protein in the serum of non-human primates, *Experientia* **24**:1277–1278.

Kueppers, F., 1968b, Genetically determined differences in the response of alpha-1-antitrypsin levels in human serum to typhoid vaccine, *Humangenetik* **6**:207–214.

Kueppers, F., 1972, Hypothesis to explain heterozygous advantage in alpha$_1$-antitrypsin deficiency, in: *Pulmonary Emphysema and Proteolysis* (C. Mittman, ed.), pp. 133–137; Academic press, New York.

Kueppers, F., 1976, α_1-antitrypsin M. A new common genetically determined variant, *Am. J. Hum. Genet.* **28**:370–377.

Kueppers, F., 1978, Inherited differences in alpha$_1$-antitrypsin, in: *Lung Biology in Health and Disease*, Vol. 13 (C. Lenfant, ed.), pp. 23–74, *Genetic Determinants of Pulmonary Disease* (S. D. Litwin, ed.), Marcel Dekker, New York.

Kueppers, F., and Bearn, A. G., 1966, A possible experimental approach to the association of hereditary α_1-antitrypsin deficiency and pulmonary emphysema, *Proc. Soc. Exp. Biol. Med.* **121**:1207–1209.

Kueppers, F., and Black, L. F., 1974, α_1-antitrypsin and its deficiency, *Am. Rev. Resp. Dis.* **110**:176–194.

Kueppers, F., and Christopherson, M. J., 1978, Alpha$_1$-antitrypsin: Further genetic heterogeneity revealed by isoelectric focusing, *Am. J. Hum. Genet.* **30**:359–365.

Kueppers, F., Brackertz, D., and Czygan, P. J., 1972, Serum α_1-antitrypsin levels during the ovarian cycle, *Clin. Chim. Acta.* **39**:131–134.

Kueppers, F., O'Brien, P., Passarge, E., and Rüdiger, H. W., 1975, Alpha$_1$-antitrypsin phenotypes in sex chromosome mosaicism, *J. Med. Genet.* **12**:263–264.

Kueppers, F., Utz, G., and Simon, B., 1977, Alpha$_1$-antitrypsin deficiency with M-like phenotype, *J. Med. Genet.* **14**:183–196.

Kühnl, P., and Spielmann, W., 1979, Pi^T: A new allele in the alpha$_1$-antitrypsin system, *Hum. Genet.* **50**:221–223.

Kurecki, I., Laskowski, M., Sr., and Kress, L. F., 1978, Purification and some properties of two proteinases from *Crotalus adamanteus* venom that inactivate human alpha-1-proteinase inhibitor, *J. Biol. Chem.* **253**:8340–8345.

Kyaw-Myint, T. O., Howell, A. M., Murphy, G. M., and Anderson, C. A., 1975, Alpha-1-antitrypsin in duodenal fluid and gallbladder bile, *Clin. Chim. Acta* **59**:51–56.

Langley, C. E., Berninger, R. W., Wolfson, S. L., and Talamo, R. C., 1979, An unusual type of α₁-antitrypsin deficiency in a child, *Johns Hopkins Med. J.* **144**:161–165.

Larsson, C., 1978, Natural history and life expectancy in severe alpha₁-antitrypsin deficiency, Pi Z, *Acta Med. Scand.* **204**:345–351.

Larsson, C., and Eriksson, S., 1977, Liver function in asymptomatic adult individuals with severe α₁-antitrypsin deficiency (Pi Z), *Scand. J. Gastroent.* **12**:543–546.

Larsson, C., Dirksen, H., Sunstrom, G., and Eriksson, S., 1976, Lung function studies in asymptomatic individuals with moderately (Pi SZ) and severely (Pi Z) reduced levels of α₁-antitrypsin, *Scand. J. Resp. Dis.* **57**:267–280.

Larsson, C., Eriksson, S., and Dirksen, H., 1977, Smoking and intermediate alpha₁-antitrypsin deficiency and lung function in middle-aged men, *Br. Med. J.* **2**:922–925.

Laskowski, M., Jr., Duran, R. W., Finkenstadt, W. R., Herbert, S., Hixson, H. F., Kowalski, D., Luthy, J. A., Mattis, J. A., McKee, R. E., and Niekamp, C. W., 1971, Kinetics and thermodynamics of interaction between soybean trypsin inhibitor (Kunitz) and bovine β-trypsin, in: *Proteinase Inhibitors* (H. Fritz and H. Tschesche, eds.), pp. 117–134, Walter de Gruyter, New York.

Laurell, C.-B., 1966, Quantitative estimation of proteins by electrophoresis in agarose-gel containing antibodies, *Anal. Biochem.* **15**:45–49.

Laurell, C.-B., 1975, Relation between structure and biological function of the protease inhibitors in the extracellular fluid, in: *Protides of the Biological Fluids* (H. Peeters, ed.), pp. 3–12, Pergamon Press, Oxford.

Laurell, C.-B., and Eriksson, S., 1963, The electrophoretic α₁-globulin pattern of serum in α₁-antitrypsin deficiency, *Scand. J. Clin. Lab. Invest.* **15**:132–140.

Laurell, C.-B., and Persson, U., 1973, Analysis of plasma α₁-antitrypsin variants and their microheterogeneity, *Biochim. Biophys. Acta* **310**:500–507.

Laurell, C.-B., and Thulin, E., 1975, Thiol-disulfide interchange in the binding of Bence Jones proteins to α₁-antitrypsin, prealbumin and albumin, *J. Exp. Med.* **141**:453–456.

Laurell, C.-B., Kullander, S., and Thorell, J., 1968, Effect of administration of a combined estrogen-progestin contraceptive on the level of individual plasma proteins, *Scand. J. Clin. Lab. Invest.* **21**:337–343.

Laurell, C.-B., Pierce, J., Persson, U., and Thulin, E., 1975, Purification of α₁-antitrypsin from plasma through thiol-disulfide interchange, *Eur. J. Biochem.* **57**:107–113.

Laurell, C.-B., Nosslin, B., and Jeppsson, J.-O., 1977, Catabolic rate of α₁-antitrypsin of Pi type M and Z in man, *Clin. Sci. Mol. Med.* **52**:457–461.

Lebas, J., Hayem, A., and Martin, J.-P., 1974, Etude des variants génétiques de l'alpha-1-antitrypsine en immunofocalisation bidimensionelle, *C.R. Acad. Sci. (Paris)* **258**:2359–2360.

Lieberman, J., and Gaidulis, L., 1976, Simplified α₁-antitrypsin phenotyping by immunofixation of acid-starch gels, *J. Lab. Clin. Med.* **87**:710–716.

Lieberman, J., Gaidulis, L., and Klotz, S. D., 1976, A new deficient variant of α₁-antitrypsin (MDuarte). Inability to detect the heterozygous state by antitrypsin phenotyping, *Am. Rev. Resp. Dis.* **113**:31–36.

Lie-Injo, L. E., 1976, α₁-Antitrypsin with unusual behaviour, *Clin. Chim. Acta* **72**:83–87.

Lipsky, J. J., Berninger, R. W., Hyman, L. R., and Talamo, R. C., 1979, Presence of alpha-1-antitrypsin on mitogen stimulated human lymphocytes, *J. Immunol.* **122**:24–26.

Mancini, G., Carbonara, A. O., and Heremans, J. E., 1965, Immunochemical quantitation of antigens by single radial immunodiffusion, *Immunochemistry* **2**:235–254.

Martin, H. B., and Boatman, E. S., 1965, Electron microscopy of human pulmonary emphysema, *Am. Rev. Resp. Dis.* **91**:206–214.

Martin, J.-P., Vandeville, D., Ropartz, C., 1973, PiB, A new allele of α_1-antitrypsin genetic variants, *Biomed. Express* **19**:395–398.

Martin, J.-P., Sesboue, R., Charlionet, R., Ropartz, C., and Pereira, M. T., 1976, Genetic variants of serum α_1-antitrypsin (Pi types) in Portuguese, *Hum. Hered.* **26**:310–314.

Martodam, R. R., Twumasi, D. Y., Liener, I. E., Powers, J. C., Nishino, N., and Krejcarek, G., 1979, Albumin microspheres as carrier of an inhibitor of leukocyte elastase: Potential therapeutic agent for emphysema, *Proc. Natl. Acad. Sci. USA* **76**:2128–2132.

Massi, G., and Vecchio, F. M., 1977, Alpha-1-antitrypsin phenotypes in a group of newborn infants in Somalia, *Hum. Gen.* **38**:265–269.

Matheson, N. R., Wong, P. S., and Travis, J., 1979, Enzymatic inactivation of human alpha-1-proteinase inhibitor by neutrophil myeloperoxidase, *Biochem. Biophys. Res. Comm* **88**:402–409.

Matthews, B. W., Sigler, P. B., Henderson, R., and Blow, D. M., 1967, Three-dimensional structure of tosyl-α-chymotrypsin, *Nature* **214**:652–656.

Mazzei, D., Novi, C., and Bazzi, C., 1966, Mitogenic action of trypsin and chymotrypsin, *Lancet* **2**:232.

Milford Ward, A., Pickering, J. D., and Shortland, J. R., 1975, The renal manifestations of Pi Z, in: L'Alpha-1-antitrypsine et le Système Pi (J.-P. Martin, ed.), pp. 131–134, INSERM, Paris.

Miller, F., and Kuschner, M., 1969, Alpha$_1$-antitrypsin deficiency, emphysema, necrotizing angiitis and glomerulonephritis, *Am. J. Med.* **46**:615–623.

Moll, F. C., Sunden, S. F., and Brown, J. R., 1958, Partial purification of the serum trypsin inhibitor, *J. Biol. Chem.* **233**:121–124.

Morin, T., Feldmann, G., Martin, J.-P., Rueff, B., Benhamou, J-P., and Ropartz, C., 1975, Heterozygous alpha$_1$-antitrypsin deficiency and cirrhosis in adults, a fortuitous association, *Lancet* **1**:250–251.

Moroz, S. P., Cutz, E., Balfe, J. W., and Sass-Kortsak, A., 1976a, Membranoproliferative glomerulonephritis in childhood cirrhosis associated with alpha$_1$-antitrypsin deficiency, *Pediatrics* **57**:232–238.

Moroz, S. P., Cutz, E., Cox, D. W., and Sass-Kortsak, A., 1976b, Liver disease associated with alpha$_1$-antitrypsin deficiency in childhood, *J. Pediat.* **88**:19–25.

Morse, J. O., 1978, Alpha$_1$-antitrypsin deficiency, *N. Eng. J. Med.* **299**:1045–1048, 1099–1105.

Morton, N. E., 1956, The detection and estimation of linkage between genes for elliptocytosis and the Rh blood type, *Am. J. Hum. Genet.* **8**:80–96.

Moskowitz, R. W., and Heinrich, G., 1972, Bacterial inactivation of human serum alpha-1-antitrypsin: A possible factor in the pathogenesis of pulmonary disease related to antitrypsin deficiency states, in: *Pulmonary Emphysema and Proteolysis* (C. Mittman, ed.), pp. 261–267, Academic Press, New York.

Musiani, P., Lauriola, L., and Piantelli, M., 1978, Inhibitory activity of alpha-1-antitrypsin bound to human IgA, *Clin. Chim. Acta* **85**:61–66.

Nachman, R. L., and Harpel, P. C., 1976, Platelet α_1-macroglobulin and α_1-antitrypsin, *J. Biol. Chem.* **251**:4514–4521.

Nalli, G., Callaneo, G., Malamani, G. D., Majolino, I., Fornassati, P. M., Almaiso, P., Piovella, F., and Ascari, E., 1977, Immunofluorescent detection of α_1-antitrypsin in platelets and megakaryocytes, *Thromb. Res.* **10**:613–617.

Nicholls, M. G., and Janus, E. D., 1973, Hashimoto's thyroiditis and homozygous alpha$_1$-antitrypsin deficiency, *Aust. N. Z. J. Med.* **3**:516–519.

Noades, J. E., and Cook, P. J. L., 1976, Family studies with the Gm:Pi linkage group, in: *Human Gene Mapping 3. Birth Defects: Original Article Series* (D. Bergsma, ed.), Vol. 12, pp. 341–344, Karger, Basel.

Novis, B. H., Young, G. O., Bank, S., and Marks, I. N., 1975, Chronic pancreatitis and alpha-1-antitrypsin, *Lancet* 2:748–749.

Odièvre, M., Martin, J.-P., Hadchouel, M., and Alagille, D., 1976, Alpha$_1$-antitrypsin deficiency and liver disease in children: Phenotypes, manifestations, and prognosis, *Pediatrics* 57:226–231.

Ohlsson, K., 1971a, Elimination of ^{125}I-trypsin-α-macroglobulin complexes from blood by reticuloendothelial cells in dog, *Acta Physiol. Scand.* 81:269–272.

Ohlsson, K., 1971b, Neutral leukocyte proteases and elastase inhibited by plasma α_1-antitrypsin, *Scand. J. Clin. Lab. Invest.* 28:251–253.

Ohlsson, K., 1975, α_1-Antitrypsin and α_2-macroglobulin: Interactions with human neutrophil collagenase and elastase, *Ann. N.Y. Acad. Sci.* 256:409–419.

Ohlsson, K., and Laurell, C.-B., 1976, The disappearance of enzyme-inhibitor complexes from the circulation in man, *Clin. Sci. Mol. Med.* 51:87–92.

Owen, M. C., Carrell, R. W., and Brennan, S. D., 1976, The abnormality of the S variant of human α_1-antitrypsin, *Biochim. Biophys. Acta* 453:257–261.

Owen, M. C., Lorier, M., and Carrell, R. W., 1978, α-1-Antitrypsin: Structural relationships of the substitutions of the S and Z variants, *FEBS Lett.* 88:234–236.

Palmer, P. E., and Wolfe, H. J., 1976, α_1-Antitrypsin deposition in primary hepatic carcinomas, *Arch. Pathol. Lab. Med.* 100:232–236.

Palmer, P. E., Wolfe, H. J., and Kostas, C.-I., 1978, Multisystem fibrosis in alpha-1-antitrypsin deficiency, *Lancet* 1:221–222.

Pannell, R. D., Johnson, D., and Travis, J., 1974, Isolation and properties of human plasma α_1-proteinase inhibitor, *Biochemistry* 13:5439–5444.

Piantelli, M., Auconi, P., and Musiani, P., 1978, Alpha-1-antitrypsin phenotype in newborns from Central and Southern Italy, *Hum. Hered.* 28:468–473.

Pierce, J. A., and Eradio, B. G., 1979, Improved identification of antitrypsin phenotypes through isoelectric focusing with dithioerythritol, *J. Lab. Clin. Med.* 94:826–831.

Pierce, J. A., Eradio, B., and Dew, T. A., 1975, Antitrypsin phenotypes in St. Louis, *J. Am. Med. Assoc.* 238:609–612.

Pierce, J., Jeppsson, J.-O., and Laurell, C.-B., 1976, Alpha$_1$-antitrypsin phenotypes determined by isoelectric focusing of the cysteine-antitrypsin mixed disulfide in serum, *Anal. Biochem.* 74:227–241.

Porter, C. A., Mowat, A. P., Cook, P. J. L., Hayner, D. W. G., Shilkin, K. B., and Williams, R., 1972, α_1-Antitrypsin deficiency and neonatal hepatitis, *Br. Med. J.* 3:435–439.

Ratnoff, O. D., 1976, Mediators of inflammation, *J. Allergy Clin. Immunol.* 58:438–446.

Rawlings, W., Jr., Kreiss, P., Levy, D., Cohen, B., Menkes, H., Brashears, S., and Permutt, S., 1976, Clinical epidemiologic, and pulmonary function studies in alpha$_1$-antitrypsin-deficient subjects of Pi Z type, *Am. Rev. Resp. Dis.* 114:945–953.

Reintoft, I., and Hagerstrand, I. E., 1979, Does the Z gene variant of alpha-1-antitrypsin predispose to hepatic carcinoma?, *Hum. Pathol.* 10:419–424.

Rennert, O. M., Kaiser, D., Sollberger, M., and Joller-Jemelka, S., 1974, Antiprotease activity in tears and nasal secretions, *Humangenetik* 23:73–77.

Rimon, A., Shamash, Y., and Shapiro, B., 1966, The plasmin inhibitor of human plasma. IV. Its action on plasmin, trypsin, chymotrypsin and thrombin, *J. Biol. Chem.* 241:5102–5107.

Robinet,-Levy, M., and Reunier, M., 1972, Techniques d'indentification des groupes Pi. *Rév. Franç. Transf.* 15:61–72.

Rodriguez, R. J., White, R. R., Senior, R. M., and Levine, E. A., 1977, Elastase release from human alveolar macrophages: Comparison between smokers and nonsmokers, *Science* 198:313–314.

Rodriguez, J. R., Seals, J. E., Radin, A., Lin, J. S., Mandl, I., and Turino, G. M., 1979,

Neutrophil lysosomal elastase activity in normal subjects and in patients with chronic obstructive pulmonary disease, *Am. Rev. Resp. Dis.* **119**:409–417.

Roll, D., Aguanno, J. J., Coffee, C. J., Glew, R. H., and Iammarino, R. M., 1978, Comparison of the carbohydrate and amino acid composition of normal and S-variant α_1-antitrypsin, *Biochem. Biophys. Acta* **532**:171–178.

Rouen Report, 1980, see Cox *et al.* (1980).

Rubinstein, H. M., Jaffer, A. M., Kudrna, J. C., Lertratanakul, Y., Chandrasekhar, A. J., Slater, D., and Schmid, F. R., 1977, Alpha$_1$-antitrypsin deficiency with severe panniculitis, *Ann. Intern. Med.* **86**:742–744.

Scharp, S., Eid, M., Cooreman, W., and Lauwers, A., 1976, α_1-Antitrypsin, An inhibitor of renin, *Biochem. J.* **153**:505–507.

Schleissner, L. A., and Cohen, A. H., 1975, Alpha-1-antitrypsin deficiency and hepatic carcinoma, *Am. Rev. Resp. Dis.* **111**:863–868.

Schuller, E., Tompe, L., Lefevre, M., and Moreno, P., 1970, Electroimmunodiffusion des protéines du liquide cephalorachidien, *Clin. Chim. Acta* **30**:73–82.

Schultze, H. E., and Heremans, J. F., 1966, *Molecular Biology of Human Proteins*, p. 354, Elsevier, Amsterdam.

Schultze, H. E., Göllner, I., Heide, K., Schönenberger, M., and Schwick, G., 1955, Zur Kenntnis der Alpha Globuline des menschlichen Normalserums, *Z. Naturforsch.* **10**:463–472.

Schultze, H. E., Heide, K., and Haupt, H., 1962, α_1-Antitrypsin aus Humanserum, Klin. Wschr. **40**:427–429.

Schumacher, G. F. B., 1970, Alpha-1-antitrypsin in genital secretions, *J. Reprod. Med.* **5**:13–19.

Schumacher, G. F. B., 1971, Inhibition of rabbit sperm acrosomal protease by human α_1-antitrypsin and other protease inhibitors, *Contraception* **4**:67.

Schumacher, G. F. B., and Pearl, M. J., 1968, Alpha-1-1antitrypsin in cervical mucus, *Fertil. Steril.* **19**:91–99.

Schwick, H. G., Heimburger, N., and Hapt, H., 1966, Antiproteasen des Humanserums, *Z. Ges. Inn. Med.* **21**:193–198.

Senior, R. M., Huebner, P. F., and Pierce, J. A., 1972, Serum elastase inhibition capacity as measure of serum α_1-antitrypsin concentrations, in: *Pulmonary Emphysema and Proteolysis* (C. Mittman, ed.), pp. 179–185, Academic Press, New York.

Sharp, H. L., 1971, Alpha-1-antitrypsin deficiency, *Hosp. Pract.* **5**:83–96.

Sharp, H. L., 1976, Relationship between α_1-antitrypsin deficiency and liver disease, in: *Liver Disease in Infancy and Childhood* (S. R. Berenberg, ed.), pp. 52–71, Williams and Wilkins Baltimore.

Sharp, H. L., Bridges, R. A., and Krivit, W., 1969, Cirrhosis associated with alpha-1-antitrypsin deficiency: A previously unrecognized inherited disorder, *J. Lab. Clin. Med.* **73**:934–939.

Sjoblom, K. G., and Wollheim, F. A., 1977, Alpha-1-antitrypsin phenotypes and rheumatic diseases, *Lancet* **2**:41–42.

Smith, M., and Hirschhorn, K., 1978, Location of the genes for human heavy chain immunoglobulin to chromosome 6, *Proc. Natl. Acad, Sci. USA* **75**:3367–3371.

Starkey, P. M., Barrett, A. J., and Burleigh, M. C., 1977, The degradation of articular collagen by neutrophil proteinases, *Biochim. Biophys. Acta* **483**:386–397.

Steven, F. S., Milsom, D. W., and Hunter, J. A. A., 1976, Human-polymorphonuclear-leukocyte neutral protease and its inhibitor. Studies with fluorescein-labelled polymeric collagen fibrils as a substrate, *Eur. J. Biochem.* **67**:165–169.

Stroud, R. M., 1974, A family of protein-cutting proteins, *Sci. Am.* **231**:74–88.

Sutcliffe, R., and Brock, D., 1973, Immunological studies on the nature and origin of the major proteins in amniotic fluid, *J. Obst. Gynec. Br. Commwlth.* **80**:721–730.

Sveger, T., 1976, Liver disease in alpha$_1$-antitrypsin deficiency detected by screening of 200,000 infants, *N. Eng. J. Med.* **294**:1316–1321.

Swedlund, M. A., Hunter, G. G., and Gleich, G. J., 1969, Alpha-1-antitrypsin increased in serum and synovial fluid in rheumatoid arthritis, *Proc. Am. Assoc. Annu. Meeting Arth. Rheum.* **12**:337A.

Szczeklik, A., Turowska, B., Czerniawska-Mysik, G., Opolska, B., and Nizankowska, E., 1974, Serum alpha$_1$-antitrypsin in bronchial asthma, *Am. Rev. Resp. Dis.* **109**:487–490.

Talamo, R. C., 1975, Basic and clinical aspects of the α_1-antitrypsin, *Pediatrics* **56**:91–99.

Talamo, R. C., Langley, C. E., and Hyslop, N. E., Jr., 1972, A comparison of functional and immunological measurements of serum α_1-antitrypsin, in: *Pulmonary Emphysema and Proteolysis* (C. Mittman, ed.), pp. 167–172, Academic Press, New York.

Talamo, R. C., Langley, C. E., Reed, C. E., and Makino, S., 1973, α_1-Antitrypsin deficiency: A variant with no detectable α_1-antitrypsin, *Science* **181**:70–71.

Talamo, R. C., Langley, C. E., and Twarog, M. D., 1975, Protease inhibitors in cord and maternal serum at birth, *Behring Inst. Res. Comm.* **58**:46–49.

Talamo, R. C., Bruce, R. M., Langley, C. E., Berninger, R. W., Pierce, J. A., Bryant, L. J., and Duncan, D. B., 1978, *Alpha$_1$-antitrypsin Laboratory Manual,* U.S. Department of Health, Education and Welfare, Publ. No. (NIH) 78–1420.

Taylor, J. C., and Kueppers, F., 1977, Electrophoretic mobility of leukocyte elastase of normal subjects and patients with chronic obstructive pulmonary disease, *Am. Rev. Resp. Dis.* **116**:531–536.

Tejler, L., and Grubb, A. O., 1976, A complex-forming glycoprotein heterogeneous in charge and present in human plasma, urine and cerebrospinal fluid, *Biochim. Biophys. Acta* **439**:82–94.

Theodoropoulos, G., Fertakis, A., Archimandritis, A., Kapordelis, C., and Angelopoulos, B., 1976, Alpha-1-antitrypsin phenotypes in cirrhosis and hepatoma, *Acta Hepat.-Gastroent.* **23**:114–117.

Tokoro, Y., Eisen. A. A., and Jeffrey, J. J., 1972, Characterization of collagenase from rat skin, *Biocheim. Biophys. Acta* **258**:289–302.

Travis, J., Baugh, R., Giles, P. J., Johnson, D., Bower, J., and Reilly, C. F., 1978, Human leukocyte elastase and cathepsin G: Isolation, characterization and interaction with plasma proteinase inhibitors, in: *Neutral Proteases of Human Polymorphonuclear Leukocytes* (H. Havemann and A. Janoff, eds.), pp. 118–128, Urban and Schwarzenberg, Baltimore-Munich.

Tullis, J. L., and Watanabe, K., 1978, Platelet anti-thrombin deficiency. A new clinical entity, *Am. J. Med.* **65**:472–478.

Twumasi, D. Y., Liener, I. E., Galdston, M., and Levytska, V., 1977, Activation of human leukocyte elastase by human α_2-macroglobulin, *Nature* **267**:61–63.

Tyler, H. M., Cheese, J. A. F., Struthers, N. W., and Dempster, W. K., 1962, A humoral change accompanying homograft rejection, *Lancet* **2**:432–434.

Van den Broek, W. G. M., Hoffmann, J. J. M. L., Dijkman, J. H., 1976, A new, high frequency variant of α_1-antitrypsin, *Hum. Genet.* **34**:17–22.

Vandeville, D., Martin, J.-P., Ropartz, C., 1974, α_1-Antitrypsin polymorphism of Bantu population: Description of a new allele PiL, *Humangenetik* **21**:33–38.

Venge, P., 1978, Polymorphonuclear leukocyte proteases and their effects on complement components and neutrophil function, in: *Neutral Proteases of Human Polymorphonuclear Leukocytes* (K. Havemann and A. Janoff, eds.) pp. 264–275, Urban and Schwarzenberg, Baltimore-Munich.

Vercaigne, D., Morcamp, C., Martin, J.-P., and Raoult, J. P., 1979, Tryptic and elastolytic inhibitory capacities of serum from various Pi phenotypes, *Clin. Chim. Acta* **93**:71–83.

Vischer, T. L., 1974, Stimulation of mouse B lymphocytes by trypsin, *J. Immunol.* **113**:58–62.

Vischer, T. L., Bretz, U., and Bagglioni, M., 1976, *In vitro* stimulation of lymphocytes by neutral proteinases from human polymorphonuclear leukocyte granules, *J. Exp. Med.* **144**:863–872.

Webb, D. R., Hyde, R. W., Schwartz, R. H., Hall, W. J., Condemi, J. J., and Townes, P. l., 1973, Serum α_1-antitrypsin variants, *Am. Rev. Resp. Dis.* **108**:918–925.

Weissmann, G., Zurier, R. B., Spieler, P. J., and Goldstein, I. M., 1971, Mechanisms of lysosomal enzyme release from leukocytes exposed to immune complexes and other particles, *J. Exp. Med.* **134**:149–165.

Weitkamp, L. R., Cox, D., Guttormsen, S., Johnston, E., and Hempfling, S., 1978, Allelic specific heterogeneity in the Pi : Gm linkage group, *Cytogenet. Cell Genet.* **22**:647–650.

Yang, S.-L., Zenveld, L. J. D., and Schumacher, G. F. B., 1976, Effect of serum proteinase inhibitors on the fertilizing capacity of rabbit spermatozoa, *Fertil. Steril.* **27**:577–581.

Yoshida, A., and Mega, T., 1979, Carbohydrate composition of normal and variant human alpha-1-protease inhibitors, *Arch. Biochem. Biophys.* **195**:591–595.

Yoshida, L., Lieberman, J., Gaidulis, L., and Ewing, C., 1976, Molecular abnormality of human α_1-antitrypsin variant (PiZ) associated with plasma activity deficiency, *Proc. Natl. Acad. Sci. USA* **73**:1324–1328.

Yoshinaga, M., Nakamura, S., and Hayashi, H., 1975, Interaction between lymphocytes and inflammatory exudate cells. I. Enhancement of thymocyte response to PHA by product(s) of polymorphonuclear leukocytes and macrophages, *J. Immunol.* **115**:533–538.

Chapter 2

Segregation Analysis

R. C. Elston

Department of Biometry
Louisiana State University Medical Center
New Orleans, Louisiana 70112

INTRODUCTION

The statistical detection of Mendelian ratios in human sibships has long
been known as segregation analysis. More generally, we can define segregation analysis as the statistical methodology used to determine from
family data the mode of inheritance of a particular phenotype, especially
with a view to elucidating single gene effects; it is thus a basic tool in
human genetics. There have been many reviews of the early literature on
this topic (see, e.g., Smith, 1956, 1959; Steinberg, 1959; Morton, 1962,
1964, 1969; Elandt-Johnson, 1974), and this will not be repeated here;
rather the present chapter will concentrate on the developments in this
area over the last decade, incidentally pointing out how the earlier methods can be considered as special cases of the more general theory that
is now available. The fast development of the subject in recent years can
be largely attributed to the increasing availability of electronic computers,
allowing the use of sophisticated analyses that would otherwise be impossible. Therefore, on the assumption that appropriate computer programs will soon be widely available, this review will pay more attention
to the statistical principles involved, and less to methods of computation.
In the first section, the various models are defined mathematically so as

A modified and abbreviated version of this article appears in *Current Developments in
Anthropological Genetics*, 1980 (J. H. Mielke and M. H. Crawford, eds.), Plenum Press,
New York.

to give precision to the assumptions that are made, and the genetic implications of these assumptions are discussed. The general likelihood method of hypothesis testing and parameter estimation is taken up in the second section, and its application to recent examples of segregation analysis is illustrated in the third section. We shall see that methods of segregation analysis are still evolving and conclude that many further refinements can be expected in the near future. Finally, in the Appendix, a few brief notes on computational methods are given.

MATHEMATICAL FORMULATION OF GENETIC MODELS

Segregation analysis is based on a mathematical model that has four major components. The first two components describe the genetic basis for the phenotypic variability of a trait in a population, the third describes how that variability is passed on from one generation to the next, while the last component describes the way in which a sample of individuals is selected from the population for study. Each of these components will be discussed in turn: the joint genotypic distribution of mating individuals, the relationship between phenotype and genotype, the mode of inheritance, and the sampling of data. In this way it will be possible to build up a general model for segregation analysis, following the papers by Elston and Stewart (1971) and Elston and Yelverton (1975).

Joint Genotypic Distribution of Mating Individuals

Random Mating, Equilibrium Conditions

When there is random mating, the joint genotypic distribution of mating individuals is simply the product of the genotypic distributions for males and females in the parental population. If, furthermore, there is no selection or mutation, and the population has reached equilibrium, these genotypic distributions depend only on the gene frequencies at the various loci.

In general we can let ψ_t be the probability that an individual in the parental population has genotype t. For two alleles at an autosomal locus, A and a say, with gene frequencies q_1 and $1 - q_1$, respectively, the genotypic distribution under random mating is given by the Hardy–Weinberg proportions $\psi_{AA} = q_1^2$, $\psi_{Aa} = 2q_1(1 - q_1)$, and $\psi_{aa} = (1 - q_1)^2$. If we consider, in addition, two alleles at a second autosomal locus, B and b say, with gene frequencies q_2 and $1 - q_2$ respectively, the genotypic distribution at equilibrium is obtained by multiplying together the probabilities for the genotypes at the individual loci and is given as follows:

Genotype, t	Probability ψ_t
AA BB	$q_1^2 q_2^2$
AA Bb	$2q_1^2 q_2(1 - q_2)$
AA bb	$q_1^2(1 - q_2)^2$
Aa BB	$2q_1(1 - q_1)q_2^2$
Aa Bb	$4q_1(1 - q_1)q_2(1 - q_2)$
Aa bb	$2q_1(1 - q_1)(1 - q_2)^2$
aa BB	$(1 - q_1)^2 q_2^2$
aa Bb	$2(1 - q_1)^2 q_2(1 - q_2)$
aa bb	$(1 - q_1)^2(1 - q_2)^2$
	1

This same procedure can be extended to obtain the genotypic distribution for any number of autosomal loci, but the number of genotypes, and hence the number of different probabilities that need to be considered, soon becomes prohibitively large. However, if we replace the genotype by the total number of "capital letters" it contains, the total being over all loci, for a large number of loci this number will tend to be normally distributed in the population. This is illustrated in Fig. 1, where the distribution of this number for two alleles at each of three autosomal loci, all gene frequencies being equal to one-half, is depicted. This gives the basis for the usual additive polygenic model in which it is assumed that there is an underlying polygenic genotype, which I shall call the *polygenotype* of an individual, that is normally distributed in the population. Without loss of generality, the polygenotype can be assumed to have a mean of zero, and its variance will be denoted σ_G^2. Thus, using the notation

$$\phi(x, \sigma^2) = \frac{1}{\sqrt{2\pi}\sigma} \exp\left[-\frac{1}{2}\left(\frac{x}{\sigma}\right)^2 \right] \tag{1}$$

and letting G denote polygenotype, the distribution of polygenotype in the population is given by $\psi_G = \phi(G, \sigma_G^2)$.

Using the symbols s and t to denote genotypes at one or a few loci

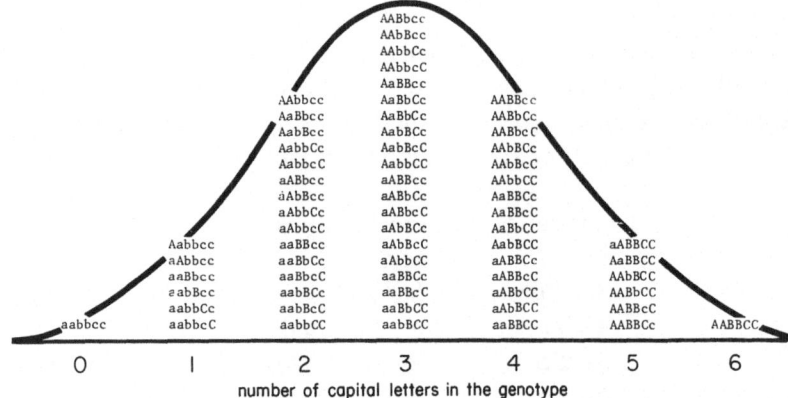

Fig. 1. Population genotypic distribution when the genotype is classified by the number of "capital letters" it contains. The histogram gives the distribution for two alleles at each of three independent loci, the smooth curve gives the distribution in the limit as the number of loci tends to infinity. (Adapted from Roberts, 1977.)

(*monogenotype* or *oligogenotype*), the distribution of the mating type $s \times t$ under random mating is $\psi_s \psi_t$, and this is easily expressible in terms of gene frequencies; using the symbols F and G to denote polygenotypes, the distribution of the mating type $F \times G$ is $\psi_F \psi_G = \phi(F, \sigma_G^2)\phi(G, \sigma_G^2)$; and if the total genotype contains both components, oligogenotype and polygenotype, the distribution of the mating type $sF \times tG$ is $\psi_s \phi(F, \sigma_G^2)\psi_t \phi(G, \sigma_G^2)$.

The distributions given above are appropriate for females whether or not X-linked loci are involved. For males, however, a modification is necessary to allow for X-linked loci. For example, denoting the male hemizygote genotypes at each locus $A\cdot$, $a\cdot$, $B\cdot$, and $b\cdot$, the male genotypic distribution at equilibrium is:

Genotype, t	Probability ψ_t
$A\cdot B\cdot$	$q_1 q_2$
$A\cdot b\cdot$	$q_1(1 - q_2)$
$a\cdot B\cdot$	$(1 - q_1)q_2$
$a\cdot b\cdot$	$\dfrac{(1 - q_1)(1 - q_2)}{1}$

It should be noted that when X-linked loci are involved the number of genotypes is different for males and females. Because of this, the limits of the summations occurring in expressions derived later (section on likelihood method) may depend upon the sex corresponding to the genotypes being summed; this is indicated explicitly by Elston and Yelverton (1975), but, for ease of notation, will remain implicit here.

Assortative Mating

Nonrandom mating, which is termed positive assortative mating if there is an increased tendency for matings between individuals with like phenotypes and negative assortative mating if there is an increased tendency for matings between unlike phenotypes, is usually ignored in segregation analysis. Nevertheless, it deserves brief mention.

In the case of polygenotypes, we can assume, as did Fisher (1918), that the distribution of mating types $F \times G$ is bivariate normal in the population, thus introducing just a single extra parameter—the marital polygenotypic correlation. For oligogenotypes, we can define $\psi_{t \mid s}$ as the probability that an individual has genotype t given that his or her spouse has genotype s, so that the distribution of mating types $s \times t$ is given by $\psi_s \psi_{t \mid s}$. Thus the only difference between this expression and the corresponding one for random mating is that $\psi_{t \mid s}$ replaces ψ_t, and this substitution can be made in all the expressions derived later for random mating. Unfortunately, however, there remain two problems: what is a reasonable way to model dependence of t on s, and what distribution of genotypes does this lead to at equilibrium? It is perhaps less necessary to solve the second problem, since the population that we study is not necessarily at equilibrium. Various one-locus models that have been investigated theoretically, but never used in segregation analysis, are given in Lange (1976), Karlin and Scudo (1969), Scudo and Karlin (1969), and Wilson (1976).

Relationship between Phenotype and Genotype

Once we know the genotypic distribution, the relationship between phenotype and genotype can be completely described by specifying, separately for each genotype, the phenotypic distribution. Let the phenotype that is being investigated be z, assuming there is a single observation on each individual, and denote the distribution of z conditional on genotype t $g_t(z)$. If the genotype is polygenic, G, we write the conditional distribution $g_G(z)$; and if both oligogenotype and polygenotype are involved we have

$g_{tG}(z)$. Classical segregation analysis was limited to qualitative phenotypes—dichotomies such as presence or absence of a disease, or polychotomies such as the patterns of agglutination possible with a set of standard antisera. Recently, however, segregation analysis has been extended to allow for quantitative traits such as the concentration of a particular component in serum. We now discuss how the distributions $g_t(z)$, $g_G(z)$, and $g_{tG}(z)$ may be mathematically specified to build up a segregation analysis model. In each case one or more of the various parameters may be sex-specific, but this point will not be belabored.

Dichotomous Phenotype

The simplest genetic model for a dichotomy, affected ($z = 1$) versus unaffected ($z = 0$), is that of two alleles at one autosomal locus. The phenotypic distributions are then completely specified by the three "penetrances" $g_{AA}(1)$, $g_{Aa}(1)$, and $g_{aa}(1)$, since $g_{AA}(0) = 1 - g_{AA}(1)$, $g_{Aa}(0) = 1 - g_{Aa}(1)$, and $g_{aa}(0) = 1 - g_{aa}(1)$. If the allele A is dominant to a we have $g_{AA}(1) = g_{Aa}(1)$; a dominant disorder with complete penetrance is equivalent to these quantities being unity, with the absence of sporadic cases corresponding to $g_{aa}(1) = 0$. Similarly, if the allele a causes a completely penetrant recessive disorder we have $g_{aa}(1) = 1$, with absence of sporadic cases corresponding to $g_{AA}(1) = g_{Aa}(1) = 0$. For an X-linked locus it is necessary to define further $g_{A\cdot}(z)$ and $g_{a\cdot}(z)$ for males; it is common to assume $g_{A\cdot}(z) = g_{AA}(z)$ and $g_{a\cdot}(z) = g_{aa}(z)$.

We do not in general wish to assume complete penetrance or absence of sporadic cases, so the assumption of a dominance relationship for two alleles at an autosomal locus reduces the number of penetrances that need to be estimated from three to two. Whether the disorder is dominant or recessive, the mathematical model is essentially the same, since one gives rise to the other by a permutation of the labels "affected" and "unaffected"; the two models are said to be images of the same phenogram (Cotterman, 1953; Hartl and Maruyama, 1968). Another two-allele one-locus phenogram is possible by assuming AA and aa, but not Aa, have the same phenotypic distribution; however, this is unlikely to occur in practice.

When we consider two-locus models many more phenograms are possible, and this may be a reason why two-locus models have rarely been used in segregation analysis. If we consider just two alleles at each of two autosomal loci, there are nine distinct genotypes (ignoring phase of linkage). The nine corresponding penetrances can be reduced to two if we assume there are basically only two kinds of individuals—those with high penetrance for the disorder and those with low penetrance. However, for just two different penetrances there are fifty phenograms (Hartl and Maruyama, 1968) possible, i.e., there are 50 genetically different ways of partitioning the nine genotypes into two sets, and in each case there are two ways of assigning the high and low penetrances to the two sets. Defrise-Gussenhoven (1962) suggested that five of these phenograms deserve special attention, and Elston and Namboodiri (1977) added a sixth to her list. These six phenograms are shown in Fig. 2.

To simplify the discussion, assume that penetrance is complete and there are no sporadic cases. Then if C is the disorder and C^* the absence of disorder, the phenogram in Fig. 2a represents a disorder that is dependent on the presence of two dominant genes (which can be symbolically represented as $Dom \cap Dom$); but if the roles of C and C^* are interchanged, the phenogram represents a disorder that can be caused by either of two recessive genes ($Rec \cup Rec$). Figure 2b represents a disorder that is due to a double recessive ($Rec \cap Rec$) if C is the disorder; and one that can be caused by either of two dominant genes ($Dom \cup Dom$) if C^* is the disorder. Figure 2c represents a disorder that is dependent on the presence of both a dominant gene and a recessive gene ($Dom \cap Rec$) if

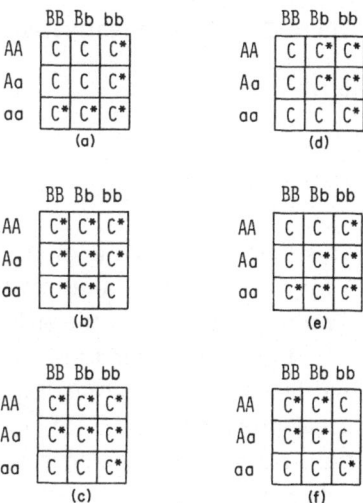

Fig. 2. Six phenograms for two phenotypes determined by two alleles at each of two loci. In each of the phenograms (a)–(f) there are nine genotypes, represented by the nine cells in each array, and the phenotype corresponding to each genotype is denoted by C or C^* in the corresponding cell. (From Elston and Namboodiri, 1977.)

C is the disorder, and conversely one that can be caused by either a dominant or a recessive gene (*Dom* ∪ *Rec*) if C^* is the disorder. It should be noted that these three phenograms begin to model what is often called "genetic heterogeneity," i.e., more than one genetic cause, when C^* is the disorder. The phenogram shown in Fig. 2d represents a disorder that is basically due to segregation at a single locus (B, b), but the genotype at a second locus (A, a) determines which allele, B or b, is dominant. In Fig. 2e, whether an individual is affected depends only on the number of capital letters in the genotype (3 or more if C is the disorder, 2 or less if C^* is the disorder), and so this phenogram is a two-locus additive "polygenic" model; it should be noted that the phenogram depicted in Fig. 2b also has this property. Lastly, the phenogram shown in Fig. 2f is a special one that has been put forward, with a simple biological explanation, to explain the genetics of handedness (Levy and Nagylaki, 1972).

When we turn to a normally distributed polygenotype, G, it would similarly be possible to assume only two penetrances, a high one and a low one, but then it would also be necessary to specify (or estimate) the value of G at which the change occurs (corresponding in the two-locus case to a choice between Fig. 2a and 2e, and whether C or C^* is the disorder). The usual model, however, is to assume that the penetrance is an increasing function of G. In particular, this function is taken to be of the form

$$g_G(1) = \Phi[(G - \theta)/\sigma_E] \tag{2}$$

where Φ is the cumulative normal function defined by

$$\Phi[(G - \theta)/\sigma] = \int_{-\infty}^{(G-\theta)/\sigma} \frac{1}{\sqrt{2\pi}} \exp\left(-\frac{1}{2}u^2\right) du$$

$$= \int_{-\infty}^{G} \frac{1}{\sqrt{2\pi}\sigma} \exp\left[-\frac{1}{2}\left(\frac{x - \theta}{\sigma}\right)^2\right] dx \tag{3}$$

This function, which is often called a risk function, is illustrated in Fig. 3 for two different values of θ: θ_1 and θ_2; the other parameter, σ_E, determines how steeply the risk rises as a function of G.

It must be understood that the form of this risk function is arbitrary, and has no biological justification. It does, however, correspond to the usual threshold model for the polygenic inheritance of a dichotomous trait. Assume the existence of a normally distributed random variable L, called liability; it is made up of two uncorrelated components: the po-

Fig. 3. Pictorial representation of two cumulative normal risk functions in relation to an underlying normally distributed polygenotype, G. (a) Distribution of G, $\psi_G = \phi(G, \sigma_G^2)$. (b) Risk function $\Phi[(G - \theta_1)/\sigma_E]$. (c) Risk function $\Phi[(G - \theta_2)/\sigma_E]$. The risk function is the probability that an individual with polygenotype G is affected; this probability may depend upon the genotype at a major locus or upon the severity of the disorder. (From Elston and Rao, 1978.)

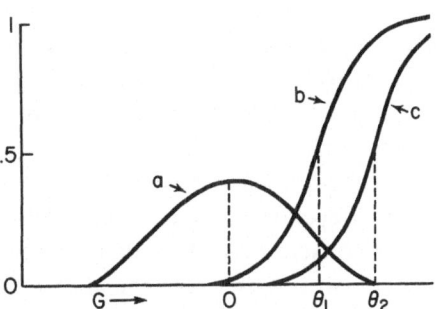

lygenotype G, normally distributed with variance σ_G^2, and an environmental component E, also assumed to be normally distributed but with variance σ_E^2. Let an individual be affected if and only if his liability is greater than θ, a parameter called the threshold. This model is illustrated in Fig. 4; the two models illustrated in Figs. 3 and 4, although seemingly different, are mathematically identical.

Since the trait we are considering is merely a dichotomous classification, and not a quantitative measure, it is impossible to give unique values separately to σ_G^2 and σ_E^2; it is only their *relative* magnitudes that are relevant. Thus, without loss of generality, some authors take $\sigma_G^2 = 1$ (e.g., Curnow, 1974), while others take $\sigma_G^2 + \sigma_E^2 = 1$ (e.g., Falconer, 1965); in the latter case σ_G^2 is the heritability of the liability to the disorder (Elston, 1973).

Morton and MacLean (1974) extended the threshold model to allow for a genotype that has both monogenic and polygenic components. Denoting the genotype tG, the "penetrance" becomes

$$g_{tG}(1) = \Phi[(G - \theta_t)/\sigma_E] \tag{4}$$

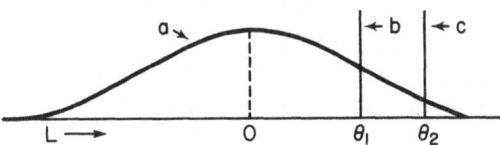

Fig. 4. Pictorial representation of the threshold approach to the polygenic model for a qualitative trait. (a) Distribution of liability L, $\phi(L, \sigma_G^2 + \sigma_E^2)$. (b) Threshold θ_1. (c) Threshold θ_2. Individuals whose total liability is greater than the threshold are affected; the threshold may depend upon the genotype at a major locus or upon the severity of the disorder. (From Elston and Rao, 1978.)

analogous to equation (2). In this model differences at the major genotype locus are reflected as shifts of the risk function (Fig. 3) or the threshold (Fig. 4) to the right or left. For two alleles at an autosomal locus there are three thresholds, θ_{AA}, θ_{Aa}, and θ_{aa}, two of which are identical if dominance is assumed.

Variable Age of Onset. Just as the phenotypic distributions may be sex-dependent, so, if we are considering a disorder that has a variable age of onset, may they depend upon the age at examination of each individual. Let a' be the age at examination of an individual, transformed (if necessary) to a scale on which the age of onset distribution for individuals with genotype t is normally distributed with mean μ_t and variance σ_E^2. Then we can write the age-specific penetrance, the probability that an individual with genotype t is affected by age a', as

$$g_t(1 \mid a') = \gamma_t \Phi[(a' - \mu_t)/\sigma_E] \tag{5}$$

where γ_t is the *susceptibility* of genotype t to the disorder: the probability that an individual with genotype t will eventually become affected *if he lives long enough*. This is a general model that allows either γ or μ to be genotype-specific; in practice only one or the other would be, because otherwise the two parameters tend to be confounded. For a polygenotype we can take the penetrance at $a' = 0$ (i.e., at birth if a' is a power of age) to be given by equation (2), and more generally at any other age of examination to be

$$g_G(1 \mid a') = \Phi[(a' + G - \theta)/\sigma_E] \tag{6}$$

and for the case of both monogenotype and polygenotype we can take

$$g_{tG}(1 \mid a') = \gamma_t \Phi[(a' + G - \theta_t)/\sigma_E] \tag{7}$$

If the ages of onset for affected individuals are known, we can consider them as part of the phenotype; then, conditional on examination at age a', the joint distributions of affection status and age of onset, a (measured on the same scale as a'), can be modeled as follows:

for an oligogenotype:

$$\left\{ \begin{array}{ll} g_t(0 \mid a') = 1 - \gamma_t \Phi[(a' - \mu_t)/\sigma_E] & \\ g_t(1, a \mid a') = \gamma_t \phi[a - \mu_t, \sigma_E^2], & a \leq a' \end{array} \right. \tag{8}$$

for a polygenotype:

$$\begin{cases} g_G(0 \mid a') = 1 - \Phi[(a' + G - \theta)/\sigma_E] \\ g_G(1, a \mid a') = \phi[a + G - \theta, \sigma_E^2], \qquad a \le a' \end{cases} \tag{9}$$

and for both components:

$$\begin{cases} g_{tG}(0 \mid a') = 1 - \gamma_t \Phi[(a' + G - \theta_t)/\sigma_E] \\ g_{tG}(1, a \mid a') = \gamma_t \phi[a + G - \theta_t, \sigma_E^2], \qquad a \le a' \end{cases} \tag{10}$$

One can, of course, use distributions other than the normal distribution to model age of onset; the important thing to note is that by making the distributions conditional on age at examination (which can be different for each individual) we eliminate the bias that is otherwise caused by the fact that ages of onset are never observed after the age at examination.

Polychotomous Phenotype

It is a simple matter to generalize all the above phenotypic distributions to allow for a polychotomy, and this will be demonstrated with particular reference to a trichotomy. As before we can let $z = 0$ for unaffected individuals, but now we have $z = 1$ or 2 for affected individuals, representing two different forms of the disorder; in particular we can let $z = 1$ represent the less severe or "intermediate" form while $z = 2$ represents the fully affected form.

For two alleles at one autosomal locus the distributions are completely specified by the six penetrances $g_t(j)$, $t = AA$, Aa or aa, and $j = 1$ or 2. By an appropriate choice of six constants θ_{jt} we can equate

$$g_t(2) = \Phi(-\theta_{2t})$$

and

$$g_t(1) = \Phi(-\theta_{1t}) - \Phi(-\theta_{2t}) \tag{11}$$

This, together with the assumption that there exist five constants μ_{AA}, μ_{Aa}, μ_{aa}, θ_1, and θ_2 such that

$$\theta_{jt} = \theta_j - \mu_t \tag{12}$$

forms the basis of the monogenic trichotomy model proposed by Reich *et al.* (1972). This assumption can be rationalized by supposing the existence of three normal liability distributions, one for each genotype: in each case the variability around the mean is environmentally caused, and the liability distribution for genotype t has mean μ_t. Furthermore, two thresholds are assumed, θ_1 and θ_2, such that individuals with liabilities above θ_2 are affected, while individuals with liabilities between θ_1 and θ_2 are intermediate. This model is illustrated in Fig. 5, and leads to penetrances satisfying (11) and (12). [It can be noted in passing that the goodness-of-fit test for a monogenic hypothesis proposed by Reich *et al.* (1972) is as much a test of whether equation (12) is true as it is of a monogenic mechanism.]

When we turn to two loci, each with two alleles segregating, there are 439 genetically different ways of partitioning the nine genotypes into three phenotypic sets (Hartl and Maruyama, 1968); as yet there have been no suggestions put forward as to which of these deserve most attention.

For a pure polygenic model we can generalize equation (2) to obtain the phenotypic distribution

$$g_G(2) = \Phi[(G - \theta_2)/\sigma_E]$$

$$g_G(1) = \Phi[(G - \theta_1)/\sigma_E] - \Phi[(G - \theta_2)/\sigma_E] \qquad (13)$$

$$g_G(0) = 1 - \Phi[(G - \theta_1)/\sigma_E]$$

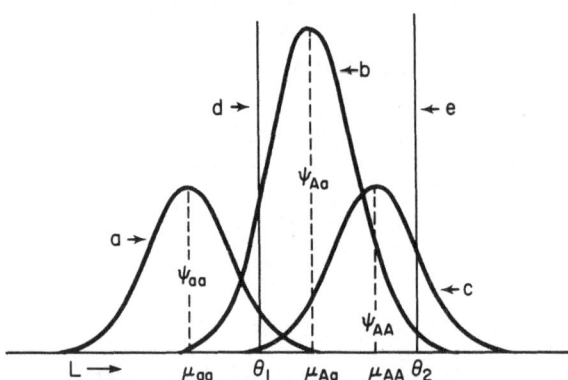

Fig. 5. Pictorial representation of a monogenic trichotomy model. There are three component liability distributions: (a) $\phi(L - \mu_{aa}, \sigma_E^2)$, for *aa* individuals, (b) $\phi(L - \mu_{Aa}, \sigma_E^2)$, for *Aa* individuals, (c) $\phi(L - \mu_{AA}, \sigma_E^2)$, for *AA* individuals; (d) threshold θ_1, below which individuals are unaffected; (e) threshold θ_2, above which individuals are fully affected. Individuals whose liabilities fall between θ_1 and θ_2 have an intermediate form of the disorder. (Adapted from Reich *et al.*, 1972.)

as illustrated in Figs. 3 and 4. This formulation assumes that the two different forms of the disorder are different degrees of the same process (Reich *et al.*, 1979). This can be further generalized into the mixed model with both monogenic and polygenic components by making θ_1 and θ_2 monogenotype-specific, i.e., by substituting θ_{1t} and θ_{2t} for them respectively, as follows:

$$g_{tG}(2) = \Phi[(G - \theta_{2t})/\sigma_E]$$

$$g_{tG}(1) = \Phi[(G - \theta_{1t})/\sigma_E] - \Phi[(G - \theta_{2t})/\sigma_E] \qquad (14)$$

$$g_{tG}(0) = 1 - \Phi[(G - \theta_{1t})/\sigma_E]$$

If we assume in addition that equation (12) holds, we arrive at the mixed model phenotypic distributions proposed by Morton and MacLean (1974) for a trichotomous trait.

Quantitative Phenotype

For a quantitative trait it is usual to assume that the distribution of z conditional on genotype is normal. Both the mean and variance of this normal distribution could in principle depend on genotype, but in order not to have too many unknown parameters, in practice the variance is taken to be the same, σ_E^2 for all genotypes. Thus for an oligogenotype we take $g_t(z) = \phi(\mu_t - z, \sigma_E^2)$, and the resulting distribution of z in the population is

$$\sum_t \psi_t \phi(\mu_t - z, \sigma_E^2) \qquad (15)$$

as depicted in Fig. 6 for the special case of two alleles at one autosomal locus. For a polygenotype we take $g_G(z) = \phi(G + \mu - z, \sigma_E^2)$, so that, analogous to (15), the distribution of z in the population is [recalling $\psi_G = \phi(G, \sigma_G^2)$ and replacing summation by integration]

$$\int_{-\infty}^{\infty} \phi(G, \sigma_G^2)\phi(G + \mu - z, \sigma_E^2)dG = \phi(\mu - z, \sigma_G^2 + \sigma_E^2) \qquad (16)$$

i.e., normal with mean μ and variance $\sigma_G^2 + \sigma_E^2$. In this situation, however, unlike the case of qualitative phenotypes, both σ_G^2 and σ_E^2 are separately estimable. The extension to the case of mixed oligogenotype and polygenotype is, as proposed by Elston and Stewart (1971), $g_{tG}(z) = \phi(G + \mu_t - z, \sigma_E^2)$; the population distribution is then again a mixture

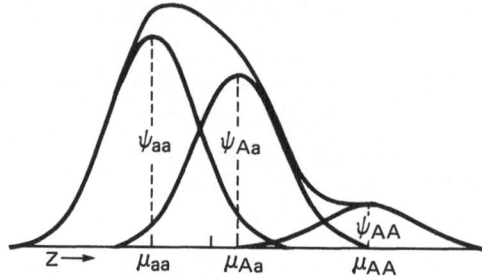

Fig. 6. Population distribution of a quantitative trait dependent on the segregation of two alleles at an autosomal locus, assuming a normal distribution for each monogenotype. The variance within each component distribution is σ_E^2 for the monogenic model, and $\sigma_G^2 + \sigma_E^2$ for the mixed monogenic and polygenic model. (Adapted from Morton and MacLean, 1974.)

of normal distributions, such as is shown in Fig. 6, but now the variance within each distribution is $\sigma_G^2 + \sigma_E^2$ rather than just σ_E^2.

All the parameters in these distributions can be made to depend on the effects of age, sex, or other measurable factors such as may be environmentally determined, but usually the effects on the means μ and μ_t are more important than the effects on the variance. If one is willing to assume that these effects are independent of genotype, approximate allowance can often be made for them by subjecting the data to initial analyses to find appropriate overall adjustments. The original phenotypes are then adjusted to a "common" age, sex, and environment prior to undergoing segregation analysis.

The assumption of normality is a critical one in one of the methods proposed for the segregation analysis of quantitative traits, and so transformation of the data may be advisable. MacLean *et al.* (1975) have suggested use of the transformation

$$z = \frac{r}{p}\left[\left(\frac{x}{r} + 1\right)^p - 1\right] \qquad (17)$$

where x is the original observation standardized to have zero mean and unit variance, r is an arbitrarily chosen scale parameter (e.g., $r = 5$) that ensures every value of $x/r + 1$ in the sample is positive, and p, which they call the "skewness parameter" is to be estimated—either on the assumption that z follows a normal distribution in the sample, or on the assumption that z follows a mixture of normal distributions. Setting $p = 1$ corresponds to no transformation ($z = x$); values of p less than unity tend to normalize positively skewed distributions, while values of p greater than unity tend to normalize negatively skewed distributions. The main advantage of this particular transformation is that it keeps the mean and variance of z approximately constant over a range of values of p. Apart from this, the same purpose can be achieved by the simpler transformation proposed by Box and Cox (1964) and given by

$$z = (y^p - 1)/p \tag{18}$$

where y is the original observation (assumed to be always greater than zero). With this transformation z is normally distributed if and only if y^p is normally distributed; and $p = 0$ corresponds to y being lognormally distributed, since $\lim_{p \to 0} (y^p - 1)/p = \ln y$.

Mode of Inheritance

The mode of inheritance from one generation to the next can be summarized mathematically by the genotypic distributions of the offspring conditional on the two parental genotypes. Let p_{stu} be the probability that an individual has genotype u given that his parents' genotypes are s and t. These conditional probabilities p_{stu} can be viewed as the elements of a three-dimensional stochastic matrix called the genetic transition matrix. Alternatively, the genetic transition matrix can be considered as a two-dimensional matrix whose elements are probability distributions. This will now be explained for the various hereditary mechanisms, in each case assuming two alleles per locus. The extension to multiple alleles is given by Elston and Stewart (1971).

Oligogenic Inheritance

Table I gives the genetic transition matrix for segregation at one autosomal locus. We order the genotypes $1 = AA$, $2 = Aa$, and $3 = aa$, so that the s-th row corresponds to genotype s for one of the parents and the t-th column corresponds to genotype t for the other parent. The (s, t)-th entry in the matrix is the vector $[p_{st1} \ p_{st2} \ p_{st3}]$, i.e., the genotypic distribution of offspring from the mating $s \times t$. Now define the transmission probability τ_t as the probability that an individual with genotype t transmits A to his offspring, so that $1 - \tau_t$ is the probability that he transmits a. Then the entries in Table I can be generated quite simply

TABLE I. Genetic Transition Matrix for Two Alleles at One Autosomal Locus[a]

		t	
s	$1 = AA$	$2 = Aa$	$3 = aa$
$1 = AA$	$[1\ 0\ 0]$	$[\frac{1}{2}\ \frac{1}{2}\ 0]$	$[0\ 1\ 0]$
$2 = Aa$	$[\frac{1}{2}\ \frac{1}{2}\ 0]$	$[\frac{1}{4}\ \frac{1}{2}\ \frac{1}{4}]$	$[0\ \frac{1}{2}\ \frac{1}{2}]$
$3 = aa$	$[0\ 1\ 0]$	$[0\ \frac{1}{2}\ \frac{1}{2}]$	$[0\ 0\ 1]$

[a] Each entry is a genotypic distribution $[p_{st1}\ p_{st2}\ p_{st3}]$; from Elston and Stewart (1971).

from

$$[p_{st1}\ p_{st2}\ p_{st3}] = [\tau_s \tau_t\ \tau_s(1 - \tau_t) + \tau_t(1 - \tau_s)\ (1 - \tau_s)(1 - \tau_t)] \quad (19)$$

substituting $\tau_{AA} = 1$, $\tau_{Aa} = \frac{1}{2}$, and $\tau_{aa} = 0$, which are the appropriate Mendelian values if there is no mutation or meiotic drive.

For an X-linked locus there are two genetic transition matrices, one for female children and one for male children. These are shown in Table II, where it is assumed that s is the male parent's genotype and t the female parent's genotype. If s is the female parent and t the male parent, then the rows and columns of the matrix must be transposed. The entries for female children can again be generated by equation (19), provided we substitute $\tau_{A\cdot} = 1$ and $\tau_{a\cdot} = 0$. For male children the entries are generated by

$$[p_{st1}\ p_{st2}] = [\tau_u\ 1 - \tau_u] \quad (20)$$

where u is s or t, whichever is the genotype of the female parent.

TABLE II. Genetic Transition Matrices for Two Alleles at One X-Linked Locus[a]

			t	
	s	$1 = AA$	$2 = Aa$	$3 = aa$
Female children	$1 = A\cdot$	$[1\ 0\ 0]$	$[\frac{1}{2}\ \frac{1}{2}\ 0]$	$[0\ 1\ 0]$
	$2 = a\cdot$	$[0\ 1\ 0]$	$[0\ \frac{1}{2}\ \frac{1}{2}]$	$[0\ 0\ 1]$
Male children	$1 = A\cdot$	$[1\ 0]$	$[\frac{1}{2}\ \frac{1}{2}]$	$[0\ 1]$
	$2 = a\cdot$	$[1\ 0]$	$[\frac{1}{2}\ \frac{1}{2}]$	$[0\ 1]$

[a] For female children each entry is a distribution for three possible genotypes; for male children each entry is a distribution for two possible genotypes. From Elston and Stewart (1971).

The extension to several unlinked loci follows in a straightforward manner from Tables I and II, since independent assortment corresponds to the multiplication of probabilities. Suppose we have two autosomal loci; the genetic transition matrix for the first is given by Table I, and that for the second by a similar matrix but with the genotypes labeled *BB*, *Bb*, and *bb*. Corresponding to any parental genotypes at the two loci we have an offspring genotypic distribution in each vector, and multiplying these together gives us the offspring genotypic distribution we want. If, for example, $s = AaBb$ and $t = AABb$, the one-locus genotypic offspring distributions are $[\frac{1}{2} \ \frac{1}{2} \ 0]$ for [*AA Aa aa*] and $[\frac{1}{4} \ \frac{1}{2} \ \frac{1}{4}]$ for [*BB Bb bb*].

Thus in this case the distribution p_{stu} is given by

$$[\tfrac{1}{8} \quad \tfrac{1}{4} \quad \tfrac{1}{8} \quad \tfrac{1}{8} \quad \tfrac{1}{4} \quad \tfrac{1}{8} \quad 0 \quad 0 \quad 0]$$

for

[*AABB AABb AAbb AaBB AaBb Aabb aaBB aaBb aabb*]

Elston and Stewart (1971) indicate how this can be represented using a Kronecker product notation, and also give the generalization for linked loci.

Polygenic and Mixed Inheritance

In the usual additive polygenic model an individual's polygenotype is the sum of the gametic values he inherits, one from each parent. The population of gametic values transmitted by any polygenotype, G, is normally distributed with mean $G/2$ and variance $\sigma_G^2/2$, and any two such gametic values produced by G have correlation $\frac{1}{2}$. It follows that under random mating the offspring genotypic distribution conditional on the parental polygenotypes F and G is normally distributed with mean $(F + G)/2$ and variance $\sigma_G^2/2$; i.e., letting H be the offspring polygenotype, we have

$$p_{FGH} = \phi(H - (F + G)/2, \ \sigma_G^2/2) \tag{21}$$

If there is assortative mating with marital polygenotypic correlation ρ, this becomes

$$p_{FGH} = \phi(H - (F + G)/2, \ (1 - \rho)\sigma_G^2/2) \tag{22}$$

The distribution for a mixed model containing both oligogenic and polygenic components is obtained by multiplying together the two corresponding distributions, just as the distribution for two independent loci is obtained by multiplying together the corresponding one-locus distributions; i.e., we can write

$$p_{sF\,tG\,uH} = p_{stu}\,p_{FGH} \qquad (23)$$

Sampling Scheme

Random Sampling

Random sampling implies the existence of a well-defined population of distinct sampling units, each of which has equal probability of being sampled. When the sampling unit is a sibship this kind of sampling is, in principle at least, feasible. All the persons living in a specified geographic area define a population of distinct sibships (including sibships of single individuals). The fact that some individuals in this population may have sibs residing outside the specified geographic area is not critical, as we can define our population of sibships as being comprised of individuals living within the specified area only. We can also define sibships such that half-sibs are automatically reckoned as being in different sibships. The important point is that we can define our sampling unit in such a way that every individual in the population belongs to exactly one sampling unit. When this is so, standard sampling methodology can be used to obtain a random sample of sampling units.

But suppose we try to make our sampling unit a larger structure, for example, the nuclear family (parents and offspring). We immediately run into the difficulty that many individuals belong to two such units, since they have both offspring and parents. In order to have a population of distinct nuclear families, and hence make random sampling possible, it becomes necessary to exclude from consideration many of the individuals living in the specified area. There will also be some individuals who have neither offspring nor parents, and so belong to "degenerate" nuclear families. These difficulties become further magnified when an attempt is made to have even larger (pedigree) structures as the sampling unit.

Fortunately, however, it is not necessary to have random sampling in the strict sense in order to make valid statistical inferences, so long as certain conditions hold. Suppose we take a random sample of individuals from the population and call these individuals "random probands." We then augment the sample of random probands by including also some of their relatives. Whether or not some of the original random probands are related to each other, we can sort the final sample out into unrelated family structures. It is then possible to treat this sample of unrelated family structures as though it is a random sample from the population, and this will lead to valid results provided the decision to include the extra relatives, and the number of such extra relatives included for each random proband, are independent of the proband's phenotype. (Strictly speaking they should also be independent of the random proband's genotype, but it is difficult to conceive of sampling independently of phenotype but not of genotype.) The simplest way to ensure that these conditions hold is to decide beforehand just what types of relatives (e.g., all first degree relatives, all first and second degree relatives) will be included in the sample, provided they exist in the population. If this is done the only possible sources of bias are the following: (1) The existence of certain types of relatives in the population may depend on the random proband's phenotype; (2) the willingness of individuals to be sampled, whether probands or relatives, may depend on their phenotypes. The first source of bias may occur because of selection, differential death rates after the age of reproduction, or differential migration from the population. If we are studying a phenotype that predisposes to early death, for example, existence of the proband's parents may well be dependent on the proband's phenotype if this is in any way genetically determined; and this may occur whether or not there is genetic selection. Biases due to differential migration can be eliminated by including in the sample those relatives who have migrated out of the population, but this may introduce other biases if these relatives have had their phenotypes altered by reason of their having lived in a different environment.

In summary, we shall use the term "random sampling" to denote those situations in which the probands are a random sample from the population and these potential biases do not exist; and we assume as a model for the actual sampling scheme random sampling of family units from a large population of similar discrete unrelated family units.

Nonrandom Sampling

Random sampling is not a very efficient way to study the genetics of rare conditions, since it leads to a sample that is largely devoid of individuals with the condition and hence largely uninformative. Typically families are selected for study because at least one member of the family has the condition, but the actual sampling scheme used is often poorly defined. Here we shall present some of the models that have been suggested to describe the kind of nonrandom sampling, or ascertainment via probands, that commonly occurs.

Ascertainment was originally considered in terms of a dichotomous trait segregating in sibships (e.g., by Fisher, 1934; Bailey, 1951; Morton 1959), using the model illustrated in Fig. 7. In this model black circles floating on the water represent affected individuals, while white circles that sink down in the water represent unaffected individuals; individuals in the same sibships are linked together. The white circles are invisible to the sampler, who picks black circles (probands) at random, each with equal probability π; and when a black circle is picked, all the circles linked to it are also drawn into the sample. This sampling procedure can be described mathematically by defining a function $\pi(z)$ which specifies the sampling probability that an individual with phenotype z in the population becomes a proband, i.e.,

$$\pi(z) = \begin{cases} \pi \text{ if } z = 1 \\ 0 \text{ if } z = 0 \end{cases} \tag{24}$$

In the limit as π tends to 0, the probability that a sibship enters the sample is proportional to the number of affected members it contains; this will be called single ascertainment. When $\pi = 1$, the probability that a sibship enters the sample is independent of the number of affected members it contains, provided it contains at least one affected member; this will be

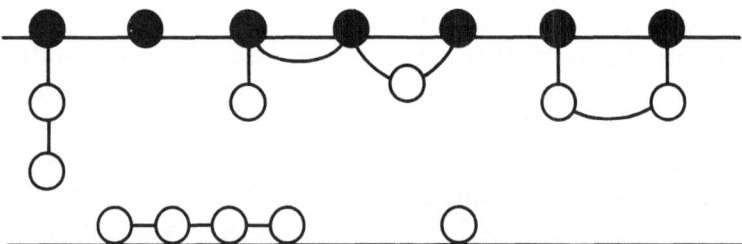

Fig. 7. Model for ascertainment of families via affected probands. Black circles represent affected individuals; white circles represent unaffected individuals. Only the floating black circles are visible to the sampler; when a black circle is sampled, so also are all the other circles linked to it. (From Batschelet, 1963.)

called complete ascertainment. The situation where π takes on any other value between 0 and 1 is called multiple ascertainment.

The function $\pi(z)$ can be generalized to polychotomous and quantitative traits. For example probands may be defined as being above some threshold, T, for a quantitative trait, so that (24) becomes

$$\pi(z) = \begin{cases} \pi \text{ if } z \geq T \\ 0 \text{ if } z < T \end{cases} \quad (25)$$

Alternatively, the probability that an individual becomes a proband may be some continuous function of phenotype; Elston and Yelverton (1975) have suggested the function

$$\pi(z) = \min[\exp(K_0 + K_1 z + K_2 z^2), 1] \quad (26)$$

This is the same as $\pi(z) = \exp(K_0 + K_1 z + K_2 z^2)$ if $K_2 < 0$ and $K_1^2 \leq 4K_0 K_2$. Similarly π might depend on both affectation status and age of onset, a, for a disorder with variable age of onset; in such a case we could model the dependency

$$\pi(z, a) = \begin{cases} \min[\exp(K_0 + K_1 a + K_2 a^2), 1] & \text{if } z = 1 \\ 0 & \text{if } z = 0 \end{cases} \quad (27)$$

In theory this same ascertainment model could be used for sampling larger pedigree structures. We could imagine Fig. 7 to include all the individuals living in a well-defined geographic area, with all related individuals linked together. However, this implies that once a proband is drawn, *all* his relatives living in the specified area, however remote the relationship, are included in the sample; otherwise (e.g., if we take his first and second degree relatives only) we have the same problem as before, namely that the population is not made up of discrete sampling units. Since it is usually impractical to sample all of a proband's relatives—and, furthermore, this would tend to defeat the original purpose of nonrandom sampling—Elston and Sobel (1979) have used the same principle as suggested above for random sampling: the family structures finally sampled for study are assumed to be taken from a similar conceptual population of discrete unrelated family units. However, it is necessary to note that the final sample will often contain individuals who, whatever their phenotype, could never be probands. For example, these could be individuals who live outside the specified geographic area, or who, when probands are chosen from the records of a particular hospital, would go to a different hospital for treatment if affected; or a sampling scheme may be used whereby only individuals of a certain age or sex could be probands. Elston and Sobel's (1979) model assumes that each individual in the sample can be unequivocally classified as to whether he could or could not be a proband, and they discuss possible ways of achieving this. Other ascertainment models have been recently discussed (Stene, 1977; Cannings and Thompson, 1977; Thompson and Cannings, 1980) but, for those cases where these models would yield theoretically different results, it is not yet clear that they have any practical advantages.

THE LIKELIHOOD METHOD

Given a mathematical basis for the four components of a segregation analysis model, an expression can be derived for the likelihood of a set of observed data: the probability (or, for continuous data, the density) of the actual observations being associated with the sample, under the assumed model. The likelihood can be thought of as summarizing the information the sample contains with respect to that model. In this section the appropriate likelihood is first derived for randomly sampled data, and then a modification is shown to be necessary when sampling is via probands of a particular phenotype. Since the likelihoods are derived in a general way, the same basic expressions can allow for all the genetic mechanisms that have been discussed so far. In the last part of this section the principles underlying the use of these likelihoods for segregation analysis are discussed.

Likelihood of Randomly Sampled Data

Nuclear Families and Sibships

First consider a nuclear family of two parents and n children, with k possible genotypes affecting the phenotype. Denote the phenotypes of the father and mother z_f and z_m, respectively, and those of their offspring $z_j (j = 1, 2, \ldots , n)$. The likelihood that the j-th offspring has phenotype z_j, conditional on his having genotype u, is $g_u(z_j)$; and the same likelihood, but conditional on his parents having genotypes s and t, is $\sum_{u=1}^{k} p_{stu} g_u(z_j)$. Now assume that, conditional on their own respective genotypes, the phenotypes of the offspring are independent of one another; then the likelihood for the whole sibship, given the parents' genotypes are s and t, is

$$\prod_{j=1}^{n} \sum_{u=1}^{k} p_{stu} g_u(z_j) \tag{28}$$

It is to be understood that the summation over u in this expression is performed separately for each offspring, and the sums are then multiplied together. From now on symbols such as Π_j and Σ_u will be written, since the limits will be clear.

Under random mating the likelihood for the two parents can be written as

$$\sum_s \psi_s g_s(z_f) \sum_t \psi_t g_t(z_m) \tag{29}$$

which is the sum of k^2 terms, each term corresponding to a particular mating type $s \times t$. Multiplying each term in this sum by the likelihood for the sibship conditional on s and t, i.e., expression (28), we arrive at the likelihood for the whole nuclear family:

$$\sum_s \psi_s g_s(z_f) \sum_t \psi_t g_t(z_m) \prod_j \sum_u p_{stu} g_u(z_j) \tag{30}$$

This expression assumes that, conditional on their genotypes, the phenotypes of all the family members are mutually independent; in other words, it assumes that there are no environmentally caused correlations among the phenotypes.

The likelihood under a completely additive polygenic model is the same, except that ψ_s and ψ_t are replaced by ψ_F and ψ_G, p_{stu} is replaced by p_{FGH}, and integration replaces summation. Thus, using the results given earlier, expressions (29) and (30) become, respectively.

$$\int_{-\infty}^{\infty} \phi(F, \sigma_G^2) g_F(z_f) \int_{-\infty}^{\infty} \phi(G, \sigma_G^2) g_G(z_m) dG \, dF \tag{31}$$

and

$$\int_{-\infty}^{\infty} \phi(F, \sigma_G^2) g_F(z_f) \int_{-\infty}^{\infty} \phi(G, \sigma_G^2) g_G(z_m)$$
$$\times \prod_j \int_{-\infty}^{\infty} \phi[H - (F + G)/2, \sigma_G^2/2] g_H(z_j) dH, \, dG, \, dF \tag{32}$$

In this expression there is a separate integration over H for each offspring, just as in expression (30) there is a separate summation over u for each offspring.

Similarly, if the model contains both oligogenic and polygenic components, we can combine expressions (29) and (31) to obtain, for the two parents,

$$\sum_s \psi_s \int_{-\infty}^{\infty} \phi(F, \sigma_G^2) g_{sF}(z_f) \sum_t \psi_t \int_{-\infty}^{\infty} \phi(G, \sigma_G^2) g_{tG}(z_m) \, dG \, dF \quad (33)$$

and we can combine expressions (30) and (32) to obtain, for the whole nuclear family,

$$\sum_s \psi_s \int_{-\infty}^{\infty} \phi(F, \sigma_G^2) g_{sF}(z_f) \sum_t \psi_t \int_{-\infty}^{\infty} \phi(G, \sigma_G^2) g_{tG}(z_m)$$

$$\times \prod_j \sum_u p_{stu} \int_{-\infty}^{\infty} \phi[H - (F + G)/2, \sigma_G^2/2] g_{uH}(z_j) dH \, dG \, dF \quad (34)$$

All these likelihoods assume that, conditional on their genotypes, the phenotypes of the individuals in the family are mutually independent. Morton and MacLean (1974) introduced a model in which, in addition to both monogenic and polygenic inheritance, there is an environmental component, C, common to all members of the same sibship; furthermore, C is assumed to be normally distributed among sibships with mean 0 and variance σ_C^2. The likelihood for the parents under this model is still (33); but that for the whole nuclear family is slightly different from (34), the factor in the second line being modified:

$$\sum_s \psi_s \int_{-\infty}^{\infty} \phi(F, \sigma_G^2) g_{sF}(z_f) \sum_t \psi_t \int_{-\infty}^{\infty} \phi(G, \sigma_G^2) g_{tG}(z_m)$$

$$\times \int_{-\infty}^{\infty} \phi(C, \sigma_C^2) \prod_j \sum_u p_{stu} \int_{-\infty}^{\infty} \phi[H - (F + G)/2, \sigma_G^2/2] \quad (35)$$

$$\times g_{uHC}(z_j) dH \, dC \, dG \, dF$$

where $g_{uHC}(z_j)$ is the probability density function of z_j conditional on oligogenotype u, polygenotype H, and common environment C. Under this model the total environmental variance σ_E^2 is partitioned into σ_C^2 and a residual σ_R^2. Thus if, for a dichotomous trait, $g_{tG}(1) = \phi[(G - \theta_t)/\sigma_E]$ for a parent, then for an offspring

$$g_{uHC}(z) = \Phi[(C + H - \theta_u)/\sigma_R] \quad (36)$$

Similarly for a quantitative trait, corresponding to $g_{tG}(z) = \phi(G + \mu_t - z, \sigma_E^2)$ for a parent, we have for an offspring

$$g_{uHC}(z) = \phi(C + H + \mu_u - z, \sigma_R^2)$$

In all of the above cases, the likelihood of a set of independent nuclear families is simply the product of the likelihoods for each of the separate families. If, however, members of two different nuclear families are in any way related, these two families should be considered as part of a single pedigree structure, including if necessary linking individuals whose phenotypes are unknown: all such individuals must be included in the likelihood that will now be derived, with $g(z)$ for them being set equal to unity wherever they occur.

Simple Pedigrees

So far an individual's phenotype has been denoted z. It will now be convenient, following Elston and Stewart (1971), to use two different symbols for phenotype, corresponding to the two different kinds of individuals who can appear in a simple pedigree—a pedigree which contains no loops and starts with a single set of original parents. Persons who are related to someone in a previous generation of the pedigree will have their phenotype denoted x, and unrelated persons "marrying into" the pedigree will have their phenotypes denoted y. In the case of the original parents, however, one will be arbitrarily denoted x and the other y.

It is also convenient to use subscripts on subscripts to denote the generation, starting with 0 for the original generation. Let the phenotypes of the original parents of the i_0-th pedigree be x_{i_0} and y_{i_0}; let the phenotype of their i_1-th child be $x_{i_0 i_1}$, and let his or her spouse's phenotype be $y_{i_0 i_1}$; similarly let the phenotype of the i_2-th child of this i_1-th child be $x_{i_0 i_1 i_2}$, and that of his or her spouse $y_{i_0 i_1 i_2}$, and so on. This notation is illustrated in Fig. 8.

If we now rewrite expression (30) in this new notation for the i_0-th set of parents, at the same time using subscripts s_0 and t_0 to indicate genotypes in the original generation, and s_1 (replacing u) for genotypes in the offspring generation, we obtain:

$$\sum_{s_0} \psi_{s_0} g_{s_0}(x_{i_0}) \sum_{t_0} \psi_{t_0} g_{t_0}(y_{i_0}) \prod_{i_1} \sum_{s_1} p_{s_0 t_0 s_1} g_{s_1}(x_{i_0 i_1})$$

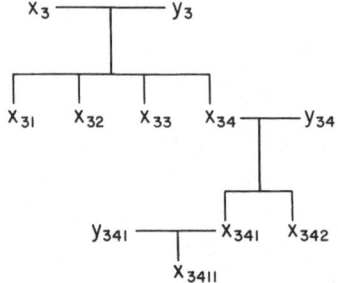

Fig. 8. Illustration of the notation for the phenotypes of the members of a pedigree: this is the third pedigree ($i_0 = 3$) in a set of pedigrees. (From Elston and Rao, 1978.)

Thus the joint likelihood of a set of independent nuclear families can be written as the product of this expression over i_0. Furthermore, under random mating the likelihood that the spouse of the i_1-th child should have phenotype $y_{i_0 i_1}$ is independent of this, and can be written

$$\sum_{t_1} \psi_{t_1} g_{t_1}(y_{i_0 i_1}) \tag{37}$$

Thus the joint likelihood for the original spouses, the children and the spouses of these children, is

$$\prod_{i_0} \psi_{s_0} g_{s_0}(x_{i_0}) \sum_{t_0} \psi_{t_0} g_{t_0}(y_{i_0})$$
$$\times \prod_{i_1} \sum_{s_1} p_{s_0 t_0 s_1} g_{s_1}(x_{i_0 i_1}) \sum_{t_1} \psi_{t_1} g_{t_1}(y_{i_0 i_1}) \tag{38}$$

Now under our model the phenotypes of the next generation depend on the previous generations only through their genotypes; and conditional on s_1 and t_1, the likelihood of the set of offspring phenotypes $x_{i_0 i_1 i_2}$ is

$$\prod_{i_2} \sum_{s_2} p_{s_1 t_1 s_2} g_{s_2}(x_{i_0 i_1 i_2}) \tag{39}$$

Thus the likelihood for a multigenerational pedigree can be expressed by (38) followed by an alternating sequence similar to (39) and (37), the subscripts on subscripts increasing by one for each new generation. Thus if we define the operator

$$\Gamma_j = \prod_{i_j} \sum_{s_j} p_{s_{j-1} t_{j-1} s_j} g_{s_j}(x_{i_0 i_1 \ldots i_j}) \sum_{t_j} \psi_{t_j} g_{t_j}(y_{i_0 i_1 \ldots i_j}) \tag{40}$$

the likelihood for a set of simple pedigrees of any number of generations can be written as the sequence of operation $\Gamma_0(\Gamma_1(\Gamma_2(\Gamma_3 \ldots .)))$, provided, corresponding to the beginning of expression (38), we define

$$p_{s_{j-1}t_{j-1}s_j} = \psi_{s_0}$$

when $j = 0$.

In the same manner, we can define Γ_j for the polygenic model as

$$\prod_{ij} \int_{F_j} \phi[F_j - (F_{j-1} + G_{j-1})/2, \sigma_G^2/2] g_{F_j}(x_{i_0 i_1 ... i_j})$$

$$\times \int_{G_j} \phi(G_j, \sigma_G^2) g_{G_j}(y_{i_1 i_2 ... i_j})$$

(41)

where the symbol $\int_F$ indicates integration of everything following it with respect to F from minus infinity to plus infinity; then the likelihood of a pedigree under the polygenic model is similarly given by the sequence of operations Γ_j, provided we now replace $\phi[F_j - (F_{j-1} + G_{j-1})/2, \sigma_G^2/2]$ by $\phi(F_0, \sigma_G^2)$ when $j = 0$.

The likelihood under a model that contains both oligogenic and polygenic components is obtained by defining Γ_j to be a combination of expressions (40) and (41), analogous to the way in which expression (34) is a combination of expressions (30) and (32); and the modification to allow for an environmental correlation within sibships can similarly be incorporated, analogous to expression (35).

Further Extensions

The approach followed here to derive the likelihood of a simple pedigree can be extended to allow for twins and half-sibships (Elston and Yelverton, 1975); it can also be extended, for an oligogenic model, to allow for arbitrary pedigree structures (Lange and Elston, 1975; Cannings *et al.*, 1978). An entirely different approach was used by Lange *et al.* (1976*b*) to derive the likelihood of an arbitrary pedigree under the polygenic model when the phenotype is normally distributed: the pedigree phenotypes are considered as a vector observation from a multivariate normal distribution. Ott (1979) has derived the likelihood of a nuclear family under the polygenic model from the same point of view, and has shown how, when thus derived, it is theoretically a simple matter to add to the likelihood a monogenic component; exactly the same method could be used to derive a theoretical expression for the mixed-model likelihood for an arbitrary pedigree, but in practice such a formulation is too cumbersome to compute for all but the smallest pedigrees.

Likelihood When Sampling via Selected Probands

Conditioning on the Parental Phenotypes

If our sample consists of independent nuclear families, each of which is ascertained via one or more probands with a particular phenotypic condition, and if in every case the proband is a parent, there is a relatively simple way of constructing a likelihood that will yield valid results when using the general maximum likelihood procedures that will be described later. All one need do is to take, instead of one of the likelihoods just derived, the likelihood for the phenotypes of the sibship conditional on the phenotypes of the two parents. For a single family this conditional likelihood is equal to expression (30) divided by expression (29) for an oligogenic model; to expression (32) divided by expression (31) for a polygenic model; and to expression (34) or (35) divided by expression (33) for mixed oligogenic and polygenic models, according to whether a sibling environmental correlation is assumed absent or present. Furthermore, if the data and the model are such that each parent can only have one genotype and the parental mating type is the same for all families in the sample, then there is only one term in the numerator of each such quotient, and the denominator is a constant that can be factored out. This is the basis for many of the early segregation analyses of data on sibships ascertained through the parents, in which it is commonly assumed that the parental genotypes are known.

Conditioning on the Sampling Model

Although there may be easier ways to obtain a likelihood that will give rise to valid results, the most informative likelihood will, in general, be the one that gives the probability (or density) of the sampled data conditional on the exact manner in which that sampling took place. This can be derived for any pedigree, if we assume the sampling model proposed by Elston and Sobel (1979), as follows.

Let L (pedigree | ≥ 1 proband) be the likelihood of the pedigree conditional on it containing at least one proband among those individuals in it who could be probands: this is thus the likelihood we need. Then we have

L(pedigree | ≥ 1 proband)

$$= L(\text{pedigree}) \cdot \frac{L(>1 \text{ proband} \mid \text{pedigree})}{L(\geq 1 \text{ proband})} \qquad (42)$$

where L(pedigree) is the likelihood of the pedigree phenotypes, assuming that the pedigree has been randomly sampled from a population of pedigrees of the same size and structure; $L(\geq 1$ proband | pedigree) is the likelihood that the pedigree, given the phenotypes of the individuals who could be probands, contains at least one proband; and $L(\geq 1$ proband) is the likelihood that an arbitrary pedigree of the same size and structure contains at least one proband among those whose particular positions are occupied by individuals who could be probands. We have already derived expressions for L(pedigree) under various models, so we now consider the correction factor by which this is multiplied on the right side of equation (42).

Suppose there are n individuals in the pedigree who could be probands, with phenotypes denoted $z_i (i = 1, 2, \ldots, n)$, and m individuals who could not, with phenotypes denoted $z_j^* (j = 1, 2, \ldots, m)$. Let the proband status of the i-th individual who could be a proband be b_i: $b_i = 1$ if that individual is a proband, $b_i = 0$ if not. Furthermore, define an ascertainment function $\alpha(z, b)$ for each individual who could be a proband as follows: $\alpha(z, 0) = 1 - \pi(z)$, $\alpha(z, 1) = \pi(z)$. Then (assuming independent ascertainments) the joint likelihood of all the proband statuses observed, conditional on the phenotypes in the pedigree, is

$$L(\text{proband statuses} \mid \text{pedigree}) = \prod_{i=1}^{n} \alpha(z_i, b_i) \qquad (43)$$

and hence the numerator of the correction factor is

$$L(\geq 1 \text{ proband} \mid \text{pedigree}) = 1 - \prod_{i=1}^{n} \alpha(z_i, 0) = 1 - \prod_{i=1}^{n} [1 - \pi(z_i)] \qquad (44)$$

The denominator is obtained by summing the whole numerator in equation (42) over all possible phenotypes for each individual, i.e., using (44),

$L(\geq 1 \text{ proband})$

$$= \sum_{z_1} \cdots \sum_{z_n} \sum_{z_1^*} \cdots \sum_{z_m^*} L(\text{pedigree})L(\geq 1 \text{ proband} \mid \text{pedigree}) \quad (45)$$

$$= 1 - \sum_{z_1} \cdots \sum_{z_n} \sum_{z_1^*} \cdots \sum_{z_m^*} L(\text{pedigree}) \prod_{i=1}^{n} [1 - \pi(z_i)]$$

(If z is continuous, integration replaces summation.) This expression can be put into a form that is very similar to that previously derived for L(pedigree), by changing the order of the summations. Looking at (40), for example, we see that the phenotype of each individual enters the likelihood in a single expression of the form $\Sigma_u p_{stu} g_u(z)$ or $\Sigma_t \psi_t g_t(z)$. Now

$$\sum_z \sum_u p_{stu} g_u(z)[1 - \pi(z)] = \sum_u p_{stu}[1 - \sum_z g_u(z)\pi(z)]$$

$$\sum_z \sum_t \psi_t g_t(z)[1 - \pi(z)] = \sum_t \psi_t[1 - \sum_z g_t(z)\pi(z)]$$

$$\sum_{z^*} \sum_u p_{stu} g_u(z^*) = \sum_u p_{stu} \quad (46)$$

and

$$\sum_{z^*} \sum_t \psi_t g_t(z^*) = \sum_t \psi_t$$

It follows that the negative of the second term of (45), i.e.,

$$\sum_{z_1} \cdots \sum_{z_n} \sum_{z_1^*} \cdots \sum_{z_m^*} L(\text{pedigree}) \prod_{i=1}^{n} [1 - \pi(z_i)]$$

can be calculated in the same way that L(pedigree) is calculated, but with the substitution of $\zeta_t = 1 - \Sigma_z g_t(z)\pi(z)$ for $g_t(z)$ in the case of an individual who could be a proband, and the substitution of unity for $g_t(z^*)$ in the case of an individual who could not be a proband. Elston and Yelverton (1975) derive, in an appendix, explicit expressions for ζ_t appropriate for various functions $g_t(z)$ and $\pi(z)$, both discrete and continuous.

Substitution of expressions (44) and (45) into (42) yields basically the same likelihood as has been commonly used, for the simple case of sibships ascertained through affected children, when the parental mating type is assumed known. For the case where the parental mating type is not known, Elandt-Johnson (1971) and Morton and MacLean (1974) have proposed that the likelihood should be further conditioned on the parental phenotypes; but this can be expected to lead to a loss of information (Go *et al.*, 1978). Prior to Elston and Sobel (1979), however, it appears never to have been suggested that account should be taken of which individuals could, and which individuals could not, be probands. Use of the likelihood (42) assumes that $\pi(z)$ (the probability that an individual with phenotype z becomes a proband) is known for all values of z, whereas in practice this is rarely the case. Usually, however, we do know (or may be prepared to assume as known) the form of $\pi(z)$ as a function of one or more unknown parameters [e.g., as in (24), (25), (26), or (27)]. These unknown parameters should thus be estimated, preferably jointly with the other parameters of the model. The appropriate likelihood for this purpose is the joint likelihood of observing the phenotypes of the pedigree *and the proband statuses,* conditional on the pedigree containing at least one proband; this is the same as (42) except that (43) replaces (44) in the numerator.

When there is complete ascertainment, i.e., $\pi(z)$ is always either 0 or 1, there is no difficulty with this formulation; in fact the calculation of (42) is simplified. When there is single ascertainment, however, i.e., $\pi(z)$ is zero or tends to zero for all values of z, this formulation of the ascertainment correction factor leads to the indefinite form 0/0. Elston and Sobel (1979) derive a general expression for the limiting form of the ascertainment correction in this situation; in the case of a dichotomy and the particular sampling function (24), their result is essentially the same as that of Thompson and Cannings (1979), derived by a different line of reasoning: the appropriate correction consists of dividing L(pedigree) by the probability that $z = 1$ for a random individual, expressed as a function of the parameters in the model.

Parameter Estimation and Testing Hypotheses

General Principles

There are different statistical methods of estimating parameters and testing hypotheses, but the method most commonly used in segregation analysis is that based on maximum likelihood. Maximum likelihood estimates of parameters are those values of the parameters that make the likelihood a maximum. They may not be unique, nor may they even exist for permissible values of the parameters. A variance, for example, must be positive; but the likelihood may be maximized by a negative value of a variance.

Provided they occur at a mathematical maximum of the likelihood, all maximum likelihood estimates enjoy the following properties asymptotically, i.e., as the size of the sample increases without bound. First, they are unbiased—the mean of many such estimates is equal to the true value of the parameter being estimated. Second, they are efficient—the variance of many such estimates is the smallest possible for any asymptotically unbiased estimate. Third, the estimates are normally distributed, and this fact can be used to obtain approximate confidence limits for the estimates.

To test a null hypothesis under a particular model we can use the likelihood ratio criterion, which is the ratio of the maximum value of the likelihood under the null hypothesis to the maximum value of the likelihood under the model. Each null hypothesis corresponds to one or more restrictions being placed on the model, and hence leads to a smaller maximum likelihood. The smaller the likelihood ratio, the less likely is the null hypothesis to be true; the test thus consists in rejecting the null hypothesis, at a specified significance level, if this ratio is smaller than a certain quantity. Asymptotically, if the null hypothesis is true, minus twice the natural logarithm of the likelihood ratio is distributed as a χ^2, the number of degrees of freedom being the difference in the number of independent parameters over which the likelihood is maximized in the denominator and numerator, respectively; i.e., the number of degrees of freedom is equal to the number of independent restrictions corresponding to the particular null hypothesis. The null hypothesis is rejected if this statistic is greater than the appropriate tabulated value of χ^2.

Maximizing the likelihood, with or without restrictions, is a computational problem that can be tackled numerically in various ways (see the Appendix). It is important to note that the likelihood surface may have several local maxima, and there is no general way of knowing how many such maxima exist. In practice, it is often necessary to search for more than one local maximum whenever a particular hypothesis is tested. When, after appropriate hypothesis testing, a decision is arrived at regarding the most appropriate genetic mechanism, the final parameter estimates quoted should be those obtained on assuming that particular mechanism as the underlying model.

Testing Genetic Hypotheses

If we assume one of the general genetic models discussed above—oligogenic, polygenic, or mixed oligogenic and polygenic—we can use these principles and the corresponding likelihoods to test various relevant hypotheses. Under a monogenic model of two alleles at one autosomal locus, for example, the hypothesis of Hardy–Weinberg equilibrium proportions, i.e., the hypothesis that the genotypic distribution is given by $\psi_{AA} = q^2$, $\psi_{Aa} = 2q(1 - q)$, and $\psi_{aa} = (1 - q)^2$, corresponds to the restriction

$$\psi_{Aa} = 2\sqrt{\psi_{AA}\psi_{aa}} \tag{47}$$

which can be tested via a χ^2 statistic with one degree of freedom.

The hypothesis of dominance corresponds to one of the two restrictions

$$g_{Aa}(z) = g_{AA}(z), \quad \text{or} \quad g_{Aa}(z) = g_{aa}(z) \tag{48}$$

which can also be tested via a χ^2 statistic with one degree of freedom if $g_{Aa}(z)$ differs from $g_{AA}(z)$ or $g_{aa}(z)$ with respect to only one parameter. Additivity corresponds to the restriction

$$g_{Aa}(z) = [g_{AA}(z) + g_{aa}(z)]/2 \tag{49}$$

In the case of two alleles at an X-linked locus, we have the extra genotypic frequencies $\psi_{A\cdot}$ and $\psi_{a\cdot} = 1 - \psi_{A\cdot}$, as well as the extra parameters involved in $g_{A\cdot}(z)$ and $g_{a\cdot}(z)$. We may wish to test the equilibrium hypothesis that the gene frequencies are the same in the two sexes, i.e.,

$$\psi_{A\cdot} = \psi_{AA} + \psi_{Aa}/2 \tag{50}$$

and we can test that the corresponding hemizygote and homozygote phenotypic distributions are equal, i.e.,

$$g_{A\cdot}(z) = g_{AA}(z), \qquad g_{a\cdot}(z) = g_{aa}(z) \tag{51}$$

The mixed model of monogenic and polygenic inheritance is a particularly interesting one, since under it one can attempt to test "what is possibly the single most important question concerning the inheritance of any character: is most of the genetic variation due to one locus, or are many gene loci necessarily involved?" (Elston and Stewart, 1971). Thus, using the likelihood (34) or (35), and assuming just three major genotypes, absence of segregation at a major locus corresponds to either of the set of restrictions

$$g_{AA\ G}(z) = g_{Aa\ G}(z) = g_{aa\ G}(z), \quad \text{or} \quad \psi_{AA} = \psi_{Aa} = 0 \qquad (52)$$

and absence of polygenic inheritance corresponds to the restriction

$$\sigma_G^2 = 0 \qquad (53)$$

If a sibling environmental correlation is incorporated into the model [using (35)], absence of such a correlation corresponds to the restriction

$$\sigma_C^2 = 0 \qquad (54)$$

Unfortunately, the application of such a mixed model to large pedigrees is not yet computationally feasible. There is, however, a way of generalizing to pedigrees the classical tests for Mendelian segregation in sibships. The usual test for a rare autosomal dominant gene is to consider the segregation ratio in sibships where one parent is affected, assuming the parental mating type is $Aa \times aa$; thus the Mendelian null hypothesis that is tested is $p_{Aa\ aa\ aa} = p_{Aa\ aa\ Aa} = \frac{1}{2}$. Similarly, for a rare autosomal recessive, assuming the parental mating type is $Aa \times Aa$, the null hypothesis that is tested as $p_{Aa\ Aa\ aa} = \frac{1}{4}$. In terms of transmission probabilities both these null hypotheses correspond to the restriction $\tau_{Aa} = \frac{1}{2}$. Thus Elston and Stewart (1971) suggest that the likelihood under a two-allele one-locus autosomal model be parametrized in terms of the three transmission probabilities, as given in equation (19), and Mendelian segregation be tested by the null hypothesis

$$\left.\begin{array}{c} \tau_{AA} = 1 \\ \tau_{Aa} = \frac{1}{2} \\ \tau_{aa} = 0 \end{array}\right\} \qquad (55)$$

against the unrestricted model in which the transmission probabilities can take on any values between zero and unity. In this model the symbols AA, Aa, and aa refer to *types* of individuals, rather than to genotypes: two individuals are of different types if they and/or their offspring come from different phenotypic distributions [Cannings *et al.* (1978) have used the term "ousiotypes" to express the same notion.] Under the genetic hypothesis (55), the types are genotypes. However, the model also includes the possibility that the offspring types are distributed independently of their parents' types, i.e., a hypothesis that corresponds to the 'two independent restrictions

$$\tau_{AA} = \tau_{Aa} = \tau_{aa} \tag{56}$$

It should be noted that, denoting this common transmission probability τ, under this "environmental" hypothesis it follows from (19) that the distribution of types among offspring (x individuals) is $[\tau^2 \ 2\tau(1 - \tau) \ (1 - \tau)^2]$ for [$AA \ Aa \ aa$]. However, this apparently unnatural restriction of having "Hardy–Weinberg" proportions under an environmental hypothesis is not a real restriction, if it can be assumed that the phenotypic distribution of Aa individuals is the same as that of AA or aa individuals [i.e., if (48) is assumed as part of the mode].

In the case of a two-allele X-linked locus, we have the further parameters $\tau_{a\cdot}$ and $\tau_{A\cdot}$, as given in equation (20), and the hypothesis of Mendelian segregation corresponds to the restrictions (55) together with

$$\left.\begin{array}{l} \tau_{A\cdot} = 1 \\ \tau_{a\cdot} = 0 \end{array}\right\} \tag{57}$$

We can also test the "environmental" hypothesis by (56) jointly with

$$\tau_{A\cdot} = \tau_{a\cdot} \tag{58}$$

on a total of three degrees of freedom, allowing the distribution of types to be different between the two sexes.

Power, Efficiency, and Robustness

The method of maximum likelihood has intuitive appeal, as well as theoretical justification asymptotically. All results based on it, however, are model-dependent; they are valid only if the assumed model does, in fact, allow as a possibility, at least approximately, the actual mechanism underlying the variability in the data. If the model is wrong, the result of any test may be meaningless: the test may be either invalid or powerless. Furthermore, there are inherent difficulties in interpreting analyses of nonexperimental data (Kempthorne, 1978). For these reasons caution is necessary in drawing conclusions from segregation analysis. Two recent simulation studies have done much to clarify some of the critical properties of the segregation analysis models that have been described above. The essential findings of these studies are now briefly summarized.

MacLean *et al.* (1975) considered models, as applied to data on nuclear families, in which any segregation is constrained to be Mendelian; and all their results are based on the use of the likelihood of the children's phenotypes conditional on their parents' phenotypes. As might be expected, the information content of a sample with respect to major locus parameters increases with the number of children per family; it is also vastly greater if quantitative data are available than if the same data are reduced to a dichotomy.

MacLean *et al.* (1975) verified that a wrong model yields inconsistent estimates and invalid results. In particular, if data are simulated to have a polygenic component but analyzed under a model that does not contain such a component, the variability due to a major locus is overestimated. A model which includes Mendelian segregation at a major locus as the only cause of familial resemblance will necessarily assign familial correlations to the major locus.

If the model includes a component for polygenic heritability but not one for sibling environmental correlation, then any data in which the sib–sib correlation is larger than the parent–offspring correlation will be interpreted as evidence of segregation at a major locus with dominance; in fact, under such a model the probability of asserting the presence of a spurious major locus approaches unity as the sibling environmental correlation increases. All these difficulties are effectively overcome by analyzing family data under a model that contains all three components: segregation at a major locus, polygenic heritability, and sibling environmental correlation. Furthermore, such a model is robust against assortative mating: although random mating and Hardy–Weinberg proportions were assumed in the likelihood formulation, even 100% genetic correlation between mates had only a small effect on the results.

However, this model is not robust in the face of departures from the assumed phenotypic distributions. If it is assumed that, conditional on genotype, the phenotype is normally distributed, then under such a model any departure from overall normality will necessarily be interpreted as due to a major locus. MacLean *et al.* (1975) specifically examine the effect of skewness, and interpret the "segregation" that it causes as being due to intrafamily variation being different in the two tails of the distribution. But, since segregation at a major locus is the only way such a model can accommodate nonnormality of any kind, nonnormality *in the absence of skewness* can also be expected to lead this model to detect a spurious major locus. If, on the other hand, the model contains a major locus segregation component parametrized in terms of transmission probabilities, as in equation (19), non-Mendelian values of these transmission probabilities allow the overall distribution of the data to take the form of a mixture of normal distributions in the absence of Mendelian segregation.

Go *et al.* (1978) demonstrated that the use of conditional likelihoods leads to less efficient parameter estimates. Morton and MacLean (1974), for example, assume the sampling units to be nuclear families, but take the appropriate likelihood to be that of the phenotypes of the offspring conditional on the phenotypes of the parents: this results in less efficient estimates. Morton and Rao (1979) have mistakenly alluded to such a statement as being an inference that the *partition of pedigrees into nuclear families* results in less-efficient estimates, correctly pointing out that the simulation studies conducted by Go *et al.* (1978) were not relevant to this question. Earlier (Morton and Rao, 1978), they expressed the opinion that partitioning pedigrees into nuclear families can lead to an exact, although less powerful, analysis.

Using the unconditional likelihood of nuclear families, Go *et al.* (1978) examined the robustness of a model that contains no polygenic or sibling environmental component, but does include the three transmission probabilities as unknown parameters. They assumed that the following three criteria must be met in order to infer the presence of a major locus: (1) the overall distribution of the data must fit a mixture of normal distributions significantly better than a single normal distribution; (2) there must be no significant departure from the Mendelian null hypothesis (55); and (3) there must be significant departure from the "environmental" null hypothesis (56). Using these criteria, the presence of neither polygenic heritability nor environmentally caused skewness or platykurtosis, with or without sibling environmental correlation, leads to the false assertion that a major locus is present. Criterion (1) effectively provides robustness against polygenic inheritance, and criteria (2) and (3) protect against the effects of nonnormality. Under the conditions studied it was only when polygenic inheritance, nonnormality, and a moderate amount of sibling environmental correlation were all present in the data together that a serious possibility of detecting a spurious major locus was found. It is quite possible, in certain types of data, for any of the criteria not to be satisfied when a major locus is segregating [see Elston *et al.* (1978) regarding criterion (3)]; but when all three are met there is strong evidence, though not proof, that a major locus is segregating.

Morton and Rao (1979) have criticized transmission probabilities as being biologically meaningless except in the Mendelian case, implying that classical segregation analysis (e.g., Morton, 1969), which, as we have seen, is based on testing the null hypothesis $\tau_{Aa} = \frac{1}{2}$, is similarly at fault. They state that "the generalized single locus model . . . rests on the assumption that non-Mendelian transmission frequencies (which are biologically meaningless) can detect and identify other phenomena not included in the model"; thus the model "shows no advantage over clinical impression, which with no greater unreliability but much less computation is also capable of 'suggesting' an untested major locus." In fact, statements such as these can be made with equal truth (or falsity) of all models for segregation analysis, though Morton and Rao (1979) do not point this out. They believe that "segregation analysis under a mixed model is capable of demonstrating a major locus," even though admitting that "it is easy to devise examples where cultural inheritance mimics Mendelian transmission." Such an example has been given by Wilson (1974). The truth is that no amount of statistical analysis of family data can prove the existence of a major locus; but, as we shall see in the examples that follow, the various methods of segregation analysis discussed above can materially influence one's confidence in the likelihood of a locus segregating in a given set of data.

EXAMPLES OF SEGREGATION ANALYSIS

In this section four recent examples of segregation analysis are summarized to illustrate the use of some of the segregation analysis models described above. All four examples concern the detection of segregation at an autosomal locus, without the complication of variable age of onset; examples of the use of two-locus models and X-linked models, and alternative approaches to the analysis of diseases with variable age of onset, can be found elsewhere (e.g., Spence *et al.*, 1974; Elandt-Johnson, 1973; Elston and Namboodiri, 1977). The first example concerns the inheritance of a quantitative trait, serum dopamine-β-hydroxylase activity, in a set of families each ascertained via a proband; the data are analyzed both as a dichotomy, using classical methods of segregation analysis, and as a quantitative trait. In the second example, another quantitative trait, serum cholesterol level, is investigated in a single large pedigree; in this pedigree, segregation analysis strongly suggests that much of the variability of this trait is due to monogenic inheritance, a finding that has more recently been confirmed by linkage analysis (Elston *et al.*, 1976). The third example reexamines the inheritance of the ability to taste phenylthiocarbamide (PTC); it shows that a single gene mode of inheritance is confirmed, with no evidence for residual polygenic variation or sibling environmental correlation. Lastly, we examine the inheritance of immunoglobulin E levels by a variety of the newer methods, and find conflicting results that are still in need of resolution.

Segregation of Dopamine-β-hydroxylase (DBH) Activity in Selected Families

On the basis of population studies, Weinshilboum *et al.* (1975) hypothesized that subjects whose serum DBH activity is less than 50 units (as defined in their paper) make up a separate subgroup of a control population. They present data on 22 families, each ascertained through a child between the ages of 6 and 12 with DBH activity less than 50 units. In 16 of these families both parents had DBH activity greater than 50

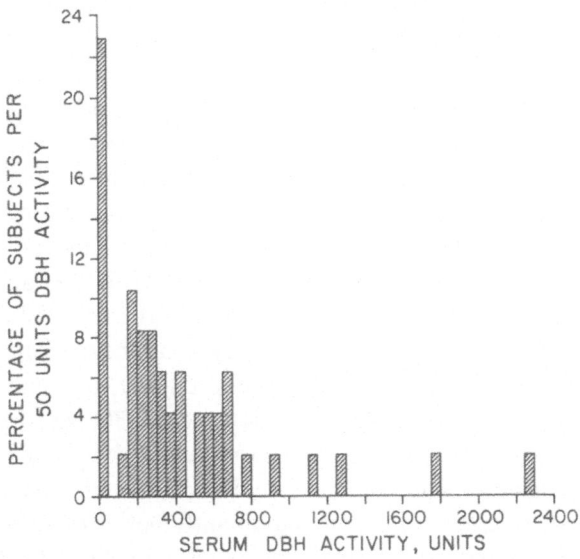

Fig. 9. Histogram of serum DBH activities in successive 50-unit increments for 48 siblings of probands with very low serum DBH activities. (From Weinshilboum *et al.*, 1975.)

units; a histogram of the DBH activities in the sibs of the probands in these families is presented in Fig. 9. Although these sibs can be divided into two groups, those with activities less than 50 units and those with activities more than 50 units, it is not at all clear from the histogram that such an underlying dichotomy represents the basic features of the distribution.

Weinshilboum et al. (1975) considered the dichotomy "low DBH" ($z = 1$) vs. "high DBH" ($z = 0$) using the 50-unit cut-off, and used classical methods of analysis to determine if the segregation in these 16 sibships is compatible with low DBH being due to a recessive autosomal locus. In these methods it is assumed that all the parents in the families are heterozygous, Aa, for the locus in question. With this assumption, the likelihood of a randomly sampled sibship is just one term of expression (30). With the further assumptions $g_{AA}(0) = g_{Aa}(0) = g_{aa}(1) = 1$, the only unknown parameter in the likelihood that is estimable is $p_{Aa\,Aa\,aa}$ $= 1 - (p_{Aa\,Aa\,AA} + p_{Aa\,Aa\,Aa})$, and in the classical method we test the null hypothesis that this "segregation parameter" is equal to 0.25. However, the likelihood is corrected for ascertainment as indicated in equation (42), assuming all the sibs could be probands, and the sampling function (24). When we set $\pi = 1$ (complete ascertainment), the maximum likelihood estimate of the segregation parameter is 0.274; this, however, is an overestimate, since it does not reflect the true mode of ascertainment. The ideal solution is to estimate π and $p_{Aa\,Aa\,aa}$ simultaneously by maximum likelihood, using an iterative method. To avoid doing this Weinshilboum et al. (1975) use Weinberg's proband method (Cavalli-Sforza and Bodmer, 1971), which approximates the maximum likelihood solution; the resulting estimate is 0.276 with a standard error of 0.075, indicating no significant departure from the Mendelian expectation of 0.25.

Elston et al. (1979) present summary results of another analysis based on the pooled data available on all members of the 22 families, for three of which information is available on three generations of individuals. They first used sex-specific quadratic regression on age to adjust the natural logarithm of each individual's DBH activity to age 30, ignoring the family structure of the data; they found that a mixture of two normal distributions fitted these adjusted values significantly better than one distribution, but that three distributions did not fit significantly better than two.

The segregation analysis was based on a likelihood that used the operation (40), with the following parameter specification of the component functions:

- p_{stu} were taken to be functions of the three transmission probabilities, as indicated in equation (19).
- ψ_t were the three parameters ψ_{AA}, ψ_{Aa}, and ψ_{aa}; but since they must add to unity, these represent only two independent parameters.
- $g_t(z)$ were taken to be $\phi(\mu_t - z, \sigma_E^2)$, where z is $\log_e$ DBH activity adjusted to age 30; this function thus depended on four independent parameters: μ_{AA}, μ_{Aa}, μ_{aa}, and σ_E^2.

The likelihood used for each family included the correction factor (43) divided by (45), assuming only children 6–12 years old could be probands. Had the DBH activities not been adjusted for age, it would have been possible to use a sampling function analogous to (25), i.e.,

$$\pi(z) = \begin{cases} \pi & \text{if } z \le 50 \\ 0 & \text{if } z > 50 \end{cases}$$

(In fact, an analysis using this function was performed, in addition to the one reported in more detail and described below: both analyses gave essentially similar results.) To allow for the fact that a single threshold was not appropriate for the age-adjusted data, the exponential function (26) was used with $K_2 = 0$, the parameters K_0 and K_1 being simultaneously estimated together with all the other parameters in the model.

In the first instance, the fit of Hardy–Weinberg proportions to the data was tested, i.e., (47). Since there was no significant departure from this hypothesis, it was assumed when conducting the later tests. The basic results are presented in Table III. Column 1 gives the estimates obtained when the likelihood is maximized for all ten parameters, the only restriction being that the estimates of probabilities were constrained to be between zero and one. The four other columns give the estimates obtained when further restrictions were placed on the model, i.e., under various hypotheses of interest; and, in the last line, the difference between the maximum likelihood obtained under that hypothesis and the maximum corresponding to the estimates in column 1 is given. As is to be expected, the estimate of K_1 is always negative, corresponding to the *lower* values of z having the higher probabilities of z leading to ascertainment.

In column 2, under the hypothesis that low levels of DBH are due to a recessive gene, the difference in log likelihood is 4.10. Twice this, or 8.2, is not significant at the 5% level if compared with the χ^2 distribution with four degrees of freedom (under this hypothesis, four independent

TABLE III. Maximum Likelihood Estimates of Pedigree Analyses on DBH Activity(log$_e$ units adjusted to age 30) in 22 Families[a]

	Unrestricted	Mendelian			Environmental
		Recessive $\mu_{AA} = \mu_{Aa}$	Dominant $\mu_{aa} = \mu_{Aa}$	Codominant $\mu_{AA} \neq \mu_{Aa} \neq \mu_{aa}$	$\mu_{AA} = \mu_{Aa}$
Transmission probability (τ_{AA})	1.0	1.0	1.0	1.0	0.571
Transmission probability (τ_{Aa})	0.431	0.5	0.5	0.5	0.571
TRansmission probability (τ_{Aa})	0.0	0.0	0.0	0.0	0.571
Gene frequency ($\sqrt{\psi_{AA}}$)	0.718	0.694	0.936	0.720	0.607
First mean (μ_{AA})	6.715	5.761	5.444	6.516	5.750
Second mean (μ_{Aa})	5.284	5.761	2.674	5.377	5.750
Third mean (μ_{aa})	1.693	1.675	2.674	1.573	1.687
Common standard deviation (σ_E)	1.198	1.298	1.806	1.281	1.284
First ascertainment parameter (K_0)	0.662	0.723	0.346	0.591	0.407
Second ascertainment parameter (K_1)	−0.909	−0.881	−0.722	−0.867	−0.623
Difference in log$_e$ likelihood	—	4.10	19.48	0.77	12.21

[a]Published by Weinshilboum et al. (1975); adapted from Elston et al. (1979).

restrictions are placed on the model: the three transmission probabilities are restricted to being Mendelian, and we also have $\mu_{AA} = \mu_{Aa}$). However it should be noted that in column 1 the estimates of τ_{AA} and τ_{aa}, i.e., 0 and 1, do not necessarily occur at a maximum on the likelihood surface, since they were constrained to be between zero and one. For this reason we do not know exactly what the asymptotic distribution of our "chi-square" statistic is, though it can be expected to be bounded by two χ^2 distributions—one with four degrees of freedom and one with two degrees of freedom (because there was free maximization with respect to just two extra independent parameters in column 1). If we were to assume either two or three degrees of freedom are appropriate, the value 8.2 would be significant at the 5% level; thus, although we may not reject this recessive hypothesis on the basis of these data, we note that it is perhaps questionable. Column 3 in Table III gives the estimates obtained at a second local maximum of the likelihood under what is mathematically the same hypothesis—but with the different genetic interpretation that low levels of DBH are due to a *dominant* gene (or, equivalently, that high levels are due to a recessive gene). This maximum is exp(19.48 − 4.10), or nearly five million, times smaller, and so can be dismissed as irrelevant. It does, however, illustrate the fact that more than one local maximum commonly occurs, and that it is important to search the likelihood surface carefully to be sure a larger likelihood has not been missed. In column 4, hypothesis (55) of Mendelian transmission is tested, without assuming a dominance relationship; twice the difference in log likelihood is 1.54, and this is not significant even when compared with the χ^2 distribution with only one degree of freedom. Finally, in column 5, the hypothesis that there are basically two groups with no transmission from one generation to the next [hypothesis (56) is tested, and the result is clearly significant].

It is also possible to test the hypothesis of a recessive gene (column 2) under the model of autosomal inheritance with codominance (column 4). The statistic 2(4.10 − 0.77) = 6.66 should be compared with the χ^2 distribution with one degree of freedom, and so is significant at the 1% level. Thus there is evidence that the means for the presumed normal homozygous and heterozygous individuals (6.516 and 5.377 in column 4) are significantly different, corresponding to the finding by Weinshilboum *et al.* (1975) that the parents of children with low DBH have significantly lower DBH than 220 randomly selected adult controls.

In summary, there is no doubt that there is transmission of this trait from one generation to the next, and that this transmission can be largely accounted for by segregation at an autosomal locus. However, this analysis in no way precludes the possibility of polygenic inheritance or environmentally caused correlations between relatives.

Segregation of Hypercholesterolemia in a Large Pedigree

Elston *et al.* (1975) investigated a 195-member pedigree, extending over five generations, for the transmission of hypercholesterolemia and hypertriglyceridemia; here we shall review the analysis they performed as it refers to elevated serum cholesterol levels. The pedigree was ascertained through four related probands, all of whom were separately referred and studied within a four-week period; two had elevated cholesterol levels, and two were referred for evaluation because of a strong family history of premature death from myocardial infarction, which is associated with elevated cholesterol levels. The 195 individuals whose cholesterol levels were analyzed included only one of these probands, who thus represented a very small fraction of 40 or so others with elevated cholesterol levels. For this reason, and to simplify the computations, the pedigree was analyzed as though it had been randomly sampled, ignoring the trivial bias in the segregation ratios that might be induced by the nonrandom sampling. It was recognized, however, that the estimate of gene frequency thus obtained is biased upward: it represents the gene frequency of the *y*-individuals *in this pedigree*, rather than in the population from which the pedigree was sampled.

The first step in the analysis ignored the pedigree structure, assuming the data to come from a random sample of individuals. Under this assumption, there was no significant sex effect, but there was a significant effect of age. Linear regression of log cholesterol on age was found to account for a larger fraction of the total variance than linear regression of cholesterol on age, and empirical cumulative plots indicated that a lognormal distribution fits the data better than a normal distribution. All further analyses were therefore conducted on the log cholesterol values. Figure 10 shows a cumulative plot of these values, adjusted to age 30, together with the best fitting single lognormal distribution and the best fitting mixture of two lognormal distributions. It is clear that a mixture of two distributions fits much better ($p < 0.01$), and the estimates of the

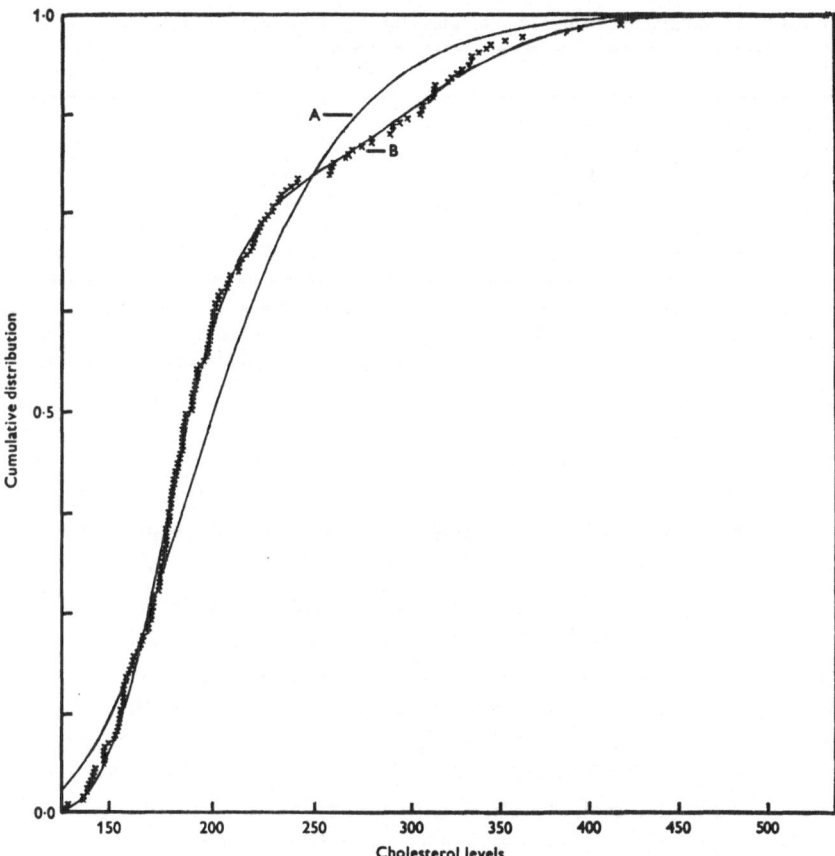

Fig. 10. Empirical and theoretical cumulative plots of the pedigree sample when one normal (A) and a mixture of two normal distributions (B) are fitted to $\log_e$ cholesterol values (after adjusting to age 30 by linear regression). The ordinate for the data points is rank/196; the original scale of cholesterol values is used as abscissa. (From Elston *et al.*, 1975.)

parameters of this mixture, together with the regression coefficient used for age adjustment, are given in column 1 of Table IV.

The segregation analysis was based on a pedigree likelihood that utilized the operation (40), but without any correction for ascertainment. The functions p_{stu} and ψ_t were specified as in the previous example, but $g_t(z)$ was taken to be ϕ ($\mu_t + \beta a - z$, σ_E^2), where a is age and z is log cholesterol; this function thus depended on five independent parameters: μ_{AA}, μ_{Aa}, μ_{aa}, σ_E^2, and β, the linear regression coefficient on age. Thus,

TABLE IV. Maximum Likelihood Estimates Obtained from 195 Log$_e$ Cholesterol Values[a]

	(1)	(2)	(3)
Mean of higher distribution (μ_{AA} and μ_{Aa})	5.748	5.755	5.754
Mean of lower distribution (μ_{aa})	5.151	5.161	5.159
Common standard deviation (σ_E)	0.170	0.167	0.167
Proportion in higher distribution ($\psi_{AA} + \psi_{Aa}$)	0.215	0.102	0.097
Proportion in lower distribution (ψ_{aa})	0.785	0.898	0.903
Linear regression coefficient on age (β)	0.005	0.005	0.005

[a] When (1) assumed to be a random sample from a mixture of two distributions; (2) assumed to be generated by dominant Mendelian inheritance; and (3) assumed to follow a model with arbitrary transmission probabilities. All mean values adjusted to age thirty; adapted from Elston et al. (1975).

in this analysis, the regression on age was estimated jointly with the other parameters.

First the fit of Hardy–Weinberg equilibrium proportions to the data was tested, and since the result was not significant, it was assumed in the later tests. Then the hypothesis of dominance, (48), was tested. Here again two local maxima were found on the likelihood surface, one corresponding to the higher levels of cholesterol being dominant, the other corresponding to their being recessive. In neither case did the likelihood become significantly larger when μ_{Aa} was estimated as a separate parameter, but here again there were two corresponding maxima. However, the hypothesis that the distribution with higher mean is composed of two dominant genotypes gave a larger likelihood (by a factor of 640), and so this was assumed in testing the hypothesis of Mendelian segregation, namely (55). Twice the difference between the two log likelihoods for this test was found to be 1.05, which, when compared to the χ^2 distribution with three degrees of freedom, is not significant. Finally the hypothesis of no genetic transmission from one generation to the next, i.e., (56) was tested: there was found to be a highly significant departure from this hypothesis ($p < 0.001$).

Columns 2 and 3 of Table IV show the maximum likelihood estimates obtained, assuming Hardy–Weinberg equilibrium and dominance, respectively, when Mendelian transmission probabilities are assumed and when the transmission probabilities are estimated to maximize the likelihood. The estimates are seen to be remarkably similar to each other, as well as to the estimates obtained when pedigree structure is ignored (column 1). The greatest discrepancy is in the proportions, which, in any case, may be expected to differ: the values in the first column are estimates of the proportions among all 195 individuals, whereas those in the last two columns (which are very similar to each other) are estimates of the proportions among the y individuals only. The similarities in this table strongly support, but do not prove, that a single dominant gene accounts for much of the variability in cholesterol levels in this family. But, as pointed out by Elston *et al.* (1975), this analysis does not deny the possibility that there may be polygenic variation or environmentally caused correlations among relatives.

Segregation of Phenylthiocarbamide (PTC) Taste Sensitivity

Rao and Morton (1977) performed segregation analysis of PTC taste sensitivity among a total of 4335 individuals distributed in nuclear families—2090 parents and 2245 children. The method of Morton and MacLean (1974) was used, i.e., the analysis was based on the likelihood of the children's phenotypes conditional on their parents' phenotypes, expression (35) divided by expression (33), with the following specifications:

- p_{stu} were fixed at their Mendelian values for two alleles at one autosomal locus.
- ψ_t were functions of the gene frequency, Hardy–Weinberg equilibrium being assumed.
- $g_{tG}(z)$ were taken to be as given in (14), together with the assumption (12); they thus involve six independent parameters: μ_{AA}, μ_{Aa}, μ_{aa}, θ_1, θ_2, and σ_E. However, it was assumed, with no loss of generality, that the underlying liability has mean zero ($\psi_{AA}\mu_{AA} + \psi_{Aa}\mu_{Aa} + \psi_{aa}\mu_{aa} = 0$) and variance 1 ($\psi_{AA}\mu_{AA}^2 + \psi_{Aa}\mu_{Aa}^2 + \psi_{aa}\mu_{aa}^2 + \sigma_G^2 + \sigma_E^2 = 1$).
- $g_{uHC}(z)$ were taken to be as given in (36), but generalized to a trichotomy as in (12) and (14): i.e.,

$$g_{uHC}(2) = \Phi[(C + H + \mu_u - \theta_2)/\sigma_R]$$

$$g_{uHC}(1) = \Phi[(C + H + \mu_u - \theta_1)/\sigma_R]$$

$$- \Phi[(C + H + \mu_u - \theta_2)/\sigma_R]$$

and

$$g_{uHC}(0) = 1 - \Phi[(C + H + \mu_u - \theta_1)/\sigma_R]$$

where $\sigma_R^2 = \sigma_E^2 - \sigma_C^2$.

Given these assumptions, there are seven independent unknown parameters in the model. However, by assuming that the population prevalences of the three phenotypes are known, there are only five independent unknown parameters; in other words, the thresholds θ_1 and θ_2 that separate the nontaster, intermediate, and taster phenotypes are implicitly defined by the other parameters and the prevalences

$$\sum_t \psi_t \int_{-\infty}^{\infty} \phi\,(G, \sigma_G^2)g_{tG}(2)\,dG \quad \text{for tasters}$$

and

$$\sum_t \psi_t \int_{-\infty}^{\infty} \phi\,(G, \sigma_G^2)g_{tG}(1)\,dG \quad \text{for intermediates}$$

Table V gives the parametrization used by Morton and MacLean (1974), together with the corresponding quantities in the notation developed here. It should be noted that Rao and Morton (1977) call B "sibling environmental correlation," a term that is better reserved for the quantity σ_C^2/σ_E^2.

When all five parameters are estimated simultaneously, it is found that $\sigma_C^2 = 0$, and so this is assumed in the following tests. Absence of segregation at a major locus is tested in the presence of polygenic inher-

TABLE V. Parameters and Notation Used by Morton and MacLean (1974), with Equivalent Notation Used Here[a]

	Notation	
Parameter	Morton and MacLean	Here
Polygenic heritability	H	σ_G^2
Relative variance due to common environment	B	σ_C^2
Gene frequency at major locus	q	$\sqrt{\psi_{AA}} = 1 - \sqrt{\psi_{aa}}$
Displacement at major locus	t	$\mu_{AA} - \mu_{aa}$
Degree of dominance at major locus	d	$(\mu_{Aa} - \mu_{aa})/(\mu_{AA} - \mu_{aa})$

[a] It is assumed that liability has mean 0 and variance 1.

itance by the hypothesis (52), i.e., $t = d = 0$ or $q = 0$, leading to a χ^2 value of 33.6. It is not clear whether this should be compared to the tabulated χ^2 with one or two degrees of freedom, but in any case it is highly significant, indicating that under this model we cannot assume absence of a major locus. On the other hand, it is found that the hypotheses $H = d = 0$ and $H = B = d = 0$ both give rise to a χ^2 value of 0.61, which is not significant whether compared to the tabulated χ^2 with two or with three degrees of freedom; this indicates that under this model there is no evidence for any polygenic inheritance or departure from complete dominance of the taster allele.

This study thus provides strong support for the monogenic hypothesis that has been previously based on the following classical evidence: bimodality of the taste threshold distribution in the population, rarity of taster children from two nontaster parents, and agreement with Snyder's ratio when the threshold distribution is dichotomized at its antimode. Rao and Morton (1977) note that such evidence is not beyond cavil, but conclude that their mixed model analysis "demonstrates that in this population skepticism about simple recessivity on the liability scale is unwarranted." However, although this analysis is far superior to any previous segregation analysis of taste-sensitivity, it must be recognized that the results do depend on certain assumptions, in particular, that conditional on genotype, the phenotypic distributions are given by equations (14) and (12). It is conceivable, but highly unlikely, that this can be a critical assumption; and as noted above, one can never be absolutely certain that in a particular situation Mendelian segregation is not being simulated by an environmental mechanism.

Segregation of Immunoglobulin E (IgE) Levels

Gerrard *et al.* (1978) studied serum IgE levels in the members of 173 nuclear families; 145 of the families were virtually a random sample, the other 28 were families with a high prevalence of atopic disease: these latter enriched the sample with families in which one or both parents had high levels of IgE, deemed necessary to determine the mode of inheritance of IgE.

Log IgE values were first adjusted for sex and age effects by regression on sex, age, age^2, age^3, sex × age, sex × age^2, and sex × age^3: the result of doing this is the same as adjusting for each sex separately for

age by using cubic regression. These adjusted values were then standardized within generations by first subtracting the mean from every observation, and then dividing by the standard deviation within each generation. These standardized values x were converted to values z using
the transformation (17), where r was taken to be 6 and p was estimated
by maximum likelihood. If z was assumed to be normally distributed, p
was estimated to be -0.639. Assuming z to be distributed as a mixture
of two normal distributions, however, led to a significantly better fit, with
p estimated to be 0.245; assuming a mixture of three distributions did not
improve the fit significantly over that for two distributions.

Using $p = 0.245$ to define the transformed observations z, the data
were subjected to a segregation analysis based on the conditional likelihood (35) divided by (33), using the following specifications of the
Morton–MacLean model:

- p_{stu} were fixed at their Mendelian values for two alleles at one
  autosomal locus.
- ψ_t were functions of the gene frequency, Hardy–Weinberg equilibrium being assumed.
- $g_{tG}(z)$ were taken to be $\phi\ (\mu_t - z,\ \sigma_E{}^2)$, thus involving the four
  parameters μ_{AA}, μ_{Aa}, μ_{aa}, and $\sigma_E{}^2$.
- $g_{uHC}(z)$ were taken to be $\phi\ (C + H + \mu_u - z,\ \sigma_R{}^2)$, where $\sigma_R{}^2$
  $= \sigma_E{}^2 - \sigma_C{}^2$.

The model is thus defined by the seven independent parameters ψ_{AA},
μ_{AA}, μ_{Aa}, μ_{aa}, $\sigma_E{}^2$, $\sigma_C{}^2$, and $\sigma_G{}^2$, parametrized by MacLean and Morton
(1974) as $q,\ t,\ d$ (defined in Table V), and the four quantities:

population mean $\qquad\qquad U = \psi_{AA}\mu_{AA} + \psi_{Aa}\mu_{Aa} + \psi_{aa}\mu_{aa}$

population variance $\qquad\quad V = \psi_{AA}\mu_{AA}^2 + \psi_{Aa}\mu_{Aa}^2 + \psi_{aa}\mu_{Aa}^2$

$$- U^2 + \sigma_G{}^2 + \sigma_E{}^2$$

polygenic heritability $\qquad H = \sigma_G{}^2/V$

relative variance due to $\quad B = \sigma_C{}^2/V$
common environment

Summary results of the analysis are presented in Table VI. Column
1 shows that under the unrestricted model d is estimated to be zero
(corresponding to high levels of IgE being due to a recessive gene) and
the relative variance due to common sibling environment, B, is also es-

TABLE VI. Maximum Likelihood Estimates of Segregation Analysis, by the Morton and MacLean (1974) Method[a]

	Unrestricted	Polygenic, sibling correlation $q = t = d = 0$	Sibling correlation and major gene $H = 0$	Polygenic, sibling correlation and dominant gene $D = 1$
	(1)	(2)	(3)	(4)
Gene frequency, q	0.489	0	0.478	0.130
Displacement, t	1.628	0	1.820	1.468
Dominance, d	0.000	0	0.197	1
Polygenic heritability, H	0.187	0.485	0	0.109
Common sibling environment, B	0.000	0.024	0.027	0.000
Overall mean, U	−0.064	−0.061	−0.063	−0.026
Overall variance, V	0.951	0.922	0.929	0.958
Difference in $\log_e$ likelihood	—	15.47	4.25	6.35

[a] Performed on transformed $\log_e$ IgE ($p = 0.245$) in 173 nuclear families; adapted from Gerrard et al. (1978).

timated to be zero. The hypothesis of no major gene effect (column 2) gives a log likelihood that is smaller by 15.47, and hence a χ^2 statistic of 30.94—highly significant whether two or three degrees of freedom are assumed to be appropriate. In column 3 the hypothesis of no polygenic component is tested; twice the difference in log likelihood is 8.49, significant at the 1% level when compared with the χ^2 distribution with one degree of freedom. Thus both major gene and polygenic effects are significant, but not sibling environmental correlation. Lastly, in column 4, we see there is highly significant departure from the null hypothesis that the major gene effect is such that high levels of IgE are dominant (χ^2 with one degree of freedom is 12.69).

Gerrard et al. (1978) further divided the data into two sets and repeated the same analysis on each set separately: in one set were the 87 families in which both parents had a value of z less than 0.39, and in the other were the remaining 86 families in which at least one parent had a value of z greater than or equal to 0.39. Since the likelihood used is conditional on the parental phenotypes, subdivision of the sample on the basis of the parental phenotypes should not affect the results; and, indeed, this was found to be the case: the estimates obtained from the two sets of data were consistent with each other and with those from all the data taken together. The purpose of this extra test was to detect any heterogeneity among mating types that might be present, a step that is advisable when the data consist of many unrelated families pooled together.

As a final check, the analysis indicated in Table VI was repeated on the observations transformed to approximate a normal distribution overall, i.e., taking $p = 0.639$. The estimates obtained were again very similar, the displacement being slightly reduced, though the tests were not as significant. The χ^2 statistic for the null hypothesis of no major locus was reduced from 30.94 to 12.25, and that for no polygenic component was reduced from 8.49 to 7.48. These are both still significant at the 1% level, and so this analysis strongly suggests that both polygenic inheritance and segregation at a major locus are important determinants of IgE levels in these families. However, Ott (1979) has reanalyzed these same data, except that one family with 10 offspring was deleted (to avoid excessive computer time), using the unconditional likelihood of both the parents and the children, i.e., (35); and when this is done the major locus component is no longer significant, even at the 10% level. In this analysis it was assumed that the families were sampled at random, whereas in fact 16% of the families were selected to increase the proportion with parents having high levels of IgE. It is difficult to believe that this or the deletion of one family would have such a large effect on the result. Since the same model was used in the analysis, the difference in result is probably due to using either the conditional likelihood or a different method of computation: whereas Ott (1979) calculated the likelihood analytically, Gerard et al. (1978) used a program that approximates the normal distribution by a polychotomy (Morton and MacLean, 1974). Until the cause of the difference has been resolved with certainty, it would be better to reserve judgement about the possibility of a major gene segregating for serum IgE levels.

CONCLUSION

Segregation analysis has seen many advances in the last decade, both in the generality of the statistical models used and in the complexity of the family structures to which it is applied. This review has been an attempt to bring these advances together into a coherent statement of the state of the art at the present time.

Comparing current methods with the classical methods of segregation analysis, there have been three major changes: the phenotype that is studied no longer need be a dichotomy, the underlying statistical model need not be dependent on just one or two parameters, and the analysis need not be restricted to data on sibships. All these changes can be correlated with the availability of increasingly powerful methods of computation, in terms of both computer hardware and computer software. Computer technology is still advancing exponentially, and we can expect newer and better methods of segregation analysis to follow quickly. Two areas in which advances may be expected in the near future are the incorporation of more environmental correlations into segregation analysis models (Boyle and Elston, 1979; Cannings *et al.*, 1980), and the use of multivariate phenotypes to define single gene effects (Weiss, 1976; Goldin *et al.*, 1980). With these elaborations we can expect segregation analysis to be even more useful in the detection of major genes. It must always be remembered, however, that segregation analysis is but one tool at the geneticist's disposal, and that its use is limited unless complemented by other methodologies such as linkage and/or biochemical analysis.

APPENDIX: NOTES ON COMPUTATIONAL METHODS

These notes are intended as a brief guide to the literature on computational methods in two areas: the calculation of pedigree likelihoods and the maximization of likelihoods.

The paper by Elston and Stewart (1971) first expressed the likelihood as the sequence of operations $\Gamma_0[\Gamma_1(\Gamma_2 \ . \ . \ .)]$, thus giving a fast algorithm for its calculation when the mode of inheritance is either oligogenic or polygenic. For the polygenic model with a quantitative trait, Elston and Stewart (1971) give the necessary general analytical result for the integral of a product of normal densities. Details of the algorithm for the oligogenic model are presented by Ott (1974), for two linked loci; this is also explained in an appendix by Conneally and Rivas (1980). The oligogenic case has been extended to pedigrees of arbitrary structure by Lange and Elston (1975) and Cannings *et al.* (1976, 1978), where the advantages and disadvantages of alternative algorithms are discussed. For the polygenic model, Lange *et al.* (1976a,b) give methods of calculation suitable for qualitative and quantitative traits, respectively, in pedigrees that are not too large. Computation of the likelihood for the mixed model is discussed briefly by Morton and MacLean (1974); Ott (1979) goes into more detail, giving an analytical method of calculation suitable for pedigrees containing up to about 10 members. Lalouel (1978) has given an algorithm for the mixed model without sibling environmental correlation, suitable for larger pedigrees, based on approximating the normal distribution by a polychotomy.

Maximizing the likelihood is usually performed by maximizing its logarithm. Bailey (1961) gives in an appendix a clear description of Fisher's scoring technique for this purpose, and its relation to Newton–Raphson iteration; see also Spence et al. (1977). Both these methods require that it be possible to express the derivatives of the log likelihood analytically. Other methods are reviewed by Fletcher (1965) and Powell (1970). Recently Ott (1977, 1979) has shown how the EM algorithm for maximum likelihood estimation (Dempster et al., 1977) can be applied to pedigree analysis. Whatever algorithm is used for maximizing the log likelihood, it is useful to be able to do so with or without constraints on the model, thus simplifying the tests of genetic hypotheses: one program package that has this general capability is MAXLIK (Kaplan and Elston, 1972).

ACKNOWLEDGMENTS. This work was supported in part by a Public Health Service Research Scientist Award (MH31732) and research grant (MH26721) from the National Institute of Mental Health, and by a research grant (GM16697) from the National Institute of General Medical Sciences.

REFERENCES

Bailey, N. T. J., 1951, A classification of methods of ascertainment and analysis in estimating the frequencies of recessives in man, *Ann. Eugen.* **16**:223–225.

Bailey, N. T. J., 1961, *Introduction to the Mathematical Theory of Genetic Linkage,* Oxford University Press, London.

Batschelet, E., 1963, Testing hypotheses and estimating parameters in human genetics if the age of onset is random, *Biometrika* **50**:265–279.

Box, A. E. P., and Cox, D. R., 1964, An analysis of transformations, *J. Roy. Stat. Soc.* B **26**:211–252.

Boyle, C. R., and Elston, R. C., 1979, Multifactorial genetic models for quantitative traits in humans, *Biometrics* **35**:55–68.

Cannings, C., and Thompson, E. A., 1977, Ascertainment in the sequential sampling of pedigrees, *Clin. Genet.* **12**:208–212.

Cannings, C., Skolnick, M. H., DeNevers, K., and Sridharan, R., 1976, Calculation of risk factors and likelihoods for familial diseases, *Comp. Biomed. Res.* **9**:393–407.

Cannings, C., Thompson, E. A., and Skolnick, M. H., 1978, Probability functions on complex pedigrees, *Adv. Appl. Prob.* **10**:26–61.

Cannings, C., Thompson, E. A., and Skolnick, M. H., 1980, Pedigree analysis of complex models, in: *Current Developments in Anthropological Genetics* (J. H. Mielke and M. H. Crawford, eds.), pp. 251–298, Plenum Press, New York.

Cavalli-Sforza, L. L., and Bodmer, W. F., 1971, *The Genetics of Human Populations*, W. H. Freeman, San Francisco.

Conneally, P. M., and Rivas, M. L., 1980, Linkage analysis in man, in: *Advances in Human Genetics*, Vol. 10 (H. Harris and H. Hirschhorn, eds.), pp. 209–266, Plenum Press, New York.

Cotterman, C. W., 1953, Regular two-allele and three-allele phenotype systems. Part 1, *Am. J. Hum. Genet.* **5**:193–235.

Curnow, R. N., 1974, The use of additional information in estimating disease risks from family histories, *Biometrics* **30**:655–665.

Defrise-Gussenhoven, E., 1962, Hypothèses de dimérie et de non-pénétrance, *Acta Genet.* **12**:65–96.

Dempster, A. P., Laird, N. M., and Rubin, D. B., 1977, Maximum likelihood from incomplete data via the EM algorithm, *J. Roy. Stat. Soc.* B **39**:1–22.

Elandt-Johnson, R. C., 1971, Complex segregation analysis. II. Multiple classification, *Am. J. Hum. Genet.* **23**:17–32.

Elandt-Johnson, R. C., 1973, Age-at-onset distribution in chronic diseases. A life table approach to analysis of family data, *J. Chron. Dis.* **26**:529–545.

Elandt-Johnson, R. C., 1974, Segregation analysis: An overview, in: *Proceedings of the 8th International Biometric Conference*, L. C. A. Corsten and T. Postelnicu, eds.), pp. 313–323, Constanta, România, Academieii Republicii Socialiste România.

Elston, R. C., 1973, Discussion to symposium on schizophrenia: Methodologies in human behavior genetics, *Soc. Biol.* **20**:276–279.

Elston, R. C., and Namboodiri, K. K., 1977, Family studies of schizophrenia, in: *Proceedings of the 41st Session of the International Statistical Institute*, New Delhi.

Elston, R. C., and Rao, D. C., 1978, Statistical modeling and analysis in human genetics, *Ann. Rev. Biophys. Bioeng.* **7**:253–286.

Elston, R. C., and Sobel, E., 1979, Sampling considerations in the gathering and analysis of pedigree data, *Am. J. Hum. Genet.* **31**:62–69.

Elston, R. C., and Stewart, J., 1971, A general model for the genetic analysis of pedigree data, *Hum. Hered.* **21**:523–542.

Elston, R. C., and Yelverton, K. C., 1975, General models for segregation analysis, *Am. J. Hum. Genet.* **27**:31–45.

Elston, R. C., Namboodiri, K. K., Glueck, C. J., Fallat, R., Tsang, R., and Leuba, V., 1975, Study of the genetic transmission of hypercholesterolemia and hypertriglyceridemia in a 195 member kindred, *Ann. Hum. Genet.* **39**:67–87.

Elston, R. C., Namboodiri, K. K., Go, R. C. P., Siervogel, R. M., and Glueck, C. J., 1976, Probable linkage between essential familial hypercholesterolemia and C3, Baltimore Conference (1975): *Third International Workshop on Human Gene Mapping*, Vol. 12, pp. 294–297, *Birth Defects: Original Article Series, 1976*, The National Foundation, N.Y.

Elston, R. C., Namboodiri, K. K., and Kaplan, E. B., 1978, Resolution of major loci for quantitative traits, *Genetic Epidemiology* (N. E. Morton and C. S. Chung, eds.), pp. 223–253, Academic Press, New York.

Elston, R. C., Namboodiri, K. K., and Hames, C. G., 1979, Segregation and linkage analyses of dopamine-β-hydroxylase activity, *Hum. Hered.* **29**:284–292.

Falconer, D. C., 1965, The inheritance of liability to certain diseases, estimated from the incidence among relatives, *Ann. Hum. Genet.* **29**:51–76.

Fisher, R. A., 1918, The correlation between relatives on the supposition of Mendelian inheritance, *Trans. Roy. Soc.* (*Edinburgh*) **52**:399–433.

Fisher, R. A., 1934, The effect of methods of ascertainment upon the estimation of frequencies, *Ann. Eugen.* **6**:13–25.

Fletcher, R., 1965, Function minimization without evaluating derivatives—a review, *Comput. J.* **8**:33–41.

Gerrard, J. W., Rao, D. C., and Morton, N. E., 1978, A genetic study of immunoglobulin E, *Am. J. Hum. Genet.* **30**:46–58.

Go, R. C. P., Elston, R. C., and Kaplan, E. B., 1978, Efficiency and robustness of pedigree segregation analysis, *Am. J. Hum. Genet.* **30**:28–37.

Goldin, L. R., Elston, R. C., Graham, J. B., and Miller, C. H., 1980, Genetic analysis of Von Willebrand's disease in two large pedigrees: A multivariate approach, *Am. J. Med. Genet.* **6**:279–293.

Hartl, D. L., and Maruyama, T., 1968, Phenogram enumeration: The number of regular genotype-phenotype correspondences in genetic systems, *J. Theor. Biol.* **20**:129–163.

Kaplan, E. B., and Elston, R. C., 1972, A subroutine package for maximum likelihood estimation (MAXLIK), Institute of Statistics Mimeo Series, No. 823, University of North Carolina.

Karlin, S., and Scudo, F. M., 1969, Assortative mating based on phenotype: II. Two autosomal alleles without dominance, *Genetics* **63**:499–510.

Kempthorne, O., 1978, Logical, epistemological and statistical aspects of nature-nurture data interpretation, *Biometrics* **34**:1–23.

Lalouel, J. M., 1978, Recurrence risks as an outcome of segregation analysis, in: *Genetic Epidemiology* (N. E. Morton and C. S. Chung, eds.), pp. 255–284, Academic Press, New York.

Lange, K., 1976, Stable gene equilibria for mixtures of random and assortative mating, *Math. Biosci.* **29**:49–57.

Lange, K., and Elston, R. C., 1975, Extensions to pedigree analysis I. Likelihood calculations for simple and complex pedigrees, *Hum. Hered.* **25**:95–105.

Lange, K., Westlake, J., and Spence, M. A., 1976a, Extensions to pedigree analysis. II. Recurrence risk calculation under the polygenic threshold model, *Hum. Hered.* **26**:337–348.

Lange, K., Westlake, J., and Spence, M. A., 1976b, Extensions to pedigree analysis. III. Variance components by the scoring method, *Ann. Hum. Genet.* **39**:485–491.

Levy, T., and Nagylaki, J., 1972, A model for the genetics of handedness, *Genetics* **72**:117–128.

MacLean, C. J., Morton, N. E., and Lew, R. 1975, Analysis of family resemblance. IV. Operational characteristics of segregation analysis, *Am. J. Hum. Genet.* **27**:365–384.

Morton, N. E., 1959, Genetic tests under incomplete ascertainment, *Am. J. Hum. Genet.* **11**:1–16.

Morton, N. E., 1962, Segregation and linkage, in: *Methodology in Human Genetics* (J. Burdette, ed.), pp. 17–52, Holden-Day, San Francisco.

Morton, N. E., 1964, Models and evidence in human population genetics, in: *Genetics Today* (S. J. Geerts, ed.), *Proceedings of the XIth International Congress of Genetics*, pp. 935–951, The Hague, The Netherlands, 1963, Pergamon Press, Oxford.

Morton, N. E., 1969, Segregation analysis, in: *Computer Applications in Genetics* (N. E. Morton, ed.), pp. 129–139, University of Hawaii Press, Honolulu.

Morton, N. E., and MacLean, C. J., 1974, Analysis of family resemblance. III. Complex segregation analysis of quantitative traits, *Am. J. Hum. Genet.* **26**:489–503.

Morton, N. E., and Rao, D. C., 1978, Quantitative inheritance in man, *Yearb. of Phys. Anthropol.* **21**:72–141.

Morton, N. E., and Rao, D. C., 1979, Causal analysis of family resemblance, in: *The Genetic Analysis of Common Diseases: Applications to Predictive Factors in Coronary Heart Disease* (C. F. Sing and M. Skolnick, eds.), pp. 431–452, Alan R. Liss, New York.

Ott, J., 1974, Estimation of the recombination fraction in human pedigrees: Efficient computation of the likelihood for human linkage studies, *Am. J. Hum. Genet.* **26**:588–597.

Ott, J., 1977, Counting methods (EM algorithm) in human pedigree analysis: Linkage and segregation analysis, *Ann. Hum. Genet.* **40**:443–454.

Ott, J., 1979, Maximum likelihood estimation by counting methods under polygenic and mixed models in human pedigrees, *Am. J. Hum. Genet.* **31**:161–175.

Powell, M. J. D., 1970, A survey of numerical methods for unconstrained optimization, *SIAM Review* **12**:79–97.

Rao, D. C., and Morton, N. E., 1977, Residual family resemblance for PTC taste sensitivity, *Hum. Genet.* **36**:317–320.

Reich, T., James, J. W., and Morris, C. A., 1972, The use of multiple thresholds in determining the mode of transmission of semi-continuous traits, *Ann. Hum. Genet.* **36**:163–184.

Reich, T., Rice, J., Cloninger, C. R., Wette, R., and James J., 1979, The use of multiple thresholds and segregation analysis in analyzing the phenotypic heterogeneity of multifactorial traits, *Ann. Hum. Genet.* **42**:371–390.

Roberts, D. F., 1977, Methods and problems in physiological genetics, in: *Physiological Variation and Its Genetic Basis* (J. S. Weiner, ed.), *Symposia of the Society for the Study of Human Biology*, Vol. 17, pp. 23–41, Taylor and Francis, London.

Scudo, F. M., and Karlin, S., 1969, Assortative mating based on phenotype. I. Two alleles with dominance, *Genetics* **63**:479–498.

Smith, C. A. B., 1956, A test for segregation ratios in family data, *Ann. Hum. Genet.* **20**:257–265.

Smith, C. A. B., 1959, A note on the effects of method of ascertainment on segregation ratios, *Ann. Hum. Genet.* **23**:311–323.

Spence, M. A., Elston, R. C., and Cederbaum, S. D., 1974, Pedigree analysis to determine the mode of inheritance in a family with retinitis pigmentosa, *Clin. Genet.* **5**:338–343.

Spence, M. A., Westlake, J., and Lange, K., 1977, Estimation of variance components for dermal ridge count, *Ann. Hum. Genet.* **41**:111–115.

Steinberg, A. G., 1959, Methodology in human genetics, *J. Med. Educ.* **34**:315–334.

Stene, J., 1977, Assumptions for different ascertainment models in human genetics, *Biometrics* **33**:523–527.

Thompson, E. A., and Cannings, C., 1979, Sampling schemes and ascertainment, in: *The Genetic Analysis of Common Diseases: Applications to Predictive Factors in Coronary Heart Disease* (C. F. Sing and M. Skolnick, eds.), pp. 363–382, Alan R. Liss, New York.

Weinshilboum, R. M., Schrott, H. G., Raymond, F. A., Weidman, W. H., and Elveback, L. R., 1975, Inheritance of very low serum dopamine-β-hydroxylase activity, *Am. J. Hum. Genet.* **27**:573–585.

Weiss, V., 1976, Die Erkennung von Hauptgenen quantitativer morphologischer Merkmale mit Familienmaterial, Geschwister und Elter-Kind-Paaren am Beispiel der Kopflange des Menschen, *Gegenbaurs morph. Jahrb., Leipzig* **122**(6):875–881.

Wilson, S. R., 1974, Simulation of Mendelism by a non-genetic Markov chain model, *Ann. Hum. Genet.* **38**:225–229.

Wilson, S. R., 1976, Two-sided assortative mating for a single locus, *Ann. Hum. Genet.* **40**:225–229.

Chapter 3

Genetic, Metabolic, and Biochemical Aspects of the Porphyrias

Shigeru Sassa and Attallah Kappas

The Rockefeller University Hospital
New York, New York 10021

INTRODUCTION

The human porphyrias comprise a group of inherited and acquired disorders characterized by aberrations in activities of specific enzymes of the heme biosynthetic pathway. The nosology, natural history, and clinical characteristics of these diseases have been well reviewed by others (Taddeini and Watson, 1968; Dean, 1971; Tschudy and Lamon, 1980; Meyer and Schmid, 1978) and will be described in this review only to the extent that the biochemistry and genetics of these disorders seem to bear on their clinical presentations.

Porphyric diseases are not limited to man and there are naturally occurring animal models of the porphyrias and experimentally inducible disturbances of porphyrin–heme biosynthesis which are of considerable interest for research purposes. Such experimental porphyrias can be evoked by a wide variety of chemicals including steroids of physiological origin. These chemically induced porphyrias have also provided valuable information concerning the host of structurally diverse drugs and other chemicals which can stimulate overactivity of the heme biosynthetic pathway—and thus might provoke exacerbations of the disease in individuals carrying the various gene lesions of the human porphyrias.

Both inherited and chemically induced porphyrias have also provided valuable insights into the regulatory mechanisms by which heme synthesis is controlled, and by inference, how heritable or environmental factors may perturb control mechanisms for heme biosynthesis. Perhaps no early paper in this field better demonstrated this point than the report of Granick and Urata (1963), in which the rate-limiting enzyme of heme synthesis in the liver was shown to be δ-aminolevulinic acid (ALA) synthase and the marked inducibility of the enzyme by foreign chemicals was recognized.

This review summarizes current information on the biochemical and enzymological aspects of heme biosynthesis and the biochemical genetics of the human porphyrias. It will be evident that, by and large, data on the physicochemical characteristics, mode of action and, where available, control mechanisms of the key enzymes of heme synthesis have been derived from animal and tissue culture studies. In contrast, knowledge concerning the specific genetic lesions of the heritable forms of the porphyrias has been generated mostly from studies in man.

TETRAPYRROLES

Tetrapyrroles provide the basic nuclear structures of the porphyrins, heme,* and the degradative products of heme, the bile pigments. They consist of four pyrrolic rings joined together by methene bridges; there may be cyclic (four methene bridges) or linearly arranged (three methene bridges) tetrapyrroles. Porphyrins are the prototypes of the cyclic tetrapyrroles, while the bile pigments represent linearly arranged tetrapyrroles which have been derived from the cleavage of the heme ring at one of the methene bridges.

It appears that tetrapyrrole compounds, at least after the formation of ALA, originate from the same biosynthetic pathway. Granick (1950) found that *Chlorella* mutants which lacked chlorophyll synthetic activity accumulated protoporphyrin IX and magnesium protoporphyrin IX. This observation was crucial in demonstrating that the biosynthesis of heme and chlorophyll shares a common pathway up to the stage of protoporphyrin IX formation. The chlorophyll biosynthetic pathway thereafter diverges by incorporating magnesium into protoporphyrin IX while iron is incorporated into protoporphyrin IX for the formation of heme. The

---

* Heme is ferroprotoporphyrin, namely the chelate of protoporphyrin IX with a ferrous ion, and is the form that exists in hemoproteins in the cell. Heme is readily autooxidized *in vitro* to ferriprotoporphyrin which is called hemin. In the text, the distinction is made between heme, the form in the cell, and hemin, the form added to experimental preparations. Hemin in aqueous medium may also be referred to as hematin solution.

three major types of cyclic tetrapyrroles of biological significance are *heme*, an iron chelate of protoporphyrin IX; *chlorophyll* and *bacteriochlorophyll*, magnesium chelates of protoporphyrin IX; and the *corrins*, cobalt chelates of uroporphyrinogen III derivatives. Linear tetrapyrroles represent metabolites of cyclic tetrapyrroles and include such compounds as biliverdin and bilirubin in animals and the phycobilins in plants. Among the cyclic tetrapyrroles, heme serves a critical metabolic function in animals as well as in certain bacterial cells as the prosthetic group of biologically important hemoproteins.

Structure of Porphyrins

The basic structures of porphyrins of biological interest are shown in Fig. 1. All naturally occurring porphyrins contain side chains at the β-carbon position of each pyrrole ring. Early porphyrin intermediates in heme biosynthesis contain more carboxylic acid side chains than do later intermediates, i.e., eight carboxylic acids in uroporphyrin, four carboxylic acids in coproporphyrin, and two carboxylic acids in protoporphyrin. There are four possible isomers for uroporphyrin and coproporphyrin, since two different substituent groups (acetic acid and propionic acid in uroporphyrin and methyl and propionic acid side chains in coproporphyrin) exist in these porphyrins; however, only the type I isomer (side chains arranged symmetrically) and the type III isomer (side chains on the D ring reversed) are known to occur in nature (Fig. 1). Protoporphyrin IX is derived from isomer III of coproporphyrinogen; although there are a total of 15 possible protoporphyrin isomers due to the three different substituents on its pyrrole rings, only the type IX isomer exists in nature. Porphyrinogens represent the reduced forms of porphyrins, and are colorless. The porphyrinogens of the isomer type III, not those of the isomer type I series, are the substrates for the heme biosynthetic enzymes subsequent to the formation of coproporphyrinogen. Protoporphyrin IX is finally converted into heme by the insertion of iron into the tetrapyrrole ring.

Physicochemical Properties of Porphyrins

Porphyrins are derivatives of porphin (Fig. 1) and they display an intense light absorption peak around the 400-nm-wavelength region; this

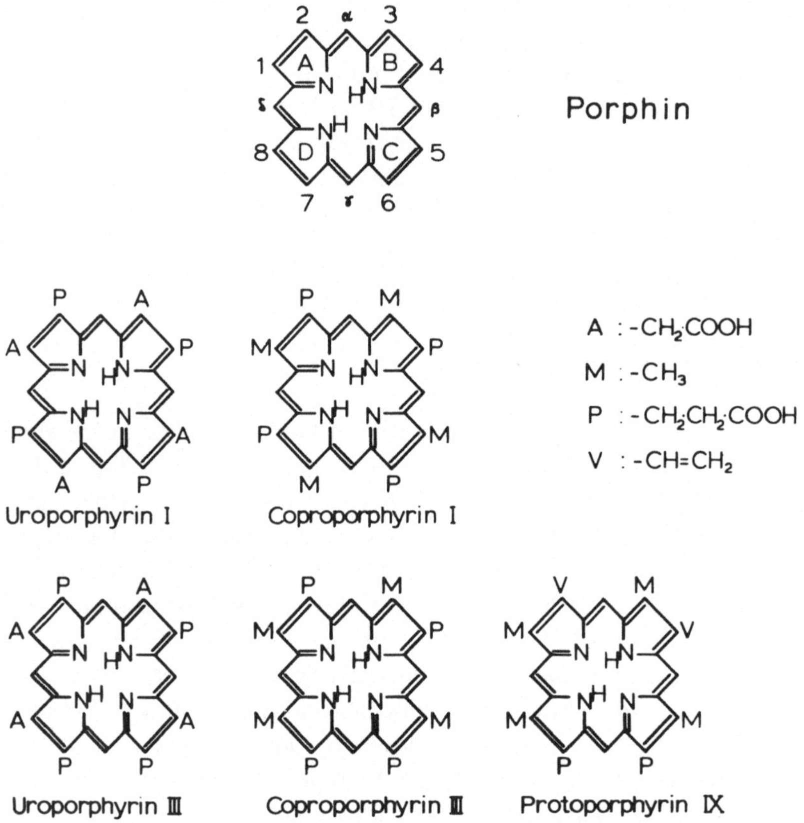

Fig. 1. Structure of porphyrins.

absorption peak is known as the Soret band. They usually have two or more, though much less intense, absorption bands in the visible light region between 500 nm and 600 nm. In addition, when these compounds are exposed to long-wavelength ultraviolet light (~400 nm), they emit an intense red fluorescence with peaks in the region of 590–660 nm (Falk, 1964; Granick and Mauzerall, 1961). These properties of porphyrins are extremely useful in detecting and quantifying them in biological samples. Water solubility of porphyrins is enhanced with increases in the number of carboxylic acid side chains. Thus, uroporphyrin is excreted in urine, coproporphyrin in urine and bile, and protoporphyrin, which is least soluble in water, is excreted only in bile.

Free porphyrins occur in nature in small concentrations and appear to have no clearly defined physiological functions. On the other hand, heme and chlorophyll, i.e., the iron and magnesium complexes of porphyrins, respectively, are in the biologic sense crucially important molecules. Heme serves as the prosthetic group for a number of hemoproteins which are essential for cellular processes involved in oxygen transport and in electron transfer. For example, heme participates in O_2 transport as the prosthetic group of hemoglobin; in transfer of electrons as the prosthetic group of cytochromes a, a_3, b, c, and c_3; in activation of oxygen molecules as the prosthetic group of mitochondrial cytochrome P-450; in the synthesis of steroid hormones and in the metabolism of hormones, drugs, and other foreign chemicals by microsomal cytochrome P-450; and in the enzymatic decomposition of H_2O_2 as the prosthetic group of catalase and peroxidase. In most of these hemoproteins, heme is associated with the apoprotein molecule noncovalently through coordination of the iron atom and a nitrogen atom of an amino acid side chain. However, heme in cytochrome c is covalently bound to the protein moiety by thioether bonds between the two vinyl groups of the heme and two cysteinyl residues of the apocytochrome.

ENZYMES AND INTERMEDIATES OF THE HEME BIOSYNTHETIC PATHWAY

The biochemical pathway for the formation of heme is illustrated in Fig. 2. The first, as well as the last three, enzymes of heme biosynthesis are localized in the mitochondria and the intermediate enzymes are found in the cytosol (Granick and Mauzerall, 1958a). The first enzyme of the pathway is ALA synthase which catalyzes the condensation of glycine and succinyl CoA to form ALA. ALA is an obligate precursor for porphyrin and heme formation. Two moles of ALA are condensed to form a monopyrrole, porphobilinogen (PBG), by a cytosolic enzyme, ALA dehydratase. Four moles of PBG are then condensed to yield the first tetrapyrrole in the pathway, uroporphyrinogen (UROgen) I; this reaction is catalyzed by a soluble enzyme, UROgen I synthase. This symmetrical isomer (type I) of UROgen is normally produced in insignificant amounts as is the next tetrapyrrole in this series, coproporphyrinogen (COPROgen) I, and there are no known heme compounds derived from the type I isomer porphyrinogens. By the combined action of UROgen I synthase and UROgen III cosynthase, four PBG molecules are converted to URO-

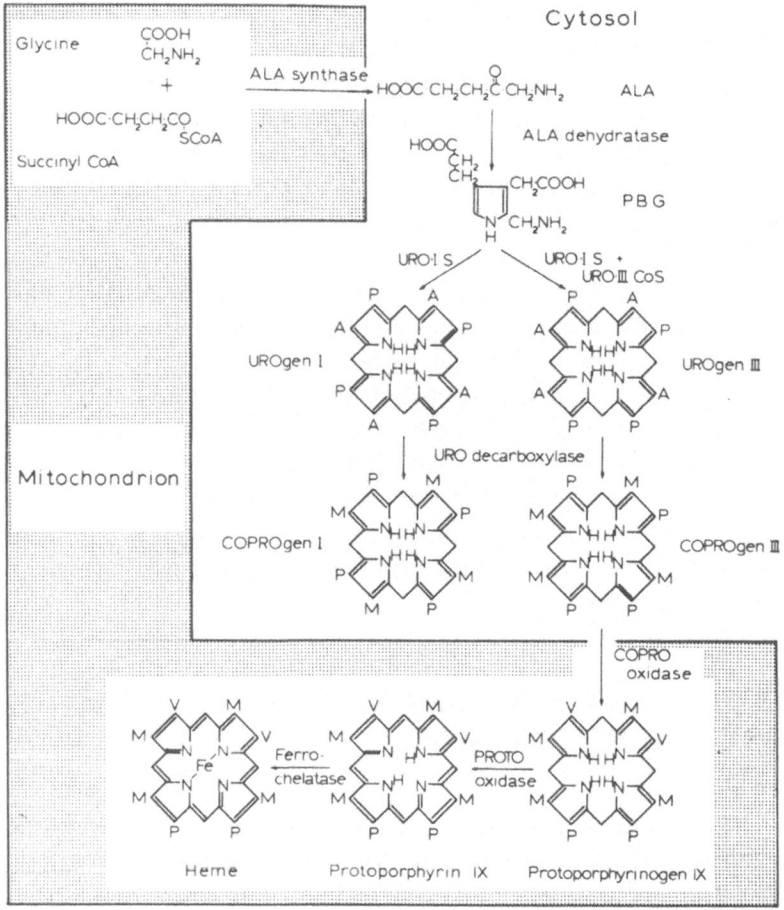

Fig. 2. Intracellular localization of enzymes and intermediates of the heme biosynthetic pathway.

gen III in which one PBG molecule (ring D in Fig. 1) is reversed. Another soluble enzyme, UROgen decarboxylase, then removes four carboxyl groups from four acetic acid side chains yielding COPROgen III which then contains four methyl groups instead of four acetic acid groups. The terminal steps in the heme biosynthetic pathway are localized again in the mitochondria. Although COPROgen I is known to occur in nature, only COPROgen III is oxidatively decarboxylated to protoporphyrinogen (PROTOgen) IX by the mitochondrial enzyme, COPROgen oxidase. Thus, PROTOgen IX contains two vinyl groups on rings A and B instead of two propionic acid groups as in COPROgen III. Six hydrogen atoms are then removed by PROTOgen oxidase to yield protoporphyrin (PROTO) IX. Finally, ferrous iron (Fe^{2+}) is inserted into PROTO IX by the action of ferrochelatase to form protoheme, i.e., heme.

Formation of δ-Aminolevulinic Acid (ALA)

ALA is formed by the condensation of glycine and succinyl CoA in the mitochondria. The enzyme which catalyzes this reaction is ALA synthase which is the first as well as the rate-limiting enzyme for heme formation, at least in liver under physiological conditions. ALA synthase requires pyridoxal 5'-phosphate as a cofactor. Vitamin B_6 deficiency is known to result in a decrease in this enzyme activity (Aoki et al., 1975).

Normally, ALA synthase activity is found only in the mitochondria. However, in rats treated with allylisopropylacetamide (AIA), a potent chemical inducer of porphyrins in liver, a substantial increase in ALA synthase activity can be detected in the cytosol (Patton and Beattie, 1973; Hayashi et al., 1969; Whiting and Elliott, 1972) by providing a succinyl-CoA-generating system in the enzyme assay mixture. There are a number of reasons to believe that both mitochondrial and cytosolic ALA synthases are synthesized on cytoplasmic polyribosomes and that the cytosolic ALA synthase probably represents a form of the enzyme in transit into the mitochondria (Hayashi et al., 1969; Granick and Sassa, 1971; Whiting and Elliott, 1972). The cytosolic ALA synthase requires monovalent or divalent cations and its apparent K_m for glycine, succinyl CoA, and pyridoxal 5'-phosphate are 1×10^{-2}, 7×10^{-5}, and 3×10^{-6} M, respectively (Scholnick et al., 1972). Utilizing an enzyme preparation purified to near homogeneity, it has been reported that the cytosolic enzyme complex has a K_m of 2.5×10^{-3} M for glycine, but an active enzyme component which is dissociated from the complex by papain treatment has a K_m of 7.5×10^{-3} M (Nakakuki et al., 1980).

The molecular weight of ALA synthase varies depending on the sources of tissues, as well as the species, and conditions of purification. One of the conditions of purification known to affect estimates of the molecular weight of this enzyme is the ionic strength of the buffer; specifically, the enzyme, particularly the cytosolic one, is readily aggregated in a buffer with a low ionic strength (Whiting and Elliott, 1972). Thus, apparent higher molecular weights of the enzyme are probably not reliable when precautions against enzyme aggregation are not taken. The purified enzyme is also known to require a sulfhydryl donor for stability. Whiting and Elliott (1972), taking account of these considerations, reported a value of 77,000 for mitochondrial ALA synthase and 178,000 for cytosolic ALA synthase. More recently, Paterniti and Beattie (1979) obtained a value of 120,000 for the mitochondrial enzyme from uninduced normal rat liver. In contrast to an apparent single molecular species of mitochondrial ALA synthase, Ohashi and Kikuchi (1979) observed three distinct bands for cytosolic ALA synthase, displaying molecular weights of 120,000, 79,000, and 51,000. These authors also demonstrated that the molecular weight of the catalytically active component of cytosolic ALA synthase had the same molecular weight as that of mitochondrial ALA synthase and that the protein exists as a dimer, with two apparently identical subunits having molecular weights of 51,000. These data suggest that cytosolic ALA synthase may be composed of a complex consisting of one catalytically active subunit of molecular weight 51,000, and two catalytically inactive protein components having molecular weights of 79,000 and 120,000. Using a specific antibody against purified ALA synthase, Nakakuki et al. (1980) have shown that both ALA synthases in the cytosol and the mitochondria are immunochemically identical. Whiting and Granick (1976a) purified mitochondrial ALA synthase from chick embryo liver preinduced with AIA and 3,5-dicarbethoxy-1,4-dihydrocollidine (DDC) and reported a molecular weight value of 87,000, consisting of two identical subunits having a molecular weight of 49,000. It is of interest that no cytosolic ALA synthase is detectable in the chick embryo liver (Tomita et al., 1974) while a significant amount of cytosolic ALA synthase is found in the chick liver after stimulation with AIA (Ohashi and Kikuchi, 1972). ALA synthase from fetal rat liver mitochondria is reported to have a molecular weight of 47,000 (Woods and Murthy, 1975) and the enzyme from rabbit reticulocytes a molecular weight of 200,000 (Aoki et al., 1971). ALA synthase from human sources has not yet been purified.

It is important to note that the half-life of ALA synthase is very short, e.g., only 34 min in fetal rat liver (Woods, 1974), 70 min in adult rat liver (Tschudy et al., 1965a; Marver et al., 1966), and 160 min in cultured chick embryo liver cells (Sassa and Granick, 1970). ALA synthase appears to be the fastest-turning-over enzyme in the heme biosynthetic pathway and its half-life is considerably shorter than the half-lives of general mitochondrial proteins, which are in the range of ~5 days (Druyan et al., 1969). The short half-life of ALA synthase permits regulation of this enzyme activity by controls on its synthetic mechanism.

Aoki (1978) described the presence of a protease in bone marrow mito-chondria specific to ALA synthase and purified it to apparent homoge-neity. The purified protease exhibited a highly specific proteolytic activity toward ALA synthase (Aoki, 1978; Aoki *et al.*, 1975). The existence of a mitochondrial protease specific for ALA synthase could explain the particularly short half-life of this enzyme.

K_m values of rat liver mitochondrial ALA synthase for succinyl CoA and glycine are 2×10^{-4} M and 1.9×10^{-2} M, respectively (Whiting and Elliott, 1972). The K_m value for pyridoxal 5'-phosphate has been estimated to be in the range of $10^{-6} \sim 10^{-5}$ M. The unusually high K_m for glycine raises the possibility that the enzyme activity may also be reg-ulated by glycine concentrations in this tissue. Increases of ALA synthase activity have, in fact, been observed with increases in glycine concen-tration in cultured chick embryo liver cells (Sinclair and Granick, 1975), and the administration of *p*-aminobenzoic acid to increase glycine excre-tion as hippurate is known to prevent the induction of hepatic porphyria by DDC (Tephly *et al.*, 1973).

The level of ALA synthase in liver cells is considered to be regulated by the end product of the pathway, heme. The enzyme activity is known to be inhibited by approximately 50% by 50 μM hemin in a partially purified ALA synthase preparation from rat liver mitochondria (Kaplan, 1971); by 10 μM hemin in a purified rat liver enzyme preparation (Whiting and Elliott, 1972); by 35 μM hemin using a purified chick embryo liver enzyme (Whiting and Granick, 1976a); and by 10 μM hemin in a partially purified enzyme preparation from rabbit reticulocytes (Aoki *et al.*, 1971). In contrast to the direct inhibition of enzyme activity by hemin in con-centrations of $\sim 10^{-5}$ M, the synthesis of ALA synthase is repressed approximately 50% by 10^{-7} M hemin in cultured chick embryo liver cells (Granick *et al.*, 1975). In the inherited hepatic porphyrias, e.g., acute intermittent porphyria, hereditary coproporphyria, and variegate por-phyria, hepatic ALA synthase activity is known to be elevated at least during acute attacks of the disease (Tschudy *et al.*, 1965b; Nakao *et al.*, 1966; Sweeney *et al.*, 1970; McIntyre *et al.*, 1971).

In plant cells ALA appears to be synthesized by a different metabolic route which does not involve succinyl CoA and glycine. Beale and Cas-telfranco (1974) demonstrated that, in greening etiolated leaves and co-tyledons, ^{14}C of glutamate or α-ketoglutarate is incorporated into ALA; however, neither ^{14}C of labeled succinyl CoA nor ^{14}C of labeled glycine is utilized significantly. In this reaction, C-1 of glutamate becomes C-5 of ALA (Beale *et al.*, 1975). γ,δ-Dioxovalerate has been proposed to be the active intermediate in the conversion of α-ketoglutarate to ALA in plants. Transamination of γ,δ-dioxovalerate to ALA has been shown to occur in extracts of *Chlorella* (Gassman *et al.*, 1968), bean leaves (Gassman *et al.*, 1966), and *Rhodopseudomonas spheroides* (Neuberger and Turner, 1963). Whether or not chlorophyll formation is exclusively associated with the γ,δ-dioxovalerate transaminase pathway is uncertain (Porra and Grimme, 1978).

Most recently, Varticovski *et al.* (1980) have isolated and purified γ,δ-dioxovalerate transaminase to apparent homogeneity from bovine liver mitochondria. The enzyme was remarkably stable with an apparent molecular weight of 240,000 and required pyridoxal phosphate as a cofactor. The enzyme had high specificity for both the amino donor (L-alanine) and the amino acceptor (γ,δ-dioxovalerate). The K_m for L-alanine was 3.7×10^{-3} M and the K_m for γ,δ-dioxovalerate was 2.4×10^{-4} M. In conformity with the classical transaminase reaction, a "ping-pong" reaction mechanism was demonstrated for γ,δ-dioxovalerate transaminase. The capacity of γ,δ-dioxovalerate transaminase to synthesize ALA appeared to be far greater than the capacity of ALA synthase from bovine liver mitochondria. It is not known as yet whether γ,δ-dioxovalerate transaminase activity changes when more or less heme synthesis is required, and this question remains to be examined.

Formation of Porphobilinogen (PBG)

ALA formed in the mitochondria is transported into the cytoplasm where two molecules of ALA are condensed to form a monopyrrole, PBG, by the cytosolic enzyme ALA dehydratase. The reaction involves the removal of two molecules of water. ALA dehydratase purified from mammalian erythrocytes (Finelli *et al.*, 1974, 1975) and liver (Cheh and Neilands, 1973; Tsukamoto *et al.*, 1979) requires a sulfhydryl donor and Zn^{2+} for maximal activity. Thus, the enzyme activity is quickly lost upon exposure to air by oxidation of SH groups of the enzyme or by inhibition of the SH groups by sulfhydryl reagents (Sassa *et al.*, 1975a), by displacement of Zn^{2+} by Pb^{2+} (Finelli *et al.*, 1975), or by chelation of Zn^{2+} with EDTA (Tsukamoto *et al.*, 1979). Lead-inhibited ALA dehydratase activity can be completely restored by the addition of dithiothreitol (Granick *et al.*, 1973) or Zn^{2+} (Finelli *et al.*, 1975). Lead does not interfere with the synthesis of human red cell ALA dehydratase, but it inhibits the enzyme activity in a noncompetitive manner, with a K_i of 0.5×10^{-6} M (Granick et al., 1973).

Purified ALA dehydratase from beef (Wilson et al., 1972), mouse liver (Coleman, 1966), and human erythrocytes (Anderson and Desnick, 1979) exhibits an apparent K_m value of approximately 4×10^{-4} M for ALA and a pH optimum in the range of 6.3–6.7. In the active site of the enzyme, there are two cysteine residues, one Zn^{2+} atom, and two histidine residues. Photooxidation of histidine residues diminishes both enzymatic activity and bound Zn^{2+}, suggesting that histidines not only occupy the catalytic site of the enzyme, but also serve as a possible ligand to Zn^{2+} (Tsukamoto et al., 1979). If Zn^{2+} is removed in the absence of oxygen and the apoenzyme is kept under strictly anaerobic conditions, the zinc-depleted enzyme remains fully active (Tsukamoto et al., 1979). These findings suggest that Zn^{2+} may protect essential SH groups from autooxidation presumably by coordination with them (Bevan et al., 1980), but that the metal is not catalytically essential if the autooxidation of the enzyme is prevented by other means. Cd^{2+} has also been shown to restore activity to the apoenzyme (Bevan et al., 1980).

The reaction mechanism of ALA dehydratase is believed to proceed as follows: one molecule of ALA forms a covalent bond to the enzyme through the formation of a Schiff's base between the ε-amino group of a lysine residue and the keto group of ALA. This stabilized carbanion then participates in an aldol condensation of a second ALA molecule with the removal of one water molecule (Nandi and Shemin, 1968). It was found that the first ALA molecule initially bound to the enzyme is the one which gives rise to the propionic acid side of PBG (Jordan and Seehra, 1980). ALA dehydratase from bovine liver has a molecular weight of 292,000 with eight apparently identical subunits (Wu et al., 1974). The purified enzymes obtained from bovine liver and human erythrocytes display a pale yellowish tint, and spectral analysis as well as kinetic studies suggest that pyridoxal 5'-phosphate may be bound to the enzyme (Anderson and Desnick, 1979).

In most mammalian tissues, ALA dehydratase activity is present at a level almost 100 times higher than that of ALA synthase. In fact ALA dehydratase is probably the enzyme whose concentration is in greatest excess as compared with the other enzymes of the heme biosynthetic pathway. Nevertheless, marked inhibition of this enzyme activity may lead to an impairment in the rate of heme formation (Ebert et al., 1979).

Lead is known to profoundly inhibit the enzyme activity *in vivo*, resulting in the excessive excretion of ALA in urine and, at least in the case of acute lead exposure, the metal can reduce the level or functional integrity of microsomal cytochrome P-450 as well (Alvares *et al.*, 1972). There is no clear evidence, however, concerning an inhibitory effect of chronic lead treatment on the synthesis or function of this hemoprotein. The level of cytochrome P-450 has been found to be decreased in rats treated with aminotriazole (Kato, 1967), an inhibitor of ALA dehydratase activity (Tschudy and Collins, 1957), suggesting that a decreased activity of ALA dehydratase may limit the supply of heme for microsomal cytochrome synthesis (Tephly, 1978).

A syndrome which presents similar symptoms to those characteristic of acute intermittent porphyria has been described in patients with hereditary tyrosinemia (Gentz *et al.*, 1965, 1969*a,b*). Because of a genetic deficiency of fumarylacetoacetase in this disorder, succinylacetone, a potent inhibitor of ALA dehydratase, accumulates and ALA dehydratase activity in erythrocytes and liver of patients with this disorder has been reported to be less than 5% and 1% of control activity, respectively (Lindblad *et al.*, 1977). Recently an inherited deficiency of erythrocyte ALA dehydratase activity (~50% of normal controls) has been reported in a family over three generations. The enzyme deficiency was inherited in an autosomal dominant pattern (Bird *et al.*, 1979). However, all subjects including those with decreased ALA dehydratase activity were clinically unaffected. In another family with hereditary deficiency of ALA dehydratase activity, clinical symptoms characteristic of AIP were observed (Doss *et al.*, 1977).

In mice, the level of ALA dehydratase in the spleen, liver, kidney (Coleman, 1966; Doyle and Schimke, 1969), and erythrocytes (Sassa and Bernstein, 1977) is regulated by at least two codominant alleles at the ALA dehydratase locus (*Lv*). Mouse strains homozygous for the *Lv^a* allele have approximately 2.5 times higher ALA-dehydratase activity than those homozygous for the *Lv^b* allele. Heterozygotes (*Lv^a/Lv^b*) display an intermediate enzyme activity. The higher enzyme activity is due to an increase in the amount of normal enzyme molecules (Doyle and Schimke, 1969). Levels of ALA dehydratase in erythrocytes from the normal human population display an approximately fourfold variation; however, the variation among siblings is considerably smaller and values are almost identical within pairs of monozygotic twins (Sassa *et al.*, 1973). The genetic control of ALA dehydratase in human populations is also compatible with an autosomal codominant inheritance (Sassa *et al.*, 1973).

ALA dehydratase activity displays marked changes during animal development. The enzyme activity is extremely high in the livers of fetal mice (Doyle and Schimke, 1969; Freshney and Paul, 1971) and fetal guinea pigs (Weissberg and Voytek, 1974), but this hepatic enzyme activity decreases to a low level immediately after birth and rises to the adult level 3 weeks later (Doyle and Schimke, 1969). In red cells of mice the enzyme activity decreases by about 75% from the third to the fifth week of post-natal life and remains approximately at that level throughout adulthood (Sassa and Bernstein, 1977). Changes in ALA dehydratase in red cells are mainly a reflection of changes in the red cell population; specifically, reticulocytes contain 10–20 times more enzyme activity compared with adult red cells (Sassa and Bernstein, 1977). Similarly, higher red cell ALA dehydratase activity is observed in human subjects with the reticulocytosis produced by hemolytic disorders (Anderson *et al.*, 1977).

Formation of Uroporphyrinogen (UROgen)

UROgen I synthase catalyzes the condensation of four molecules of PBG to form a symmetrical cyclic tetrapyrrole UROgen I. A second enzyme, UROgen III cosynthase, acting together with UROgen I synthase leads to the formation of UROgen III, in which one of the four PBG molecules (ring D, Fig. 1) is reversed. Assembly of 4 pyrrole rings starts from ring A with sequential addition of rings B, C and ring D. Ring D is one which undergoes intramolecular rearrangement leading to the formation of the asymmetrical type III tetrapyrrole (Battersby *et al.*, 1979). UROgen I synthase and UROgen III cosynthase are also referred to as PBG deaminase and PBG isomerase, respectively. UROgen I synthase is relatively heat-stable (up to ~55°C) while UROgen III cosynthase is heat-labile.

The activity of UROgen I synthase is usually very low, in fact close to that of ALA synthase (30–100 nmol ALA formed/hr/g mouse liver) (Hutton and Gross, 1970) which is considered to be the rate-limiting enzyme of heme synthesis, at least in the liver. In the mouse liver, UROgen I synthase activity has been reported to be 35–60 nmol PBG consumed/hr/g liver (Hutton and Gross, 1970) or 22 nmol UROgen I formed/hr/g liver (Miyagi *et al.*, 1971). Apparently erythroid tissue contains a higher enzyme activity than the liver. For example, a value of 35–36 nmol UROgen I formed/hr/ml erythrocytes (Sassa *et al.,* 1974*a*) has been reported for circulating erythrocytes—a value which is only about $\frac{1}{10}$ the activity found in reticulocytes (Sassa and Bernstein, 1977). The enzyme activity in spleen is also approximately 10-fold greater than that in liver (Hutton and Gross, 1970).

UROgen I synthase activity in tissues, for example, erythrocytes, fibroblasts, and liver cells, from patients with acute intermittent porphyria is about one-half the normal value (Miyagi *et al.,* 1971; Strand *et al.,* 1970, 1972; Sassa *et al.,* 1974*a*). The low level of hepatic UROgen I synthase activity associated with the elevation of ALA synthase activity probably explains why biochemical disturbances of porphyrin metabolism, i.e., accumulation of the porphyrin precursors, ALA and PBG, occur in the liver of patients with acute intermittent porphyria, but not in erythroid tissues where a 50% reduction of the enzyme would still leave sufficient residual activity to catalyze formation of UROgen to sustain a normal rate of heme synthesis.

Recently, human UROgen I synthase has been purified from erythrocytes more than 42,000-fold with a 25% yield (Anderson, 1979; Anderson and Desnick, 1980). The purification procedures included anion-exchange chromatography, hydrophobic chromatography using octyl- and phenyl-Sepharose, gel filtration, and a second anion-exchange chromatography. In the final step, using DEAE cellulose, the enzyme separated into five forms which were designated as UROgen I synthase A, B, C, D, and E. These five forms were also separated by polyacrylamide gel electrophoresis and by isoelectric focusing. The five forms of the enzyme, however, showed only one band in SDS-polyacrylamide gel electrophoresis suggesting that the purified UROgen I synthase was homogeneous. The mol. wt. of UROgen I synthase was determined to be in the range of 36,000–42,000 by gel filtration and the mol. wt. of a single polypeptide of UROgen I synthase was also determined to be 37,000 by SDS-polyacrylamide gel electrophoresis, suggesting that the enzyme is monomeric.

The purified enzymes A and B had the following properties; a pH optimum of 8.2; a K_m value of about 6×10^{-6} M for PBG; a specific activity of about 2300 nmol UROgen formed/mg protein/hr; a turnover number of 1.42, and marked temperature dependence. The enzymes were inhibited by sulfhydryl reagents and heavy metals. Amino acid analysis revealed an absence of tryptophan and methionine. The enzymes were found to be thermostable. The multiple forms of UROgen I synthase A, B, C, D, and E were thought to represent different degrees of enzyme–substrate (i.e., none, mono-, di-, tri-, and tetrapyrrole) complexes. When homogeneous UROgen I synthase A and B were incubated with [^{3}H]-PBG and then subjected to electrophoresis, all five forms of the enzyme were observed; the more anodal bands contained proportionally more radiolabel, consistent with the above hypothesis. The multiple forms of UROgen I synthase were also observed when enzyme preparations from liver or brain were applied to DEAE cellulose column chromatography. The presence of multiple forms of UROgen I synthase in human erythrocytes has been also reported by Miyagi et al. (1979). These authors proposed that human erythrocyte UROgen I synthase is a group of six isoenzymes with different electric charges.

The elution patterns of the multiple forms of UROgen I synthase in erythrocytes from patients with acute intermittent porphyria were examined, and two types were observed (Anderson, 1979; Anderson and Desnick, 1980). One pattern was similar to the normal but was characterized by approximately a 50% reduction in each form. The second pattern was characterized by elevation of the C enzyme with respect to the enzyme forms A and B. Patients with the latter pattern of enzyme–substrate intermediates has high urinary PBG levels and/or were experiencing neurologic symptoms of the disease. Immunochemical studies utilizing a specific antibody against purified human UROgen I synthase indicated that several individuals with acute intermittent porphyria possessed material in erythrocyte lysates that was noncatalytic but immunologically cross-reactive.

UROgen I synthase activity in the liver, erythrocytes, cultured skin fibroblasts, amniotic cells, and mitogen-stimulated lymphocytes from patients with acute intermittent porphyria is about one-half the control level, thus characterizing this enzyme deficiency as an expression of the primary gene defect of this disorder. As noted earlier, UROgen I synthase activity is considerably higher in reticulocytes (10–20 times) than in adult red cells (Sassa and Bernstein, 1977). This fact has to be taken into consideration if screening for acute intermittent porphyria gene carriers is made solely on the basis of the erythrocyte UROgen I synthase assay (Anderson et al., 1977). Blood samples with increased reticulocytes or equivocal erythrocyte UROgen I synthase values require further confirmatory testing using a UROgen I synthase assay which is independent of cell age, as in cultured skin fibroblasts (Bonkowsky *et al.*, 1975*b*; Sassa *et al.*, 1975*b*) or mitogen-stimulated lymphocytes (Sassa *et al.*, 1978*b*).

A microsomal-membrane-bound PBG oxygenase, another enzyme which utilizes PBG, has been described in rat liver and brain (Frydman *et al.*, 1973*a*; Tomaro *et al.*, 1973; Frydman *et al.*, 1975). The enzyme activity has also been found in wheat germ (Frydman *et al.*, 1973*b*). This enzyme converts PBG to 2-hydroxy-5-oxo PBG. Normally, this enzyme activity appears to be suppressed to a very low level by the presence of an inhibitor, but the enzyme activity is known to increase upon removal of the inhibitor by storage, by dialysis, or by other purification procedures. PBG oxygenase activity is also reported to increase in rat liver after administration of progesterone or phenobarbital (Frydman *et al.*, 1973*a*).

UROgen III cosynthase has proved to be much more difficult to purify than UROgen I synthase mainly because of its instability. This cytosolic enzyme was first separated from UROgen I synthase and partially purified from spinach (Bogorad, 1958*a,b*) and *Rhodopseudomonas spheroides* (Hoare and Heath, 1958). The enzyme activity is generally much greater ($\sim 10\times$) than that of UROgen I synthase (Frydman and Feinstein, 1974), thus ensuring the formation of the type III UROgen. UROgen III cosynthase purified from wheat germ has a molecular weight of about 62,000 (Higuchi and Bogorad, 1975). When UROgen I synthase and UROgen III cosynthase are incubated together in the presence of PBG, the two enzymes form a complex and catalyze a reaction which converts PBG to UROgen III (Higuchi and Bogorad, 1975). A UROgen I synthase–UROgen III cosynthase complex has also been isolated from bovine liver (Sancovich *et al.*, 1969) and avian erythrocytes (Llambias and Batlle, 1971). The two enzymes from human erythrocytes also reassociate *in vitro* in the absence of PBG to form a stable complex (Frydman and Feinstein, 1974). These findings suggest that the two enzymes may interact physically and that the two-enzyme complex isolated from various tissues may represent the functional form of an enzyme complex involved in UROgen III formation from PBG. A genetic deficiency of UROgen III cosynthase has been described in human and bovine erythropoietic porphyria (Romeo and Levin, 1969; Levin, 1968) which is accompanied by an excessive excretion of URO I and COPRO I. UROgen III cosynthase activity in the fox squirrel (*Sciurus niger*) is only 10% of that in the grey squirrel and the former, not the latter, is known to excrete a large amount of URO I in urine (Levin and Flyger, 1973).

Formation of Coproporphyrinogen (COPROgen)

A cytosolic enzyme, UROgen decarboxylase, catalyzes the decarboxylation of the four carboxylic acid groups of the acetic acid side chains of UROgen to yield COPROgen (Granick and Mauzerall, 1958b). The enzyme decarboxylates all four isomers of UROgen, the III isomer being decarboxylated most rapidly followed by the IV, II, and I in decreasing order (Granick and Mauzerall, 1958b; Cornford, 1964). Although some earlier studies suggested that UROgen III may be decarboxylated at the same rate as UROgen I (Romeo and Levin, 1971; Kushner et al., 1975; Rasmussen and Kushner, 1979), more recent studies appear to have established that the decarboxylation of the III isomer is faster than that of the I isomer (Smith and Francis, 1979). There is no evidence that there is more than one enzyme for the four successive decarboxylation reactions (Tomio et al., 1970). Thus, the reaction catalyzed by UROgen decarboxylase yields from UROgen hepta-, hexa-, penta-, and tetracarboxylic porphyrinogens.

Partially purified UROgen decarboxylase from mouse spleen displays a K_m of about 1×10^{-6} M for UROgen I and UROgen III (Romeo and Levin, 1971). Unlike other decarboxylases, UROgen decarboxylase does not require any cofactor, such as pyridoxal 5'-phosphate (Romeo and Levin, 1971). Jackson et al. (1975) demonstrated that the order of removal of carboxyl carbons from each pyrrole ring proceeds in a clockwise fashion with UROgen III starting from ring D (Fig. 1), but in a random manner with UROgen I (Jackson et al., 1975). An apparent K_m value for rat liver UROgen decarboxylase was reported to be about 8×10^{-6} M using a pentacarboxylic substrate and a range of approximately $0.5-1.5 \times 10^{-6}$ M was estimated for the remaining porphyrinogens (Smith and Francis, 1979). The affinity of the enzyme appears to decrease with each decline in the number of carboxyl groups in the substrate. Another recent study found that the K_m for the III isomer was always higher than that for the corresponding I isomer (DeVerneuil et al., 1980). It has also been suggested that both types I and III porphyrinogens with the same number of carboxylic groups are decarboxylated at the same active center, while the decarboxylation of porphyrinogen substrates with different numbers of carboxylic groups occurs at four different active centers (DeVerneuil et al., 1980).

The activity of UROgen decarboxylase in mouse spleen (0.35 nmol COPROgen formed/mg protein in the 25,000g supernatant) is increased about four-fold after treatment of mice with phenylhydrazine for 6 days, suggesting that the enzyme activity is much higher in erythroid precursor cells. The hepatic enzyme activity, however, is not affected appreciably by this treatment (Romeo and Levin, 1971). The enzyme activity is also detectable in brain, heart, lung, testis, and kidney, with specific activities ranging from 0.13 to 0.28 nmol COPROgen formed/mg protein in the 25,000g supernatant (Romeo and Levin, 1971).

This enzyme, which is known to be inhibited by iron (Kushner *et al.*, 1975), is decreased in erythrocytes as well as in the liver (Elder and Tovey, 1977; Elder *et al.*, 1978) of patients with porphyria cutanea tarda (Kushner *et al.*, 1975, 1976; Felsher *et al.*, 1978; Tiepermann and Doss, 1978), and decreased to undetectable levels in liver after induction of experimental porphyria in rats by feeding iron and hexachlorobenzene (Taljaard *et al.*, 1971, 1972). It is interesting to note that iron removal by phlebotomy alleviates the clinical symptoms of porphyria cutanea tarda (Danby *et al.*, 1966; Epstein and Redeker, 1965; Hickman *et al.*, 1967, Ippen, 1961, 1977; Saunders, 1963; Sweeney and Jones, 1979), and that an experimental porphyria normally inducible with 2,3,7,8-tetrachloro-dibenzo-*p*-dioxin (TCDD) in mice does not occur in iron-deficient animals (Sweeney *et al.*, 1979). A similar synergism with hepatic siderosis in the development of porphyria has also been observed in rats treated with hexachlorobenzene (Louw *et al.*, 1977; Smith *et al.*, 1979). Susceptibility of hepatic UROgen decarboxylase activity to TCDD treatment is also known to be under genetic influence in mice (Jones and Sweeney, 1977)—namely, the enzyme is markedly inhibited in genetically responsive mice for cytochrome P-448 induction (C57Bl/6J) but not in nonresponsive mice (DBA/2J).

The enzyme activity is inhibited *in vitro* by Hg^{2+}, Cu^{2+}, iodoace-tamide, and *p*-chloromercuribenzoate; and these inhibitions can be reversed by reduced glutathione, suggesting that UROgen decarboxylase is an SH enzyme. Oxygen and Mn^{2+} are also inhibitory to the enzyme activity (Mauzerall and Granick, 1958).

Formation of Protoporphyrinogen (PROTOgen)

The two propionic acid groups on pyrrole rings A and B of COPRO-gen III are oxidatively decarboxylated to two vinyl groups yielding PRO-TOgen IX; this reaction is catalyzed by the mitochondrial enzyme, CO-PROgen oxidase. The enzyme requires the presence of molecular oxygen

(Sano and Granick, 1961) and does not utilize either COPROgen I or COPROgen II as substrate (Porra and Falk, 1964).

Recently, COPROgen oxidase has been purified to homogeneity from bovine liver (Yoshinaga and Sano, 1980). The purified enzyme has a molecular weight of 74,000 by gel filtration and 71,600 by SDS-polyacrylamide gel electrophoresis, suggesting that the enzyme is a monomeric protein. Partially purified enzymes from yeast (Poulson and Polgase, 1974a) and rat liver (Batlle et al., 1965) also have similar molecular weights. The purified bovine liver enzyme has a specific activity of 6920 nmol PROTOgen formed/hr/mg protein. The K_m value of the purified enzyme is 4.8×10^{-5} M while it is 2.5–3.0×10^{-5} M in a crude preparation of liver mitochondria (Yoshinaga and Sano, 1980). The enzyme is rich in aromatic amino acid residues. Although 9.5 residues of free SH groups are present per mole of enzyme, the enzyme activity is not significantly inhibited by sulfhydryl reagents. It should be noted that, unlike many oxidases, no metal is found in this purified enzyme, nor do metal chelators have any effect on its activity. Using the purified enzyme and synthetic porphyrinogen intermediates, Yoshinaga and Sano (1980) demonstrated that the enzymatic conversion of the propionate groups of COPROgen III to the two vinyl groups of PROTOgen IX occurs in a stepwise fashion from positions 2–4 through the formation of a β-hydroxypropionate porphyrinogen as an intermediate. The enzyme activity is known to be decreased in patients with hereditary coproporphyria (Nordmann et al., 1977; Elder et al., 1976b). No chemicals which directly affect the enzyme activity in vivo are presently known.

Formation of Protoporphyrin IX (PROTO IX)

PROTOgen IX is oxidized in mitochondria to PROTO IX. This reaction can proceed nonenzymatically in the presence of molecular oxygen (Sano and Granick, 1961), and it was believed that a nonenzymatic mechanism is responsible for the formation of PROTO IX in vivo. However, recently a distinct mitochondrial enzyme, which catalyzes the oxidation of PROTOgen IX to PROTO IX, has been isolated and characterized by Poulson and Polgase (1974a,b, 1975). PROTOgen oxidase purified from yeast (Poulson and Polgase, 1974a) and rat liver (Poulson, 1976) requires molecular oxygen. The rat liver enzyme has a molecular weight of 35,000, an optimal pH of 8.6–8.7 and a K_m of 1.1×10^{-5} M (Poulson, 1976). A possible deficiency of PROTOgen oxidase has been proposed by Brenner and Bloomer (1979) in cultured skin fibroblasts from patients with variegate porphyria.

Formation of Heme

The final step in heme biosynthesis is the insertion of iron into PROTO IX by the mitochondrial enzyme ferrochelatase. Unlike other enzymatic steps that require a reduced porphyrinogen as substrate, this enzymatic step proceeds with oxidized PROTO IX as substrate. In rat liver cells, the enzyme is localized in the inner membrane of mitochondria (Jones and Jones, 1969; McKay *et al.*, 1969; Bugany *et al.*, 1971). In addition to Fe^{2+}, Zn^{2+} (Johnson and Jones, 1964) and Co^{2+} (Labbe and Hubbard, 1961) are good substrates for the reaction. Fe^{3+}, however, is not a substrate. Mesoporphyrin or deuteroporphyrin are also known to substitute for PROTO IX (Johnson and Jones, 1964; Jones and Jones, 1969; Porra and Jones, 1963). Removal of lipids inactivates the enzyme activity and the addition of phospholipids restores activity to normal (Yoshikawa and Yoneyama, 1964; Sawada *et al.*, 1969).

Ferrochelatase activity has been assayed by the incorporation of $^{59}Fe^{2+}$ into heme; by the formation of heme as determined by the pyridine hemochromogen assay; or by the disappearance of PROTO IX from the reaction mixture. As mentioned earlier, Co^{2+} for Fe^{2+} and meso- or deuteroporphyrin for PROTO IX have been used as alternative substrates. The assumption is made that the enzyme activity obtained by these assays is generally proportional, but it should be noted that the activity as determined with one assay may change without changing the activity determined by another assay (Tephly, 1978). For example, phenobarbital treatment causes induction of ferrochelatase while it does not affect chelatase activity for Co^{2+} (Tephly *et al.*, 1971; Hasegawa *et al.*, 1970). DDC also inhibits ferrochelatase activity (Onisawa and Labbe, 1963) but not the chelatase activity for Co^{2+} (Tephly *et al.*, 1971). When Co^{2+} and mesoporphyrin are used as substrates, however, DDC inhibits Co^{2+}-mesoheme formation (DeMatteis *et al.*, 1973).

Ferrochelatase deficiency has been described in tissues from patients with erythropoietic protoporphyria (Bonkowsky *et al.*, 1975a; Bottomley *et al.*, 1975), and in the bovine form of erythropoietic protoporhyria (Ruth *et al.*, 1977). The enzyme activity was reported to be normal in patients with variegate porphyria by Langelaan *et al.* (1970); however, another group reported a decrease in ferrochelatase activity in bone marrow (Becker *et al.*, 1977) and in cultured skin fibroblasts (Viljoen *et al.*, 1979) from patients with variegate porphyria.

REGULATION OF ENZYMES OF HEME BIOSYNTHESIS

Liver

Normally, the liver synthesizes about 15% of the heme made in the whole body (Granick and Sassa, 1971). The remainder of heme synthesis is almost entirely accounted for by the formation of hemoglobin in bone marrow erythroblasts. The amount of ALA necessary for maintaining normal levels of hemoproteins in the rat is summarized in Table I. In normal animals, ALA required for the synthesis of hepatic microsomal cytochrome P-450 (respresenting 40% of total hepatic heme) is approximately 65% of all of the ALA synthesized in the liver. The hemoprotein requiring the next greatest concentration of ALA in the liver is catalase (10% of total hepatic heme) which utilizes approximately 15% of the ALA synthesized in this organ. Hemoproteins of much lower concentrations include mitochondrial cytochromes ($\sim$30%), cytochrome b_5 ($\sim$20%), and tryptophan pyrrolase ($\sim$0.25%). Hepatic microsomal hemoproteins, particularly of the group collectively known as cytochrome P-450, turn over rapidly and their synthetic rates can be increased greatly under various conditions. The short half-lives of these hemoproteins contrast with the long life-span of hemoglobin which is essentially the same as that of the life-span of red cells. Only a few chemicals are known which induce heme synthesis in erythroid cells, while a great number of chemicals are now known to stimulate heme synthesis in the liver. Thus, under steady state conditions, heme synthetic rates are higher in erythroid cells than in the liver, but the high heme synthetic rates in erythroid cells are relatively constant in normal animals and cannot readily be stimulated further by chemicals. In contrast, heme synthesis in the liver is readily stimulated by a wide variety of drugs, chemicals, and hormones and such induction may increase hepatic heme synthesis to rates which occur normally in erythroid cells.

The rate of heme synthesis in the liver is thought to be primarily controlled by ALA synthase activity. The major reasons for this conclusion can be summarized as follows. (1) ALA synthase activity of unin-

TABLE I. Relative Concentrations and Turnover Rates of Hemoproteins in Rat Tissues

Hemoprotein	Site of production	Function	Concentration (nmol/g)	$t_{1/2}$ (hr)	ALA utilized for synthesis (nmol/g/hr)
Hemoglobin	Bone marrow	O_2 Transport	4470	600–720	99
Catalase	Liver	H_2O_2 decomposition	5.3	29	3.5
Cytochrome b, c, c_1, a + a_3		Electron transport	16.4	140	1.3
Cytochrome b_5			12.0	55	1.6
Cytochrome P-450			22.5	biphasic $\{8, 47\}$	14.4
Tryptophan oxygenase		Trytophan oxidation	0.14	2.2	0.3
Peroxidase		Reduction of H_2O_2 by electron donors	Trace	—	—
Myoglobin	Muscle	O_2 transport	40–70	biphasic $\{480, 2000\}$	0.14

duced normal liver is very low, e.g., 30–100 nmol of ALA/g mouse liver/ hr at 37°C (Hutton and Gross, 1970). This level is sufficient to maintain the steady-state concentrations of hepatic hemoproteins, but not enough to provide ALA for increased cytochrome P-450 formation under various treatments. However, in response to treatment with various chemicals, ALA synthase in the liver can be markedly induced. For example, hepatic ALA synthase activity in the rat treated with DDC plus phenylbutazone is about 1140 nmoles of ALA/g liver/hr at 37°C (DeMatteis, 1973) and in the chick embryo treated with DDC and AIA, about 6000 nmol of ALA/ g liver/hr at 37°C (Whiting and Granick, 1976a,b); these synthetic rates are at least 50–100 times greater than those observed in untreated control animals or untreated avian livers. (2) Enzymes that convert ALA to heme are considered to be present at nonlimiting activities in liver. For example, the addition of ALA to rat hepatocyte cultures increases the level of cytochrome P-450 (Bissell and Hammaker, 1976a,b). ALA administration also blocks ALA synthase induction, presumably as a result of ALA conversion to heme since labeled ALA can be readily incorporated into hepatic heme (Levin et al., 1972; Bissell and Hammaker, 1976a,b). ALA administration also increases bilirubin production (Song et al., 1971), presumably by ALA conversion to heme.

Whether ALA administration can enhance hemoprotein concentrations in vivo in the liver, is not known with certainty; the findings of Song et al. (1971) indicate that ALA administered to neonatal rats, though increasing bilirubin production, does not increase hepatic cytochrome P-450 levels per se, a finding confirmed in adult rats (Padmanaban et al., 1973), mice, and chick embryos in recent studies from our laboratory. There are variable experimental data concerning the effect of ALA on heme formation but such data may in part reflect species or tissue specificities or differences in experimental conditions. For example, ALA addition to cultured chick blastoderms is known to stimulate an early formation of hemoglobin when incubated in a plain buffer–salt–glucose medium (Levere and Granick, 1967), but ALA does not stimulate early hemoglobin synthesis when a more enriched medium is used (Wilt, 1968). ALA is known to increase cytochrome c synthesis in the developing Polyphemus silk moth (Soslau et al., 1971). Hemin added to cultured mouse Friend erythroleukemia cells increases hemoglobin formation in a dose-dependent manner, but ALA does not increase hemoglobin synthesis (Sassa, 1976; Ross and Sautner, 1976).

The rate of turnover of ALA synthase is very short which is a suitable characteristic for a rate-limiting enzyme (Goldberg and St. John, 1976). The half-life of this enzyme is about 70 min in the rat liver (Tschudy *et al.*, 1965*a*) and about 3 hr in the mouse liver (Gayathri *et al.*, 1973) and in cultured chick embryo liver cells (Sassa and Granick, 1970). The half-life of ALA synthase is considerably shorter than that of general mito-chondrial proteins (~5 days) (Druyan *et al.*, 1969). Aoki (1978) described a mitochondrial protease (mol. wt. 18,000) specific for ALA synthase, as noted earlier. This protease may be potentially important in the regulation of heme formation by affecting the level of ALA synthase in mitochondria.

Hepatic ALA synthase production can be regulated by heme, the end product of the pathway. Partially purified ALA synthase preparations from most mammalian sources can be inhibited by hemin at concentra-tions equal to or greater than 5×10^{-6} M. Repression of synthesis of ALA synthase by hemin, however, occurs at 10^{-7} M (Granick *et al.*, 1975). Thus, it appears that enzyme repression rather than inhibition of activity of preformed enzyme may be the principal regulatory action of hemin on ALA synthase in cells; however, enzyme inhibition by hemin cannot be totally excluded as a control on the enzyme because of the fact that the site of formation of heme in mitochondria is in the close vicinity of ALA synthase. Another possible mechanism of heme action on ALA synthase is the inhibition of intracellular transfer of the cytosolic form of the enzyme into mitochondria, a finding observed in the rat liver treated with AIA and hemin (Hayashi *et al.*, 1969; Yamauchi *et al.*, 1980). The most potent induction of ALA synthase is observed when animals or chick embryos are treated with chemicals which in some way reduce the ''free'' heme concentration. Free heme can be considered as a cellular fraction of heme which is rapidly turning over and which is not present as the prosthetic group of any known hemoprotein. This free heme con-centration, as well as the balance of heme supply and demand, is illus-trated in the hypothetical scheme shown in Fig. 3. The free heme pool exists at a concentration lower than the practical detection level; thus, this heme fraction remains hypothetical. However, considerable infer-ential evidence suggests that such a heme pool does exist as a metabol-ically functional entity which turns over very rapidly and in significant amounts. Free heme can be labeled as a rapidly labeled fraction of heme 1–2 hr after a brief pulse of [^{14}C]glycine in the rat (Yannoni and Robinson, 1975). The rapidity of the appearance of the labeled heme fraction does not conform to the known turnover rates of hepatic hemoproteins and

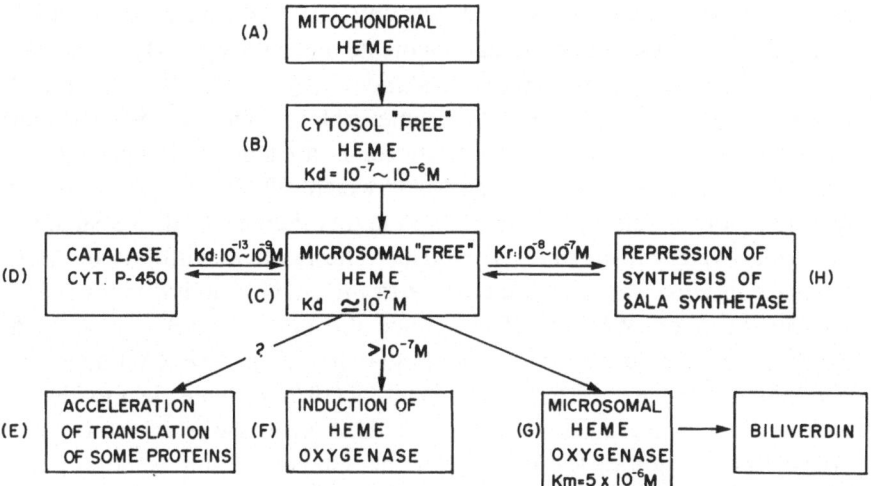

Fig. 3. A schematic model indicating the affinities of cell constituents and hemoproteins for heme and the control of heme metabolism (Granick *et al.*, *J. Biol. Chem.* **250**:9215–9225, 1975).

strongly suggests the existence of a rapidly synthesized heme fraction unassociated with specific apohemoproteins.

An important question concerning the regulation of ALA synthase is whether inducing compounds increase the amount of ALA synthase or enhance the catalytic activity of the preformed enzyme. Patton and Beattie (1973) suggested that the increased rate of heme synthesis in the livers of AIA-treated rats may be due to an enhancement in the enzyme activity rather than an actual increase in the enzyme protein. On the other hand, a more recent study by Whiting and Granick (1976*b*) utilizing a monospecific antibody raised against ALA synthase demonstrated that there are proportional increases in the amount of enzyme protein and the enzyme activity in the chick embryo liver treated with AIA or DDC. Conversely, decreases of ALA synthase activity by hemin treatment were accompanied by decreases in the immunoprecipitable enzyme protein. When the relative rate of synthesis of ALA synthase was determined by pulse labeling of the liver enzyme with [³H]leucine, followed by quantitative immunoprecipitation and SDS-polyacrylamide gel electrophoresis of the enzyme protein (Whiting and Granick, 1976*b*), it was demonstrated that there was a 500-fold increase in relative synthesis of the enzyme when a 360-fold increase in enzyme activity was produced *in vivo* in chick embryo liver. These findings clearly indicate that enhanced ALA synthase activity during the induction process elicited by chemicals is the result of an increased synthesis of the enzyme protein. Whiting (1976) further demonstrated that cytoplasmic polyribosomes are able to synthesize ALA synthase as a polypeptide of molecular weight 70,000, which *in vivo* is rapidly incorporated into mitochondria and is then excised to a smaller-molecular-weight form (ml. wt. 49,000). The relative synthesis of ALA synthase *in vitro* was also comparable to the enzyme activity induced *in vivo* in the liver (Whiting, 1976). Since the rate of synthesis of ALA synthase *in vitro* depends on the availability of its mRNA under the conditions of the assay, the data by Whiting (1976) provide clear evidence to indicate that increased synthesis of ALA synthase is a result of increased production of the mRNA for this enzyme. Using isolated poly-adenylated mRNA from chick embryo liver following induction of hepatic porphyria, Brooker *et al.* (1980) confirmed that induction of hepatic ALA synthase is due to an increase in the level of ALA synthase mRNA.

The most potent chemical inducers of hepatic ALA synthase are AIA and DDC. These two chemicals are distinct from other less-potent inducers of ALA synthase in that, in addition to their direct enzyme-inducing action, AIA degrades free heme as well as the heme moiety of cytochrome P-450; and DDC inhibits ferrochelatase activity markedly. Thus, both chemicals can be expected to reduce the cellular free heme concentration considerably, resulting in the potentiation of their direct inducing effects on ALA synthase. Potentiation of ALA synthase induction can be produced by the combination of AIA and desferrioxamine (Sinclair and Granick, 1975) or AIA and CaMgEDTA (Sassa and Kappas, 1977). Both desferrioxamine and CaMgEDTA, though having no inducing capacity of their own, chelate iron thus inhibiting ferrochelatase activity and reducing the free heme concentration even further than AIA treatment alone. This type of treatment would not be expected to potentiate the action of an inducer which directly inhibits ferrochelatase activity, and, in fact, neither CaMgEDTA nor desferrioxamine has a significant potentiating effect on the induction of ALA synthase activity following treatment with DDC, a known inhibitor of ferrochelatase (Rifkind, 1979). On the other hand, phenobarbitone, a compound which induces apocyto-chrome P-450, thereby drawing upon the pool of cellular free heme for use in cytochrome P-450 formation, can cause a greater induction of ALA synthase when combined with DDC than when administered alone

(DeMatteis, 1973). The most potent induction of hepatic ALA synthase results from the combined administration of AIA and DDC, which can lead to increases in ALA synthase ranging up to 500 times greater than normal (Whiting and Granick, 1976a,b).

Recently, it has been shown that treatment of mice with DDC (DeMatteis et al., 1980) or griseofulvin (DeMatteis and Gibbs, 1980) gives rise to green porphyrins in the liver with profound inhibitory activity toward ferrochelatase. The inhibitor can be labeled with [^{14}C]-ALA, suggesting that it is derived from prelabeled hepatic heme. Spectrally similar green porphyrins can be isolated from the livers of animals treated with ALA, secobarbital, or l-ethynylcyclohexanol (DeMatteis and Cantoni, 1979); however, they are devoid of the inhibitory activity on ferrochelatase. The site of production and the structure of these green porphyrins remains to be elucidated.

ALA synthase of rat kidney (Barnes et al., 1971), chick embryo (Yoda et al., 1974), rat adrenal (Condie et al., 1976a,b), and chick blastoderm (Levere et al., 1967) is also inducible with certain chemicals, but to a much lesser extent compared with the enzyme in liver. ALA synthase in the mouse spleen (Wada et al., 1967), mouse Harderian gland (Margolis, 1971), rat heart (Briggs et al., 1976), rat testes (Tofilon and Piper, 1980), and cultured human amniotic cells (Sassa et al., 1974b) is not inducible by chemicals that are known to induce ALA synthase in the liver. The regulatory mechanism of ALA synthase in rat fetal liver is also different from that in the adult liver. In fetal rodent liver, the level of ALA synthase activity is elevated above normal adult levels (Woods and Dixon, 1970, 1972) and hemin has no inhibitory effect on the level of the enzyme activity (Woods and Murthy, 1975). This may be partly due to the fact that prenatal rodent liver is predominantly erythropoietic and erythroid ALA synthase does not appear to be suppressible by hemin (Sassa, 1980).

Erythroid Cells

Control of heme biosynthesis in erythroid cells appears to be under a different control mechanism than that which operates in the liver. Three major erythroid systems, i.e., embryonic chick blastoderms, mammalian bone marrow cultures, and mouse Friend erythroleukemia cells have been extensively investigated. In chick blastoderms incubated in vitro, hemoglobin appears 48 hr after the initiation of incubation. However, hemoglobin appears about 12 hr earlier when cultures are incubated with ALA (Levere and Granick, 1965, 1967; Levere et al., 1967; Wainwright and Wainwright, 1966, 1967). By using inhibitors such as actinomycin D, cycloheximide, or puromycin, it has been suggested that globin mRNA is already present in the blastoderm but not translated until a sufficient amount of heme is formed. In the chick blastoderm, naturally occurring steroids stimulate the formation of ALA synthase (Irving et al., 1976) and hemoglobin (Levere et al., 1967), but chemicals like AIA and DDC which induce ALA synthase in liver do not induce this enzyme in the chick blastoderm.

Mouse Friend-virus-transformed erythroleukemia cells normally continue to divide in culture without demonstrating substantial amounts of heme or hemoglobin formation (benzidine-positive hemoglobinized cells are less than 1%). When these cells are treated with dimethylsulfoxide (DMSO) or a variety of other chemicals, they undergo erythroid differentiation and finally become hemoglobinized. It takes about 3–4 days after DMSO treatment in culture for cells to display significant increases of hemoglobin. However, ALA synthase, ALA dehydratase, UROgen I synthase (Sassa, 1976), and globin mRNA (Ross et al., 1972) are induced earlier than the increase of heme. It has been shown that a sequential induction of enzymes in the heme biosynthetic pathway takes place and that this induction occurs in the order that the enzymes are arranged in the biosynthetic sequence of reactions (Sassa, 1976; Rutherford et al., 1979). An earlier induction of ALA synthase than of hemoglobin is also observed in the chick blastoderm treated with the steroid, etiocholanolone (Irving et al., 1976). In mouse Friend cells, the fact that essentially all heme pathway enzymes are induced while cells undergo erythroid differentiation suggests that ALA synthase is not rate-limiting. Indeed, some mutants of Friend erythroleukemia cells (e.g., F4N + 2, Ma, U99) show normal induction of ALA synthase but without heme formation (Eisen et al., 1978; Sassa et al., 1978a). Other evidence indicates that in this system the terminal step of heme synthesis, ferrochelatase, may be the rate-limiting one for the following reasons: (1) in contrast to the early induction of ALA synthase, ALA dehydratase, and UROgen synthase (within 48 hr), induction of ferrochelatase, as determined by the incorporation of ^{59}Fe into heme, occurs on the fourth day after DMSO treatment when the heme concentration in the cultures also starts to increase (Sassa, 1976); (2) the time course of ^{59}Fe incorporation into heme in culture and the increase of heme concentration are very similar (Sassa, 1976); (3) hemin alone, but not ALA, PBG, or PROTO, in Friend erythroleukemia cells causes erythroid differentiation (Ross and Sautner, 1976; Granick and Sassa, 1978); and (4) DMSO plus hemin can correct defects of heme formation in certain hemoglobin-noninducible mutants of Friend cells (Sassa et al., 1978a). These findings strongly suggest that, at least in mouse Friend erythroleukemia cells, the regulation of heme biosynthesis is not at the level of ALA synthase but probably at the level of ferrochelatase (Rutherford et al., 1979; Sassa, 1976, 1980). A similar conclusion has also been reached in studies of normal human bone marrow cultures treated with erythropoietin (Sassa and Urabe, 1979). It is also known that normal mouse-bone-marrow-derived erythroid colonies are stimulated to make hemoglobin by treatment with hemin but not with ALA (Porter et al., 1979).

Indirect evidence by Karibian and London (1965) suggested, however, that heme may inhibit ALA synthase activity in erythroid cells because the in vitro addition of hemin to rabbit reticulocytes inhibited labeled glycine incorporation into heme but not the incorporation of ALA. This type of indirect experiment must be interpreted with caution, however, since hemin treatment also markedly decreases ^{59}Fe incorporation into heme (Neuwirt and Ponka, 1977; Granick and Sassa, 1978). Ibrahim et al. (1978) confirmed the findings of Karibian and London (1965) in reticulocytes, but more recently they suggested that there is a regulatory step for heme synthesis beyond ALA dehydratase in developing mouse erythroid colonies in culture (Ibrahim et al., 1979). Neuwirt and Ponka (1977) suggested that hemin's principal role in heme synthesis may be to exert a control on the entry of iron into reticulocytes and mitochondria. At any rate, treatment of Friend cells in culture with hemin shows an induction, rather than suppression or decrease of ALA synthase, followed by an increased formation of heme (Granick and Sassa, 1978). Also in fetal rodent liver, which is known to be erythropoietic, ALA synthase is refractory to the action of hemin (Woods, 1974).

There is suggestive evidence for tissue-specific isozymes of ALA synthase in the guinea pig (Bishop, 1976). Utilizing AMP (6-amino) and CoA (sulfhydryl) carboxyhexyl-Sepharose affinity column chromatography, a similarity of the erythropoietic fetal liver form to the erythropoietic adult form, but not to the adult liver form of the enzyme, has been demonstrated (Bishop, 1976). Clinical evidence which suggests that hemin might suppress ALA synthase activity in bone marrow derives from the effect of hemin treatment (Watson et al., 1974) in a patient with congenital erythropoietic porphyria. In this study, hemin infusion was reported to suppress the excretion of ALA and diminish the photosensitivity of the patient. The possible extent to which ALA and porphyrins may be produced by the liver in this patient, however, is not known.

Other Cell Types

Very little is known about the control of ALA synthase and heme biosynthesis in cell types other than liver and erythroid cells. However, available observations suggest that in most nonhepatic tissues, the regulation of ALA synthase may be different from that occurring in the liver. Hemin does not appear to inhibit ALA synthase in Harderian gland in mice (Margolis, 1971), heart (Briggs et al., 1976), adrenal gland (Condie et al., 1976a,b), and testis in the rat (Tofilon and Piper, 1979). Fetal rat liver ALA synthase is also refractory to treatment with hemin (Woods and Murthy, 1975). AIA and DDC, potent inducers of hepatic ALA synthase in adult rats, fail to induce ALA synthase in mouse Harderian gland (Margolis, 1971), rat testis (Tofilon and Piper, 1980), adrenal gland (Condie et al., 1976a,b), heart (Briggs et al., 1976), spleen (Wada et al., 1967), brain (Paterniti et al., 1978), or in cultured human amniotic cells (Sassa et al., 1974b).

The effects of hormones and nutritional factors on ALA synthase activity vary among different tissues. Fasting increases ALA synthase activity in the adrenal gland (Condie et al., 1976a,b), but decreases the enzyme activity in the heart (Briggs et al., 1976; Sedman and Tephly, 1980), and has no effect on brain ALA synthase activity (Paterniti et al., 1978). It is known that experimental hepatic porphyria elicited by chemicals is not well induced in well-fed rats compared with starved animals. Adrenal ALA synthase activity is affected by ACTH (Condie et al., 1976a,b), testicular ALA synthase is increased considerably by human chorionic gonadotropin (Tofilon and Piper, 1980), and splenic ALA synthase by erythropoietin (Wada et al., 1967). Cobaltous chloride treatment caused a marked decrease in cardiac ALA synthase activity in rats that remained suppressed for a considerably longer duration of time than in rat liver (Sedman and Tephly, 1980).

In the liver, several lines of evidence suggest that ALA synthase induction is potentiated by a decrease in free heme concentration (see above). Induction of hepatic ALA synthase in patients with acute intermittent porphyria was considered to be the result of derepression of ALA synthase formation by a presumed (though still unproved) heme deficiency in this organ. However, cultured skin fibroblasts (Bonkowsky *et al.*, 1975*b*) or mitogen-stimulated lymphocytes (Sassa *et al.*, 1978*b*) from these porphyric patients do not show elevated ALA synthase activity in spite of a deficiency of UROgen I synthase activity comparable to that found in the liver in AIP. Though indirect, such data suggest that ALA synthase is not the rate-limiting enzyme for heme formation in nonhepatic cells and it is therefore probably risky to assume that the same controls on heme synthesis which exist in liver also exist in nonhepatic cells.

CLASSIFICATION OF THE HUMAN PORPHYRIAS

Porphyrias are classified as erythropoietic or hepatic in type depending on the primary organ in which excess production of porphyrins or precursors takes place (Table II). All erythropoietic porphyrias and the majority of the hepatic porphyrias are inherited and generally rare disorders. The relatively common hepatic porphyria, known as porphyria cutanea tarda, appears to occur primarily as an acquired disorder. All porphyrias were originally classified on the basis of clinical characteristics and patterns of excretion of porphyrins and their precursors. Recent biochemical findings, particularly those related to the identification of the genetic deficiencies of enzymes of the heme biosynthetic pathway have contributed greatly to an understanding of the pathogenesis of many porphyrias, but some complexities in terms of the classification of these diseases remain. For example, the clinical syndrome of acute intermittent porphyria, with neurovisceral attacks, and the excessive urinary excretion of ALA and PBG occur at a far lesser frequency than does the decreased UROgen I synthase activity, which is found in approximately 50% of individuals in patients' families. Thus, acute intermittent porphyria is completely expressed at its enzymological or molecular level, but in its clinical phenotypes is a genetic disease having a low and variable degree of expressivity. A precise classification of the porphyrias at the molecular level will be possible only when the enzymological defects of all of the porphyrias have been elucidated; until such information becomes available it remains necessary to rely to some extent on the clinical characteristics and the biochemical findings in blood, urine, and feces of some patients for classification purposes.

TABLE II. Classification of the Human Porphyrias

| Classification | Symptoms | Biochemical findings[a] | | | Inheritance |
		Urine	Stool	Red cells	
I. Erythropoietic porphyria					
A. Congenital erythropoietic porphyria	Severe photosensitivity + hemolysis	URO I, COPRO I	—	URO I, COPRO I	Autosomal recessive
B. Erythropoietic protoporphyria	Photosensitivity	—	PROTO	PROTO	Autosomal dominant
II. Hepatic porphyria					
A. Acute intermittent porphyria	Neurovisceral	ALA, PBG	—	—	Autosomal dominant
B. Hereditary coproporphyria	Neurovisceral ± photosensitivity	ALA, PBG, COPRO	COPRO	—	Autosomal dominant
C. Variegate porphyria	Neurovisceral ± photosensitivity	ALA, PBG, COPRO	COPRO, PROTO ISOCOPRO	—	Autosomal dominant
D. Porphyria cutanea tarda	Photosensitivity	URO	ISOCOPRO	—	—
III. Toxic porphyria					
Hexachlorobenzene porphyria	Photosensitivity	URO	ISOCOPRO	—	—

[a] The major porphyrin or heme intermediate found in excess is underlined.

Congenital Erythropoietic Porphyria (CEP)

Incidence and Mode of Inheritance

CEP in man is characterized biochemically by the excessive production and excretion of type I isomers of URO and COPRO. CEP is an
extremely rare disorder transmitted in an autosomal recessive fashion.
Meyer and Schmid (1978) summarized 60 authentic cases of CEP, 32
being male and 28 being female. No chromosomal abnormalities have
been described (Heilmeyer et al., 1963; Gross, 1964). The earliest sign
of the disorder is a massive excretion of URO in the urine which is
noticeable in early childhood or in the neonatal period. A recent report
indicates that CEP can be recognized in utero by identifying elevated
amounts of porphyrins in amniotic fluid (Nitowsky et al., 1978). Unlike
acute intermittent porphyria or variegate porphyria which display a preferential occurrence in certain populations, CEP appears to have no specific population distribution.

Clinical Symptoms

The major clinical symptoms of CEP are photosensitivity and hemolytic anemia. Autonomic nervous symptoms, e.g., hypertension, neurological disturbances, and visceral pain, which characterize the inherited
hepatic porphyrias are not observed with this disorder. Skin lesions due
to photosensitivity are extensive in CEP, and are usually recognized by
the fifth year of age, though they may be lacking in the neonatal period
(Meyer and Schmid, 1978). Only two cases have been reported to have
the onset of symptoms in late life, i.e., >50 years (Kramer et al., 1965;
Pain et al., 1975). Red or pinkish urine is also noted as the first sign of
the disease. In a case where the prenatal diagnosis of CEP was made,
amniotic fluid had a dark port wine appearance (Nitowsky et al., 1978).
Skin lesions are vesiculous to bullous in nature and are usually located
on the exposed portions of the body. Blisters occur more frequently in
the summer than in other seasons. Bullae, occuring after exposure to
sunlight, may become infected and ulcerated leaving fibrous scars or
tissue deformity. Severe tissue deformity includes the loss of nails or
terminal phalanges, the tip of the nose and/or the ears, and loss of eyelids.
Hypertrichosis and generalized pigmentation and depigmentation are also
common. Both deciduous and permanent teeth may show brownish stains

("erythrodontia") which, under ultraviolet-light illumination, emit the intense red fluorescence characteristic of porphyrins. The porphyrin deposited in the teeth and bone is mainly URO, probably due to its affinity for calcium phosphate (Sveinsson et al., 1949).

Mild to severe hemolytic anemia may be present (Grinstein et al., 1949; Gross, 1964; Kench et al., 1955; Schmid et al., 1955; Watson et al., 1958); but, in some cases, is lacking (Canivet and Pelhard-Considère, 1958; Chatterjea, 1964). Splenomegaly is almost always present. One case required multiple transfusions (Gross, 1964), and two early deaths have been reported as a result of anemia (Simard et al., 1972; Sato and Takahasi, 1926). The life span of autologous red cells has been shown to be decreased in patients with CEP by using [^{15}N]glycine as a tracer (Grinstein et al., 1949; Watson et al., 1958). Splenectomy may improve the erythrocyte life span dramatically in some patients (Heilmeyer et al., 1963; Taddeini and Watson, 1968), but may be without effect in others (Gray and Neuberger, 1952).

Erythrocytes from patients with CEP, however, appear to show heterogeneous life-spans (Gray, 1952; Gray et al., 1950a,b). Approximately one-half the erythrocytes display a life-span of 20 days or shorter while the rest show an almost normal survival. Reticulocytosis, erythroid hyperplasia of bone marrow, circulating normoblasts, and increased fecal urobilinogen are common findings (Schmid et al., 1955).

The anemia of CEP appears to consist of two components, i.e., peripheral hemolysis and ineffective erythropoiesis in the bone marrow. Hemolysis in the circulation is evident from the shortened erythrocyte life-span (Kramer et al., 1965). A positive direct Coombs' test has been reported in a case of CEP, suggesting that an autoimmune hemolytic anemia may develop in some patients with CEP (Chatterjea, 1964). Ineffective erythropoiesis in the bone marrow is suggested by the fact that there is increased incorporation of [^{15}N]glycine into early-labeled bile pigment in CEP patients (Gray et al., 1950b). This finding implies that there is an increased turnover of bone marrow heme, but it does not necessarily exclude the possibility that certain "early-labeled" peaks may arise from hepatic free heme. Ferrokinetic studies in two patients with CEP also showed data compatible with either hemolysis or ineffective erythropoiesis (Kramer et al., 1965).

The bone marrow of patients with CEP shows the presence of many fluorescent erythroblasts (Stich, 1958). Fluorescence is principally localized in the nuclei of these cells (Varadi, 1958; Watson et al., 1958;

Schmid *et al.*, 1954, 1955; Larriza, 1962; Heilmeyer *et al.*, 1963; Gross, 1964). Excess heme has also been demonstrated in the nuclei of fluorescent erythroblasts by benzidine stain (Schmid *et al.*, 1955). The extrusion of nuclei containing heme is considered to contribute to the major portion of increased early-labeled bile pigments (Watson, 1966). The mechanism of hemolysis in this disorder is not yet clearly understood. Photohemolysis of red cells can be demonstrated *in vitro*; however, the extent of hemolysis *in vivo* is not correlated with exposure to light. Splenic destruction of porphyrin-laden erythrocytes, however, is clearly an important factor in the hemolytic process in this disorder. No other enzymatic defect apart from that involving the heme pathway has been demonstrated in the erythrocytes from CEP patients (Zail *et al.*, 1967; Haining *et al.*, 1968). In the bovine form of CEP, it has been suggested that porphyrin-induced decreases of phosphofructokinase activity may explain the shortened erythrocyte survival (Zinkl and Kaneko, 1973).

Biochemical Findings

Large amounts of URO and COPRO are excreted in urine (Varadi, 1958; Heilmeyer *et al.*, 1963; Taddeini and Watson, 1968; Meyer and Schmid, 1978) without increases of ALA and PBG (Watson and Schwartz, 1941). The color of urine varies from faint pink to dark reddish depending on the porphyrin content. The amount of COPRO is usually less than that of URO which has been reported to reach 500 mg/day in some patients (Varadi, 1958; Stich, 1958; Meyer and Schmid, 1978). Smaller amounts of less-carboxylated porphyrins are also excreted. The major fraction of urinary URO and COPRO is of isomer type I (Aldrich *et al.*, 1951; Taddeini and Watson, 1968; Meyer and Schmid, 1978); however, smaller increases of type III URO and COPRO are also found (Rimington and Miles, 1951; Larriza, 1962; Heilmeyer *et al.*, 1963; Watson *et al.*, 1964*a*; Heilmeyer and Clotten, 1965; Taddeini and Watson, 1968; Meyer and Schmid, 1978). Feces contain large amounts of COPRO, less URO, but not increased PROTO (Meyer and Schmid, 1978). Fecal COPRO is mainly the type I isomer. Plasma also contains URO and COPRO, which are thought to derive from porphyrins eluted from circulating red cells and from extruded nuclei of erythroblasts in the bone marrow (Aldrich *et al.*, 1951; Schmid *et al.*, 1954; Rosenthal *et al.*, 1955; Watson *et al.*, 1958; Heilmeyer *et al.*, 1963). Circulating erythrocytes contain large amounts of URO I and lesser concentrations of COPRO I. Usually, PROTO in the erythrocytes is not increased; however, in some atypical cases of CEP, PROTO may increase and become the major porphyrin in erythrocytes (Hofstad et al., 1973).

Genetic Abnormality

The genetic abnormality of CEP must account for the excessive production and excretion of the type I isomeric porphyrins as well as at least a normal rate of synthesis of type III porphyrins and normal heme formation in affected individuals (Gray, 1952). There must, therefore, be an imbalance between the type I isomer and the type III isomer production with an overall increase in porphyrin synthetic activity. Determinations of enzymes of the heme biosynthetic pathway in the bone marrow and in erythrocytes from CEP patients have shown 1.5- to 2-fold increases of ALA synthase (Masuya, 1969), ALA dehydratase (Heilmeyer and Clotten, 1965; Masuya, 1969), UROgen I synthase (Heilmeyer and Clotten, 1965), and ferrochelatase (Masuya, 1969). In contrast, UROgen III cosynthase was shown to be decreased to about one-tenth to one-third the normal level according to Romeo and Levin (1969). Based on these findings in human CEP and in the bovine form of CEP (Levin, 1968), Romeo and Levin proposed that the basic abnormality of the disease is a structural gene defect for UROgen III cosynthase (Levin, 1968, 1971; Romeo and Levin, 1969). The UROgen III cosynthase reaction was found to be reduced in vitro as the UROgen I synthase reaction proceeded. Thus, these investigators suggested that if the inactivation of the cosynthase also occurs in vivo as the synthase reaction proceeds, the result would be the imbalance between the synthase and cosynthase reaction which was observed in vitro. Although this hypothesis could explain the unusual isomer distribution pattern in CEP, it does not offer an explanation for the hyperactivity of other enzymes of the heme biosynthetic pathway. There is also an unusual case of CEP with a predominant increase of erythrocyte PROTO described by Eriksen and his co-workers (Hofstad et al., 1973; Eriksen and Eriksen, 1974) which cannot be explained by a UROgen III cosynthase deficiency. Marked elevation of erythrocyte PROTO has been observed in bovine CEP as well (Watson et al., 1958).

To account for overactivity of the heme pathway with an imbalance between the type I and III isomer production, Watson et al. (1964a) proposed a constitutive regulator mutation for certain steps in the heme biosynthetic pathway. Another possibility has been suggested by the same group (Miyagi *et al.*, 1976), namely, that a structural gene mutation leading to hyperactive ALA synthase or UROgen I synthase could provide a reasonable explanation for the biochemical findings in CEP. Although either proposal could account for the observed porphyrin abnormalities in CEP, no direct proof for either mechanism has been obtained. Recently, Moore *et al.* (1978) determined ALA synthase, COPROgen oxidase, and ferrochelatase in a buffy coat of blood consisting mainly of granulocytes in two CEP patients; in addition ALA dehydratase, UROgen I synthase, UROgen III cosynthase, and UROgen decarboxylase were determined in the erythrocytes from the same two individuals. Following the rate of conversion of UROgen through the UROgen III cosynthase reaction to the corresponding COPRO isomers, the ratio of series I isomer to series III isomer was found to be 3.6 and 4.2 in two cases of CEP, while in normal controls such small quantities of the series I isomer were formed that quantitation of the ratio was precluded. Since ALA synthase (~4.8 times), ALA dehydratase (1.4–2.2 times), UROgen I synthase (~1.5 times), UROgen decarboxylase (1.3–1.7 times), COPROgen oxidase (~1.2 times), and ferrochelatase (0.9–1.5 times) were all found to be generally increased, the data were interpreted to indicate that the activity of UROgen III cosynthase is proportionally less than the activity of the UROgen I synthase (Moore *et al.*, 1978). The authors concluded that the primary control of heme biosynthesis in CEP is at the level of ALA synthase, with a secondary control point at the level of UROgen III cosynthase, although the genetic basis for these enzymatic abnormalities remains unclear.

CEP with an Unusual Pattern of Porphyrin Distribution

In 1973, Hofstad *et al.* described a case of CEP whose porphyrin excretion pattern was quite unusual compared with the patterns described for other CEP patients. The patient was a boy who had reddish urine from the first day of life and was known to have a hemolytic anemia with splenomegaly and a 6.2–7.5% reticulocytosis when he was 7 weeks old. However, the diagnosis of CEP was not made until he was 2 years old, when he started to display a typical rash after exposure to sunlight. Porphyrin analyses of urine, feces, plasma, and erythrocytes revealed a large amount of porphyrins in these tissues. Urinary porphyrins consisted primarily of 7-carboxylic porphyrin, and fecal porphyrins consisted of

COPRO and smaller amounts of PROTO. Erythrocytes contained primarily PROTO. These features are considerably different from those observed in other cases of CEP, but the marked elevation of PROTO in the erythrocytes is similar to the findings in bovine CEP (Watson *et al.*, 1959).

A study of the excretion pattern and porphyrin distribution in this case was published subsequently by Eriksen and Eriksen (1974). Unlike the predominance of URO I in the urine of most CEP patients, more than 50% of the porphyrins excreted in the urine in this case were the type III isomers, most of which were hepta-, hexa-, and pentacarboxylic porphyrins. The activity of porphyrin synthesizing enzymes in erythrocytes appeared within normal limits except for a moderate increase of UROgen I synthase activity (Eriksen and Seip, 1973). An apparent isocoproporphyrin (tris-[2-carboxyethyl]-carboxymethyl ethyl trimethyl porphyrin) was found in feces, urine, bone marrow, buffy coat, plasma, and liver. Isocoproporphyrin is a four-carboxylic porphyrin in which the propionic acid side chain on position 2 has already been decarboxylated, but the acetic acid side chain on position 5 of ring C is still intact. The patient's liver contained a large amount of hexacarboxylic porphyrin, while the spleen contained relatively large amounts of URO. These findings were subsequently confirmed by Rimington and With (1973) in the same patient. Desethylisocoproporphyrin and dehydroisocoproporphyrin, known to be excreted in symptomatic porphyria cutanea tarda (Elder, 1972), were not detected (Eriksen and Eriksen, 1974), nor was hydroxyisocoproporphyrin found (Eriksen and Eriksen, 1974; Rimington and With, 1973) in the patient. The excretion of large amounts of heptacarboxylic porphyrins in this patient and the presence of the porphyrin fraction called "P1" probably containing one acetate and three propionate β-substituents (Elder, 1972), was suggested to be indicative of a disorder in hepatic porphyrin metabolism (Rimington and With, 1973). Although two liver biopsies were performed, the amounts of tissues taken were too small for porphyrin analysis (Eriksen and Eriksen, 1974). The same was true for plasma studies in the patient. Eriksen and Eriksen (1974) believe that the presence of trace amounts of an apparent isocoproporphyrin and relatively large amounts of hepta- and hexacarboxylic porphyrins both in the liver and in bone marrow suggest an erythroid origin of porphyrins in this subject. The finding that more than 50% of the excreted porphyrins were type III in this case cannot be explained by a deficiency of UROgen III cosynthase alone (Romeo and Levin, 1969). Thus, if classical CEP is due to UROgen

III cosynthase deficiency, this unique CEP case originally reported by Hofstad *et al.* (1973) may represent a distinct genetic variant.

Two apparently similar cases of porphyria have also been reported by Piñol Aguadé *et al.* (1969, 1975). These cases developed blisters, hypertrichosis, and scarring of the exposed skin in early childhood. Porphyrin examination showed an abnormal urinary excretion pattern with predominantly pentacarboxylic porphyrins and elevated erythrocyte PROTO. The fecal porphyrin pattern was similar to that of porphyria cutanea tarda.

Erythropoietic Coproporphyria

Two cases of another possible variant of CEP have been described by Heilmeyer and Clotten (1964, 1966). The first was a 22-year old woman who suffered only a mild cutaneous photosensitivity from the age of 10. Her erythrocytes and normoblasts exhibited prominent red fluorescence upon exposure to ultraviolet light and contained a large amount of COPRO III (200 times normal); feces contained a large amount of COPRO and PROTO. These findings indicate that there is an overproduction mainly of type III isomer porphyrins in the erythroid cells of this patient. A first cousin (daughter of her father's brother) was said to suffer from a similar light sensitivity. Her asymptomatic mother also displayed a similar porphyrin abnormality, suggesting that the disorder was transmitted in an autosomal dominant fashion. In contrast to the marked elevation of COPRO in erythrocytes and in feces, urinary COPRO excretion remained within normal limits.

Erythropoietic Protoporphyria (EPP)

EPP is an inherited disorder of porphyrin metabolism which was originally described by Magnus *et al.* (1961). The disease is characterized clinically by cutaneous photosensitivity to wavelengths of light in the long-ultraviolet range. Patients with EPP have marked increases of PROTO in their erythrocytes, plasma, and feces but not in their urine. Photosensitivity is not related to the PROTO levels in their erythrocytes but probably to the PROTO in plasma (Redeker *et al.*, 1963). In contrast to CEP which is rare, EPP is much more common; more than 300 cases have been recorded (DeLeo *et al.*, 1976) in the nearly 20 years since its first description. EPP is an autosomal dominant disorder occurring in both sexes, with a variable degree of penetrance (Reed *et al.*, 1970). Identical twins with EPP have been described (Redeker and Bronow, 1964). The disease can also occur in a completely latent form (Haeger-Aronsen and Krook, 1966; Sassa *et al.*, 1979*b*). In contrast to the hepatic porphyrias, in which latent subjects may be vulnerable to the chemical provocation of their disease, no latent EPP subjects have been shown to be activated clinically by exposures to chemicals (Lynch and Miedler, 1965).

Clinical Findings

DeLeo *et al.* (1976) reported their ten-years' experience with 32 cases of EPP. They recorded the onset in childhood of burning (97%) and itching (88%) of the skin on exposure to sunlight, accompanying edema (94%) and erythema (69%), with a less-frequent incidence of vesicles, petechiae, and residual scarring. Cholelithiasis (12%), anemia (27%), and abnormal liver function tests (4%) were also noted as associated abnormalities. Certain of the symptoms that characterize CEP do not occur with EPP thus allowing the distinction of these two disorders. These symptoms are erythrodontia and fluorescence of the teeth and skin, hirsutism, and hyperpigmentation in CEP (Haeger-Aronsen and Krook, 1966). Hemolytic anemia in EPP occurs only occasionally and usually is insignificant. Ferrokinetics are generally within normal limits (Turnbull *et al.*, 1973).

Biochemical Findings

EPP is a unique porphyria in that there is no abnormality in porphyrin pattern in the urine. The most remarkable findings in EPP are the striking increases of PROTO in erythrocytes and feces and lesser, but significant, amounts of PROTO in plasma. Red fluorescence characteristic of porphyrins is detectable in most reticulocytes (Clark and Nicholson, 1971) but only in small numbers of mature erythrocytes (Kaplowitz *et al.*, 1968; Magnus *et al.*, 1961; Cripps *et al.*, 1966). Only 9–33% of fluorocytes are found in the peripheral blood (Langhof *et al.*, 1961, 1964). Curiously, no fluorescing erythroblasts are found in the bone marrow according to Porter and Lowe (1963), or fluorescence is only present in the cytoplasm (Haeger-Aronsen, 1963) which is in contrast to the findings in CEP. Apparently, PROTO accumulation occurs in late normoblasts and reticulocytes but the porphyrin quickly diffuses out of red cells into the blood stream.

The leakage of porphyrin out of circulating erythrocytes has been examined by Piomelli *et al.* (1975) using age-dependent fractionation of erythrocytes on discontinuous density gradients. They described that the PROTO concentration declined rapidly with erythrocyte age in EPP patients—indeed the bulk of PROTO and of fluorocytes was lost in less than 3 days. In the bone marrow, fluorescence was observed only in occasional late normoblasts, but in all reticulocytes. These findings suggest that PROTO accumulation occurs at the late stage of erythroid maturation and that all the erythroid cells uniformly carry the gene defect of this disorder. In contrast, in erythrocytes from patients with lead poisoning, PROTO declined only slightly with cell age and erythrocytes of essentially all ages fluoresced. Incubation of cells in plasma demonstrated rapid diffusion of PROTO from erythrocytes derived from patients with EPP but not from erythrocytes derived from patients with lead poisoning. Likewise, plasma PROTO was found elevated only in EPP. The daily loss of erythrocyte PROTO in EPP was estimated to be 40% of the total PROTO content of the circulating red-cell mass. These workers concluded that this high daily loss, combined with the additional loss from bone-marrow reticulocytes, could readily account for most of the daily fecal PROTO excretion in this disorder without the need to postulate a preponderant extraerythropoietic source, i.e., liver, of PROTO formation (Scholnick *et al.*, 1971).

Genetic Defect

Earlier observations in EPP raised the possibility of dual sources for PROTO production, i.e., erythrocytes and the liver. First, the daily excretion of PROTO in feces is far greater (~50 times) than the total amount of PROTO in circulating erythrocytes. Second, several affected individuals with elevated fecal PROTO lacked PROTO increases in erythrocytes (Redeker and Bryan, 1964; Haeger-Aronsen, 1963). Third, isotopic studies with [^{15}N]glycine in an EPP patient demonstrated that the specific activity of erythrocyte PROTO was much lower than that of fecal stercobilin and fecal PROTO (Gray *et al.*, 1964). Fourth, elevated hepatic ALA synthase activity (Cripps and MacEachern, 1971; Masuya, 1969) as well as an increase in PROTO formation by caloric restriction (Redeker and Sterling, 1968), an effect thought to be specific for hepatic ALA synthase, have been described in several patients with EPP. These findings suggested that a certain amount of PROTO in EPP might derive from the liver. Scholnick *et al.* (1971) labeled PROTO with [^{14}C]glycine and [^{3}H]-ALA and followed the radioactivity in PROTO in blood and feces. This experimental protocol was based on the fact that the labeling pattern of porphyrins in the liver and in erythroid cells by the two isotopes is markedly different—namely, tracer ALA labels liver porphyrins and heme over 1,000 times more intensely than erythrocyte heme, whereas tracer glycine labels these fractions more uniformly (Shemin, 1955; Robinson *et al.*, 1966). Analysis of the isotope kinetic data suggested that the major fraction of PROTO appeared to be derived from liver and the rest from erythrocytes. Based on these findings, Scholnick *et al.* (1971) proposed that EPP be renamed "erythrohepatic porphyria." Their conclusion is in apparent conflict with the more recent study by Piomelli *et al.* (1975) who advocated a strict erythropoietic origin of PROTO in EPP. Schwartz *et al.* (1971), using two labeled precursors as did Scholnick *et al.* (1971), proposed earlier a conclusion similar to that of Piomelli *et al.* (1975), i.e., that the excess PROTO in EPP derived only from erythropoietic cells. The possible heterogeneity of EPP, however, should not be overlooked.

The original findings by Magnus *et al.* (1961) suggest that there may be a deficiency of ferrochelatase in the peripheral blood of patients with EPP. More recently, direct proof for a ferrochelatase deficiency has been obtained by enzyme assay in bone marrow (Bottomley *et al.*, 1975), in liver and in cultured skin fibroblasts (Bonkowsky *et al.*, 1975a), and in nucleated blood cells in the circulation (De Goeij *et al.*, 1975; Brodie *et al.*, 1978). These results clearly indicate that ferrochelatase deficiency is present, probably in all tissues, and that this enzyme deficiency then becomes rate-limiting for heme formation in erythroid cells in the marrow and possibly in the liver as well. ALA synthase activity in bone marrow cells has been reported to be increased in some EPP patients (Takaku *et al.*, 1972), but not in others (Bottomley *et al.*, 1975).

A recent study from our laboratory has shown that a partial deficiency of ferrochelatase activity in EPP can be functionally demonstrated in mitogen-stimulated human lymphocytes by incubating cells with ALA and determining the amounts of PROTO formed (Sassa et al., 1979b). Mitogen-stimulated EPP lymphocytes accumulated substantially greater amounts of PROTO than did mitogen-stimulated normal lymphocytes when incubated with ALA (Table III). Since the stimulation of DNA synthesis by mitogens of EPP lymphocytes is similar to that of normal lymphocytes, these data suggest that ferrochelatase activity in EPP cells is less than that in normal cells. Moreover, the fraction of ferrochelatase in EPP lymphocytes which could be inhibited by CaMgEDTA was less than that in normal lymphocytes (ratio B/A in Table III) and iron supplementation facilitated PROTO utilization for heme synthesis in normal lymphocytes but not in EPP cells (ratio C/A in Table III). These findings provide good evidence for a functional deficiency of ferrochelatase activity in EPP lymphocytes.

Utilizing this method, we have been able to identify patients with clinically manifested EPP as well as completely latent gene carriers with the EPP defect and also to distinguish these two populations from normals. For example the father (I-1) and a daughter (II-1) in the family lineage shown in Fig. 4 were normal in all respects and thus considered to be normal controls in this family. The mother (I-2) and two daughters (II-3, II-4) were clinically normal and did not have photosensitivity or elevated erythrocyte PROTO levels. PROTO formation in lymphocytes from these subjects was also normal. However, plasma from these three

TABLE III. PROTO Formation from Added ALA in Human Lymphocytes[a,b]

Subjects	Protoporphyrin IX (pmol/mg protein/24 hr)				
	(A) ALA	(B) ALA + CaMgEDTA	(C) ALA + Iron	Ratio: B/A	Ratio: C/A
Normal	$269 \pm 24(7)^{b}$	$776 \pm 81(7)$	$201 \pm 11(3)$	$2.58 \pm 0.23(7)$	$0.74 \pm 0.08(3)$
EPP	$638 \pm 54(11)$	$1043 \pm 77(11)$	$638 \pm 103(2)$	$1.71 \pm 0.16(11)$	$1.02 \pm 0.06(2)$
P value	<0.001	<0.001	<0.001	<0.001	<0.001

[a] From Sassa et al., 1979, Trans. Assoc. Am. Physicians **92**:268–276.
[b] Lymphocytes were preincubated with the mitogens for 3 days and PROTO formation was studied during the following 24-hr period in the presence of (A) 0.6 mM ALA, (B) 0.6 mM ALA + 5 mM CaMgEDTA, and (C) 0.6 mM ALA + 25 μM ferrous ammonium sulfate.
[c] Numbers in parentheses indicate the number of cultures studied.

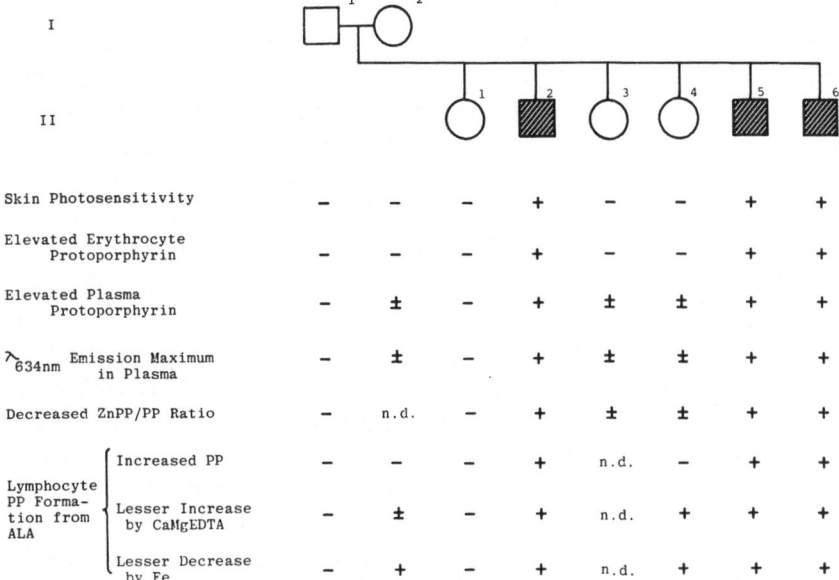

	1	2	3	4	5	6
Skin Photosensitivity	−	−	−	+	−	−	+	+
Elevated Erythrocyte Protoporphyrin	−	−	−	+	−	−	+	+
Elevated Plasma Protoporphyrin	−	±	−	+	±	±	+	+
λ_{634nm} Emission Maximum in Plasma	−	±	−	+	±	±	+	+
Decreased ZnPP/PP Ratio	−	n.d.	−	+	±	±	+	+

Lymphocyte PP Formation from ALA	Increased PP	−	−	−	+	n.d.	−	+	+
	Lesser Increase by CaMgEDTA	−	±	−	+	n.d.	+	+	+
	Lesser Decrease by Fe	−	+	−	+	n.d.	+	+	+

Fig. 4. A family pedigree of EPP. The father (I-1) and one daughter (II-1) were normal clinically as well as biochemically. The mother (I-2) and two daughters (II-3, II-4) were clinically unaffected but displayed biochemical characteristics of EPP. Three sons (II-2, II-5, II-6) displayed both clinical and biochemical characteristics of EPP. PP) protoporphyrin IX; ZnPP) zinc protoporphyrin IX; n.d.) not determined (Sassa *et al., Trans. Assoc. Am. Physicians* **92:**268–276, 1979).

subjects contained a small but distinctly elevated amount of PROTO, thus distinguishing them from the normal controls. Moreover, the responses of their lymphocytes in culture to the presence of CaMgEDTA or iron were clearly abnormal, characterizing them biochemically as similar to EPP patients (Fig. 4). Therefore, these three subjects (I-2, II-3, and II-4) can be considered to be latent gene carriers of EPP, since they displayed no clinical symptoms of this disease but clearly showed some of its biochemical characteristics.

These technique has also been applied to cultured skin fibroblasts to diagnose the bovine form of EPP by incubating cells with ALA, ALA plus CaMgEDTA, and ALA plus iron. The bovine form of EPP is an autosomal recessive disorder (Ruth *et al.,* 1977) in contrast to the autosomal dominant nature of human EPP. It was possible with this methodology to diagnose clinically expressed homozygotes, clinically latent heterozygotes, and normal controls among animals of a bovine EPP lineage.

The lymphocyte culture technique used to study human EPP is simple and allows a relatively rapid diagnosis of the disease as well as the identification of latent gene carriers of EPP over a four-day culture period. The short culture time for lymphocytes is a considerable advantage over the long time period required for identification of the EPP gene defect utilizing skin fibroblast cultures. Mitogen-stimulated lymphocytes were first utilized for detection of the gene defect of URO I synthase activity in patients with AIP (Sassa *et al.*, 1978*b*). Using CaMgEDTA, which blocks the utilization of PROTO for heme formation, the level of PROTO formation from added ALA in AIP lymphocytes was approximately one-half that observed in normal controls, thus reflecting the partial deficiency of UROgen I synthase characteristic of this disorder.

Ferrochelatase activity directly determined in cultured skin fibroblasts from EPP patients has been found to be approximately 10% of control (Bonkowsky *et al.*, 1975*a*; Bloomer *et al.*, 1977). This is a rather unexpected finding since EPP subjects, who are all considered to represent heterozygotes for this enzymatic deficiency, are expected to display a 50% deficiency of ferrochelatase. In fact, in other dominant forms of the human porphyrias, e.g., AIP and hereditary coproporphyria, an enzymatic deficiency of 50% has been described (Strand *et al.*, 1972; Sassa *et al.*, 1974*a,b*; Elder *et al.*, 1976*b*; Grandchamp *et al.*, 1977). Only one case of homozygous hereditary coproporphyria (Grandchamp *et al.*, 1977) has been reported; this individual showed a level of COPROgen oxidase activity only 5% of that of control cells. In the recessive bovine form of EPP, ferrochelatase activity in the affected homozygotes was approximately 10% of controls; and was about 50% below normal in the clinically latent heterozygotes (Ruth *et al.*, 1977). Thus, an extremely low ferrochelatase activity (~10% of controls) in cultured skin fibroblasts from EPP patients, who presumably represent heterozygotes for the gene defect, is most unusual. Bloomer (1980) proposed that ferrochelatase in EPP may be structurally different from normal, with greatly reduced catalytic activity. No direct proof for this hypothesis, however, has yet been provided. On the other hand, assessment of ferrochelatase activity indirectly in intact cells by the determination of the rate of PROTO accumulation after incubation of cultured fibroblasts (Bloomer *et al.*, 1977) or mitogen-stimulated lymphocytes (Sassa *et al.*, 1979*b*) with ALA suggests that functional ferrochelatase activity in EPP cells is approximately

50%, not 10%, of that of control cells, and this level of enzyme deficiency is in keeping with the heterozygous state of patients with this disease. The possibility needs to be considered that attempts to measure ferrochelatase activity directly in isolated mitochondria or after release of the enzyme from the mitochondrial membranes in tissues with a very low enzymatic activity, may impair the enzyme's catalytic activity. If so, an indirect assessment of the functional activity of ferrochelatase in whole cells, i.e., as in mitogen-stimulated lymphocytes or skin fibroblasts treated with ALA may provide a more accurate assessment of ferrochelatase activity (Sassa *et al.*, 1979*b*).

Acute Intermittent Porphyria (AIP)

AIP is an autosomal dominant disease affecting porphyrin–heme metabolism in the liver. It is the most common form of inherited hepatic porphyria and is unique in that the clinical syndrome does not include cutaneous photosensitivity. Porphyrin excretion is not usually excessive. AIP is often termed the Swedish type of porphyria (vs. variegate porphyria, as the South African type). In Sweden, there are more than 600 well-documented cases of AIP (Waldenström, 1957; Wetterberg, 1967). Clinical manifestations of this disorder have been very rarely observed before puberty. One case of putative AIP occurring at the age of four months has been reported (Beauvais *et al.*, 1976). Although this case was associated with a somewhat low erythrocyte UROgen I synthase activity, urinary PBG was slightly elevated only on two out of eight determinations, thus leaving considerable doubt about the diagnosis. Erythrocyte UROgen I synthase in children in this age range is also known to be lower than that of adults (Nordmann *et al.*, 1976). The incidence of AIP was originally estimated to be 1.5/100,000 in Sweden based on urinary ALA and PBG determinations (Goldberg and Rimington, 1962), but this is clearly an underestimate from more recent experience, since urinary ALA and PBG are not necessarily elevated in gene carriers of this disorder. Wetterberg (1967) estimates that the incidence of AIP in the population 15 years of age or older in Sweden is 7.7/100,000. The incidence of AIP is clearly affected by its occurrence in large families (Meyer and Schmid, 1978). The highest incidence has been observed in Lapland (100/100,000) (Goldberg and Rimington, 1962) and all AIP subjects there appear to be related to a single family. In Finland, Mustajoki and Koskelo (1976) recorded 107 patients with AIP and 45 patients with variegate porphyria in 42 families during a period of nine years and the prevalence of hereditary hepatic porphyria (including both AIP and variegate porphyria) was calculated to be 3.4/100,000. This figure is more or less similar to the one for the rest of the world but quite different from the particularly high value for the Swedish population. With the use of the more sensitive UROgen I synthase assay which detects all gene carriers for this disorder, including clinically manifest as well as completely latent cases in adults and children, the incidence of the gene carrier state for AIP must be expected to increase, probably at least by a factor of 10 (Strand *et al.*, 1972; Sassa *et al.*, 1974*a*).

Clinical Findings

The clinical symptoms of the three inherited hepatic porphyrias, AIP, hereditary coproporphyria, and variegate porphyria, are very similar with the exception that the latter two porphyrias may also be associated with skin photosensitivity. The incidence of specific symptoms and physical signs in patients with acute attacks have been summarized in Table IV. All of the symptoms can be related to the neurological disturbances of this disorder. The most characteristic neurological symptoms are those involving the autonomic nervous system, i.e., abdominal pain, vomiting, and constipation (Goldberg and Rimington, 1962; Taddeini and Watson, 1968; Tschudy *et al.*, 1975; Tschudy and Lamon, 1980; Meyer and Schmid, 1978).

The severity of clinical manifestations, incidence of acute episodes, and the date of onset of the clinical expression of AIP are quite variable. In addition, the great majority of gene carriers of the disorder, i.e., those characterized by a 50% deficiency of UROgen I synthase, remain clinically asymptomatic throughout their lives. These facts make it evident that additional factors including dietary, chemical, and hormonal influences must be important in the clinical expression of the gene defect of AIP. Many of the chemicals which are potentially hazardous for patients with the inherited forms of hepatic porphyrias have been listed in the reports of Wetterberg (1975) and Rifkind (1976). Most of these substances have been shown to cause the induction of ALA synthase and porphyrins in cultured chick embryo liver cells or the chick embryo liver *in ovo*, an

TABLE IV. Incidence of Clinical Symptoms and Signs in Acute Porphyric Attacks in
Patients with AIP

Symptom	Waldenström (1957) ($N = 321$)	Mustajoki and Koskelo (1976)[a] ($N = 88$)	Goldberg (1959) ($N = 50$)
Abdominal pain	85%	95%	90%
Vomiting	59	80	75
Constipation	48	80	70
Diarrhea	9	5	10
Paresis/paralysis	42	50	66
Muscle pain	—	70	—
Convulsions	10	20	16
Sensory loss	9	25	38
Transient amaurosis	4	—	6
Diplopia	3	—	—
Bulbar paralysis	14	20	—
Delirium	28	25	—
Hysteria, depression	27	—	12
Psychosis	—	—	12
Hypertension	40	55	55
Tachycardia	28	85	60
Pigmentation	7	—	16
Fever	37	—	12
Azotemia	10	—	6
Jaundice	3	—	—
Oliguria	7	—	—
Proteinuria	9	—	—
Urinary PBG	100	100	100

[a] This study includes both acute intermittent porphyria and variegate porphyria.

experimental model system which is quite useful for screening chemicals
for their ability to activate the porphyrin–heme pathway in liver.

Biochemical Findings

AIP is characterized by the urinary excretion of large amounts of
PBG and ALA. PBG can be detected by reacting it with Ehrlich's aldehyde reagent which forms a salt with a reddish color. The Ehrlich's
aldehyde–PBG complex displays in intense absorption band at 553 nm
and a weaker band at 525 nm, the ratio of the absorptions at 525 nm and

553 nm being 0.85 (Mauzerall and Granick, 1956). The concentration of PBG in normal urine is less than the detection level of the original method described by Watson and Schwartz (1941); when PBG excretion is markedly elevated, however, as in acute attacks of AIP, this test produces an intense reddish color. The Watson–Schwartz test may, as the authors have noted (Watson and Schwartz, 1941), give false positive tests. Therefore, the modified Watson–Schwartz test is recommended to distinguish PBG from urobilinogen; the modification takes advantage of the fact that the Ehrlich's aldehyde–PBG complex is insoluble in chloroform and butanol while the complex with urobilinogen can be extracted from urine into these organic solvents (Watson et al., 1961, 1964b; Taddeini and Watson, 1968). Quantitative and specific analysis of PBG, however, requires the column chromatography method of Mauzerall and Granick (1956) using DOWEX ion-exchange resin. This method is sensitive enough to quantitate PBG in normal urine (normal levels <4 mg/day).

The amount of ALA in urine usually parallels the excretion of PBG. Urinary PBG, however, is more specific than ALA in the diagnosis of the hereditary hepatic porphyrias, since PBG does not increase in other conditions in which ALA excretion may be elevated, e.g., lead poisoning. Urinary PBG may decrease during clinical remissions of AIP, but urine PBG rarely becomes completely normal during remission in AIP patients who have had only one acute episode of the disease. On the other hand, many gene carriers of AIP, characterized by low UROgen I synthase activity in erythrocytes or other cells, remain PBG negative throughout their lives and never develop symptoms.

Porphyrin concentrations in freshly voided urine from AIP patients in remission and in relapse are usually not elevated significantly (Cookson and Rimington, 1954; Tschudy, 1965; Meyer and Schmid, 1978). Upon standing, however, urinary porphyrin values may increase, presumably because of spontaneous cyclization of PBG to form porphyrins (Cookson and Rimington, 1954; Watson, 1954). This cyclization is facilitated by the acidic pH of urine and by light (Cookson and Rimington, 1954). Thus, most of the urinary porphyrins in AIP appear to be formed artifactually in urine after voiding (Meyer and Schmid, 1978). In agreement with this view, COPRO and PROTO in the feces are only slightly elevated.

Both ALA and PBG have been detected in the plasma and cerebrospinal fluid of AIP patients (Sweeney et al., 1970; Bonkowsky et al., 1971). During acute attacks, plasma ALA concentration may be as high as 75 μg/100 ml (Sweeney et al., 1970; Bonkowsky et al., 1971). ALA, probably because of its chemical similarity to GABA, is able to compete for synaptic γ-isobutyric acid (GABA) receptor binding (Müller and Snyder, 1977); inhibit potassium-stimulated release of GABA from preloaded synaptosomes (Brennen and Cantrell, 1979); mimic the neurophysiological effects of GABA in eliciting presynaptic and postsynaptic inhibition in spinal cord (Nicoll, 1976); inhibit muscular responses to nerve stimulation (Cutler *et al.*, 1978), and inhibit the release of acetylcholine evoked by a nerve impulse (Bornstein *et al.*, 1979). These effects occur only at very high concentrations of ALA (at least 10^{-4} M), except for the inhibition of potassium-stimulated release of GABA at 10^{-6} M (Brennan and Cantrell, 1979). It is not clear, therefore, that porphyric symptoms are caused by a direct effect of ALA on central nervous system cells. It is also known that levels of ALA and PBG in cerebrospinal fluid are considerably lower than those in plasma in the acute porphyrias (Bonkowsky *et al.*, 1971; Percy and Shanley, 1977). Intraperitoneal or intrathecal administration of ALA to rats has been reported to induce neurological and behavioral abnormalities (Moore and Meredith, 1976; Pierach and Edwards, 1978); however, Pierach and Edwards (1978) concluded that ALA (or PBG) is unlikely to cause the neurological disturbances of the porphyrias because relatively large amounts are required to cause neurological aberrations in rats. Moreover, seizures are the major symptoms in animals and, while these may occur in AIP patients, they are not common.

The concentration of PBG is high in the livers of AIP subjects, but this compound is not detected in other tissues (Schmid *et al.*, 1954; Smith, 1960; Goldberg and Rimington, 1962).

Gene Defect

The fact that urinary ALA and PBG are excreted in excess and that hepatic ALA synthase activity is increased during acute episodes of the disease in AIP patients (Tschudy *et al.*, 1965*b*; Nakao *et al.*, 1966; Dowdle *et al.*, 1967; Sweeney *et al.*, 1970), led to the original hypothesis that AIP might be the first genetic disease characterized by the over-production of an enzyme—specifically, ALA synthase (Tschudy *et al.*, 1965*b*). Though later studies established a different enzymatic expression of the primary gene defect of AIP, this hypothesis (Tschudy *et al.*, 1965*b*), together with the earlier demonstration by Granick and Urata (1963) that ALA synthase was inducible, stimulated the interest of many investigators in the regulation of this enzyme by endogenous and exogenous chemicals.

Recent studies indicate that there may be combined abnormalities of the heme pathway in this disorder, namely, UROgen I synthase deficiency and ALA synthase elevation, but that the URO I synthase deficiency reflects the primary gene defect of AIP. Decreased UROgen I synthase activity has been described not only in the livers of patients with AIP (Strand *et al.*, 1970; Miyagi *et al.*, 1971), but also in erythrocytes (Fig. 5) (Meyer *et al.*, 1972; Strand *et al.*, 1972; Sassa *et al.*, 1973, 1974a; Magnusson *et al.*, 1974; Schumaker *et al.*, 1976; Peterson *et al.*, 1976; Nordmann *et al.*, 1976; Grandchamp *et al.*, 1976; Mustajoki, 1976; Astrup, 1978; Doss and Tiepermann, 1978; Brocklehurst, 1978; Piepkorn *et al.*, 1978), cultured fibroblasts (Meyer, 1973; Sassa *et al.*, 1975a,b; Bonkowsky *et al.*, 1975b), cultured amniotic cells (Sassa *et al.*, 1975a,b), and mitogen-stimulated lymphocytes (Sassa *et al.*, 1978b). The UROgen I synthase deficiency (~50% of the normal level) has been observed invariably in these tissues regardless of the presence or absence of clinical symptoms of AIP or the presence or absence of a concomitantly elevated

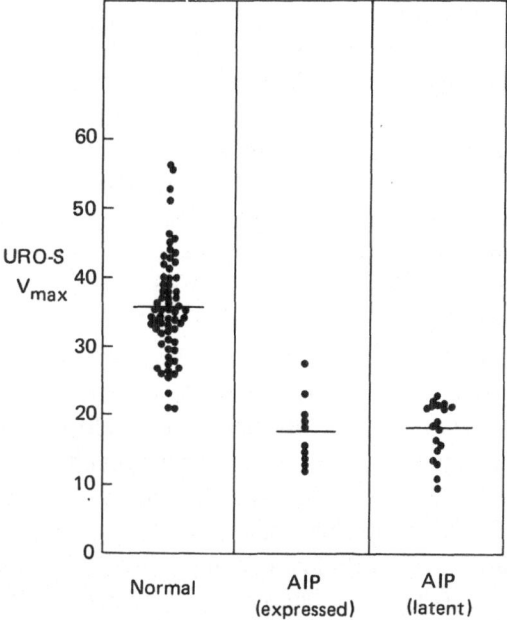

Fig. 5. Erythrocyte UROgen I synthase activities in 37 normal subjects, 10 patients with expressed AIP, and 18 subjects with completely latent AIP. The enzyme activity (V_{max}) is expressed as nmol of UROgen per ml of erythrocytes per hr at 37°C. The horizontal line represents the mean value for each group (Sassa *et al.*, *Proc. Natl. Acad. Sci. USA* **71**:732–736, 1974a).

ALA synthase. In fact, ALA synthase activity has been found to be elevated only in the liver cells of AIP patients during acute attacks of the disease. These findings establish that the UROgen I synthase deficiency is the primary inherited defect of AIP; that this enzymic deficiency characterizes all AIP gene carriers including those who remain clinically completely asymptomatic; and that the elevation in ALA synthase activity is probably an associated phenomenon which may result from a number of different contributing factors.

It has been postulated that the hepatic UROgen I synthase deficiency in AIP leads to a deficiency of hepatic heme formation, which then results in derepression of the synthetic mechanism controlling ALA synthase production (Strand *et al.*, 1970). A deficiency in hepatic heme formation in AIP has not yet been directly demonstrated, however, and it is important to note that completely latent AIP gene carriers have a URO I synthase deficiency in their cells (erythrocytes, fibroblasts, lymphocytes, etc.) equivalent to that found in AIP individuals with the expressed clinical syndrome. It is possible, though improbable, that these two populations of gene carriers have differing levels of URO I synthase activity in their livers; until information on this point becomes available it is important to emphasize that decreased UROgen I synthase does not of itself determine the clinical status of AIP and that additional metabolic factors must be involved in the expression of the clinical syndrome of AIP.

The biochemical data in Fig. 5 affirm these points, and specifically indicate that (1) levels of erythrocyte UROgen I synthase activity extend over an approximate threefold range in the normal population (left column in Fig. 5); (2) enzyme levels in active AIP patients average one-half the normal mean and also extend over a threefold range (middle column in Fig. 5); and (3) erythrocyte UROgen I synthase activity in individuals in whom the disease has remained completely latent exactly parallels that of individuals in whom the disease has become clinically expressed. In the family study depicted in Fig. 6, UROgen I synthase activities were determined in a female propositus (case 20), her parents (cases 4 and 5), and the maternal lineage (all paternal relatives were normal). Of the 15 individuals in this lineage with low erythrocyte UROgen I synthase, only in the propositus were any AIP symptoms expressed; and although several of the adult carriers of the gene defect excreted slightly higher than normal levels of porphyrin precursors in urine, most did not. Of the seven prepubertal children in the lineage (cases 37, 39, 40, 41, 42, 44, and 45), none manifested any abnormalities of porphyrin-precursor excretion or AIP

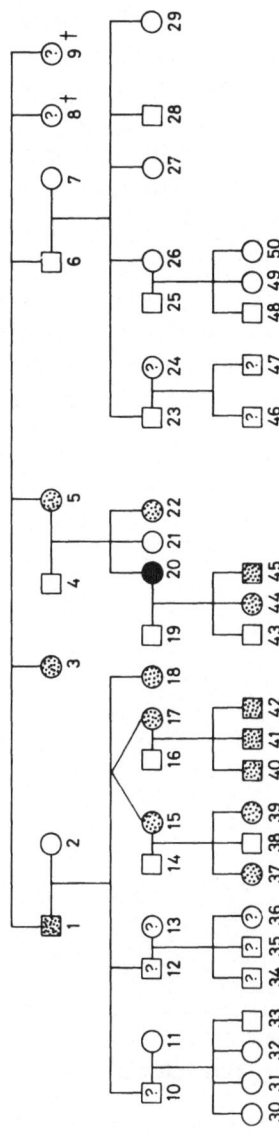

Fig. 6. Study of erythrocyte UROgen I synthase activity in a family with AIP. The propositus is case 20. The UROgen I synthase defect was transmitted in the maternal (case 5) lineage. The study confirms the autosomal dominant mode of transmittance of the UROgen I synthase defect. Cases 37, 39, 40, 41, 42, 44, and 45 were all prepubertal children (Sassa *et al.*, *Proc. Natl. Acad. Sci. USA* **71**:732–736, 1974*a*).

symptoms. It is of special interest that one of the prepubertal children (case 37, 7 years old at the time of the study) had been treated with Dilantin and phenobarbital for 2.5 years for epileptic seizures. Nevertheless, she was completely asymptomatic and her urinary excretion of ALA and PBG were within normal limits. This child remained on Dilantin and phenobarbital medication through puberty without developing clinical or biochemical expressions of her AIP gene defect.

More recent studies have demonstrated that the UROgen I synthase deficiency of AIP can be detected in lymphocytes undergoing mitogen-induced transformation, and the activity of this induced enzyme serves to clearly distinguish AIP from normal cells (Sassa *et al.*, 1978*b*) (Fig. 7). The deficient induction of UROgen I synthase is not due to a difference in the extent of mitogen-induced transformation of AIP lymphocytes because their DNA and RNA synthetic rates are equivalent to those of normal lymphocytes. Moreover, the activities of ALA synthase, ALA

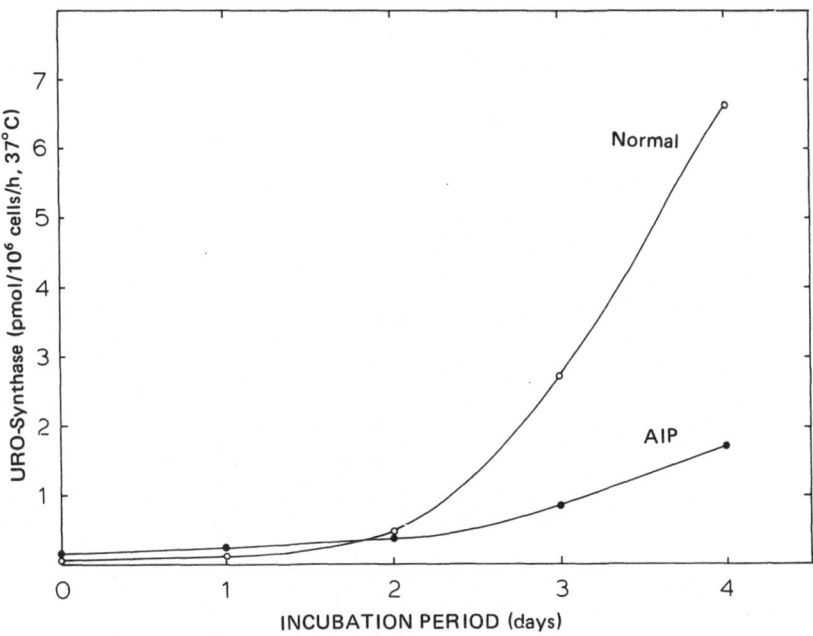

Fig. 7. Induction of UROgen I synthase in mitogen-stimulated lymphocytes. UROgen I synthase activity was determined fluorometrically using 5×10^5 cells per assay and PBG as the substrate. Lymphocytes were incubated with the mitogens before the assay, and the enzyme activity is displayed as a function of incubation period with the mitogens. Points are the mean of triplicate determinations (Sassa *et al.*, *J. Clin. Invest.* **61**:499–508, 1978).

dehydratase, catalase, and the heme content of AIP lymphocytes are similar to those of normal cells. These findings thus confirm that deficient induction of UROgen I synthase in AIP lymphocytes is the expression of a specific gene defect which characterizes these cells.

It has been generally assumed that the AIP gene defect is equivalent to the UROgen I synthase deficiency. However, a recent study by Anderson (1979) utilizing an antibody against UROgen I synthase, showed that some AIP individuals contained positive cross-reactive material (CRM) in their erythrocytes whereas one subject with AIP did not contain CRM. These findings raise the possibility that UROgen I synthase deficiency in AIP subjects may not be due to a single mechanism. This important observation, however, needs to be confirmed in more AIP individuals and in other tissue types as well. Amino acid analysis of the isolated purified positive CRM also needs to be done.

As noted above, it has been proposed (Strand et al., 1970) that the low levels of UROgen I synthase activity in the livers of AIP patients may impair hepatic heme synthesis and that, as a consequence, heme repression of ALA synthase production is lessened, thus resulting in enhanced production of this enzyme. Recent studies on avian embryo hepatic cells in culture indicate that it is necessary to decrease concentrations of liver heme considerably below normal before the enhanced synthesis of ALA synthase will occur (Sinclair and Granick, 1975; Sassa and Kappas, 1977). It is not clear how the 50% deficiency of UROgen I synthase activity in the liver cells of AIP patients can alone depress hepatic heme synthesis sufficiently to lead to secondary stimulation of ALA synthase production. To elicit the overproduction of this enzyme in individuals carrying the low UROgen I synthase trait, it is therefore probable that other metabolic factors are also required.

The metabolic factors known to be associated with the activation of the AIP clinical syndrome include the onset of puberty; variations in hormone production during the menstrual cycle; exposure to environmental chemicals or to a wide variety of drugs; changes in dietary composition and reduced caloric intake (Wellend et al., 1964; Tschudy et al., 1964; Kappas et al., 1974; Meyer and Schmid, 1978; Anderson et al., 1979; Tschudy and Lamon, 1980). Among the endogenous factors which may enhance the susceptibility of some AIP individuals to activation of their disease, steroid hormones and their derivatives have been the subject of particular study. It is now clear that a wide spectrum of metabolites derived from the biotransformation of gonadal and adrenal steroid hormones or their precursors have potent abilities to induce ALA synthase and to enhance porphyrin production in cultured avian embryonic liver cells (Granick, 1966; Granick and Kappas, 1967; Kappas and Granick, 1968; Kappas et al., 1968; Sassa et al., 1979a). Examples of the induction responses to several of these compounds are shown in Fig. 8. Phenolic estrogens and their metabolites have little or no porphyrinogenic properties; adrenal steroids of the hydrocortisone type are devoid of such activity. The metabolites of C19 and C21 neutral steroid hormones such as progesterone, testosterone, Δ^4-androstenedione and related compounds, on the other hand, have ALA-synthase-inducing activity which on a molar basis is equivalent to or greater than that of potent drug inducers of this liver enzyme. Steroid metabolites of both the 5β and 5α type are porphyrinogenic, but within many pairs of 5β and 5α isomers, it has been shown that the 5β (A:B cis) ring structure confers more inducing potency than the 5α (A:B trans) structure in cultured avian liver cells (Granick and Kappas, 1967; Kappas and Granick 1968; Sassa et al., 1979a) and in the chick embryo liver in ovo (Kappas et al., 1968; Sassa et al., 1979a). A comparable structure–activity relationship has been described in certain other steroid responsive experimental systems as well (Singer et al., 1976; Irving et al., 1976; Urabe et al., 1979); it is also of interest that 5β steroids have been shown to bind more tightly than 5α steroids to cytochrome P-450 (Estabrook et al., 1975) and that 5β steroids inhibit P-450-dependent oxidations to a greater extent than do 5α compounds (Soyka and Long, 1972; Soyka and Deckert, 1974).

The actions of steroid inducers of ALA synthase in liver are essentially analogous to those of drug and other chemical inducers of this enzyme. Thus, chelators such as CaMgEDTA greatly enhance the steroid induction response; hemin in concentrations of $\sim 10^{-7}$ M inhibits it; and insulin alone, or insulin with hydrocortisone and triiodothyronine exerts a "permissive" effect on steroid-induced porphyrinogenesis. Steroids are presently the only major class of compounds natural to man which have been shown to possess substantial ALA synthase and porphyrin-inducing capacity in liver (Granick and Kappas, 1967; Kappas and Granick, 1968; Sassa et al., 1979a) and in erythroid cells (Levere et al., 1967; Irving et al., 1976; Singer et al., 1976; Urabe et al., 1979). Thus their potential role in enhancing the susceptibility of AIP gene carriers to provocation of their disease is of considerable interest.

In exploring the role which the endocrine system may play in the pathogenesis of AIP, it has been shown that clinically expressed AIP

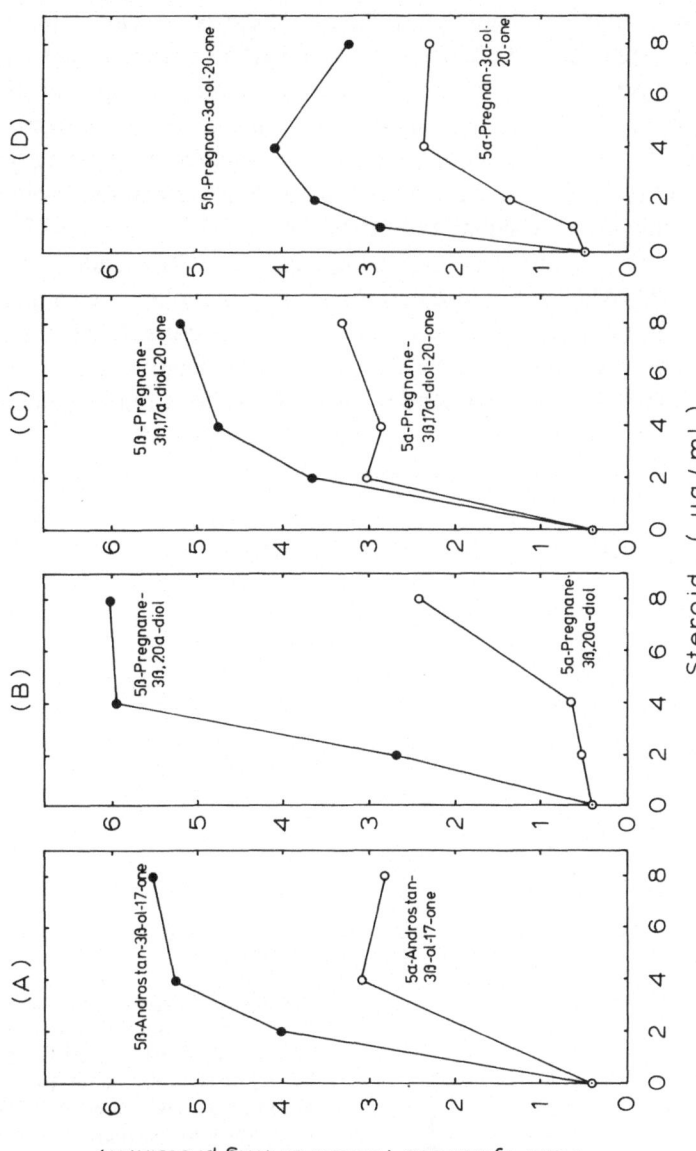

Fig. 8. Effects of 5α and 5β epimers of C-19 and C-21 steroids on the induction of ALA synthase in cultured chick embryo liver cells. The cells were cultivated in a serum-free modified F12 medium supplemented with insulin (1 μg/ml) and hydrocortisone (0.05 μg/ml) for 24 hr. The medium was then replaced with fresh medium containing the steroid. ALA synthase activity was determined after incubation for 20 hr (Sassa et al., J. Biol. Chem. 254:10011–10020, 1979).

patients display a major aberration of steroid hormone metabolism in comparison with normal subjects or AIP gene carriers in whom the porphyric trait has remained completely latent. There is, in clinically expressed AIP, a significant deficiency of steroid Δ^4-5α reductase activity in liver as assessed with radiolabeled tracer hormones (Kappas *et al.*, 1972; Bradlow *et al.*, 1973); this impairment is manifest by the disproportionate generation of 5β, as compared with 5α metabolites, during the reductive metabolism of precursor hormones such as testosterone which are normally converted to equal extents to 5α and 5β compounds (Fig. 9). With steroids such as the adrenal compound 11-hydroxyandrostenedione (11-OHAD) whose structure determines that reductive metabolism occurs primarily via the 5α pathway, the steroid reductive deficiency is manifest by an approximately 50% decrease in the relative amount of the principal metabolite formed (Anderson *et al.*, 1979). Comparisons of

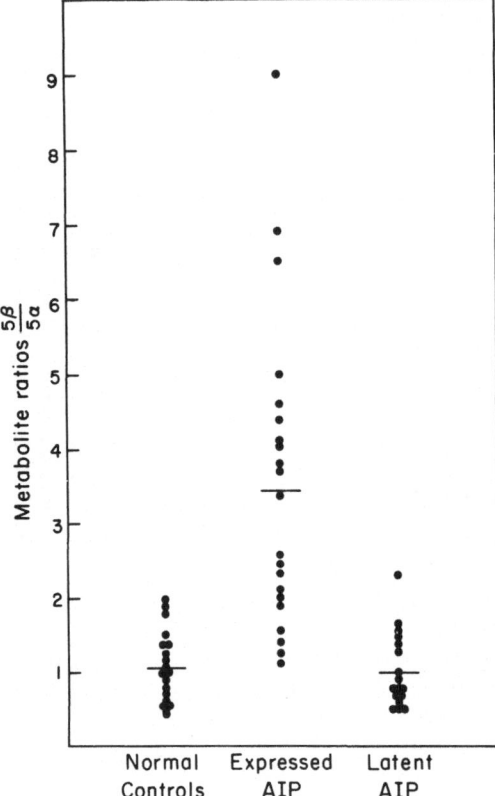

Fig. 9. Ratio of 5β to 5α metabolites formed from radioactive testosterone in 20 normal subjects, 21 patients with clinically expressed AIP, and 18 subjects with latent AIP. The mean value for each group is indicated and was significantly higher than normal ($P < 0.001$) only in the group with clinically expressed AIP (Anderson *et al.*, *Am. J. Med.* **66**:644–650, 1979).

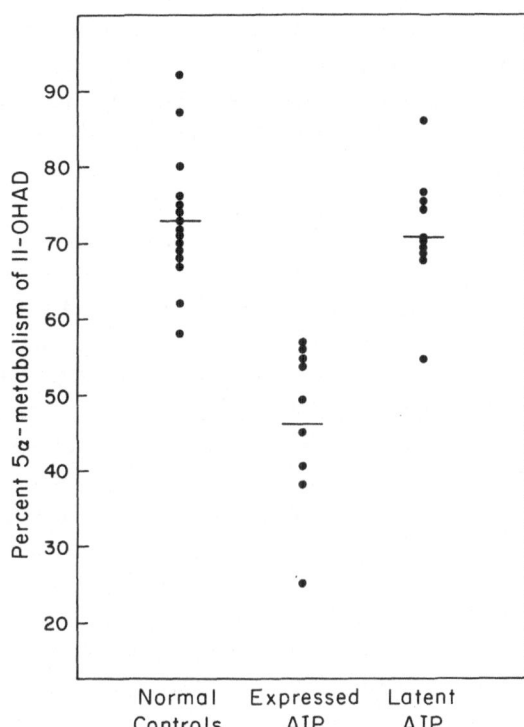

Fig. 10. Percent 5α-metabolism of radioactive 11-hydroxyandrostene-dione (11-OHAD) in 12 normal subjects, 9 patients with clinically expressed AIP, and 10 subjects with completely latent AIP. The mean value was reduced ($P < 0.001$) in the group with clinically expressed AIP but not in the group with completely latent AIP (Anderson *et al.*, *Am. J. Med.* **66:**644–650, 1979).

steroid metabolism patterns using radioactive hormone substrates in normals, clinically expressed AIP patients, and completely latent AIP subjects are shown in Fig. 10. The impairment of steroid Δ^4-5α reductase activity in clinically expressed AIP is substantial, particularly when the enzyme activity is tested with the specific steroid substrate 11-OHAD (Anderson *et al.*, 1979); and it is extremely interesting that this aberrant pattern of steroid metabolism distinguishes clinically expressed AIP patients not only from normal subjects but also from completely latent AIP gene carriers. Thus, although individuals with expressed and latent AIP are similar with respect to the extent of their erythrocyte UROgen I synthase deficiency, the former group displays an abnormality in activity of a steroid-metabolizing enzyme in liver, even in clinical remission, which is not observed in the gene carrier group in whom AIP has remained completely latent. It has also been demonstrated that there are significant elevations of several porphyrinogenic steroids in plasma and urine of patients with expressed AIP (Goldberg *et al.*, 1969; Paxton *et al.*, 1974).

The basis for the steroid Δ^4-5α reductase deficiency in clinically expressed AIP is not known; however, the occurrence of this endocrine abnormality in clinically expressed AIP subjects may enhance the susceptibility of such individuals to activation of their disease. It is of interest that phenobarbital, a precipitating drug in AIP, can produce, in normal subjects, those aberrations of steroid metabolism which characterize patients with expressed AIP (Kappas *et al.*, 1977).

Agents which are capable of stimulating porphyrin formation in cultured chick embryo liver cells have also been described in the sera of patients with AIP (Kappas *et al.*, 1969; Strand and Marver, 1970; Rifkind *et al.*, 1974). The agents in sera responsible for porphyrin stimulation are heat-stable; insensitive to trypsin; present in the supernatants after ethanol precipitation of plasma proteins; extractable in ethyl acetate; nondialyzable; and migrate with the albumin fraction of serum during electrophoresis (Rifkind *et al.*, 1974). The chemical nature of these agents is not known. In addition to the porphyrin-inducing substances found in certain human sera, there are also humoral agents which inhibit porphyrinogenesis. These are present in the ethanol precipitates of sera and are found consistently in high concentrations in pregnancy and postpartum sera (Rifkind *et al.*, 1974). This finding is of interest in view of the fact that hepatic ALA synthase becomes highly refractory to induction by AIA during late pregnancy in the rat (Paul *et al.*, 1974).

Acute exacerbations of the hereditary hepatic porphyrias may be associated with curtailment of dietary intake (Knudsen *et al.*, 1977) and may be ameliorated by increases in carbohydrate ingestion or by infusions of large amounts of glucose (Welland *et al.*, 1964). Levulose has also been shown to be effective in treating acute porphyric attacks (Brodie *et al.*, 1977*a*). The salutary effects of glucose in the treatment of hepatic porphyria has an experimental analogy ("glucose effect") in the prevention of chemically induced experimental porphyria—specifically, glucose administration can block the chemical induction of hepatic ALA synthase (Rose *et al.*, 1961; Tschudy *et al.*, 1964; Marver *et al.*, 1966). The precise mechanism of this effect is not known; moreover, glucose infusions in acute attacks of human porphyrias may not be consistently beneficial (Brodie *et al.*, 1977*a*), although such treatment is an important mode of possible therapy in a group of diseases for which highly specific and consistently effective therapeutic interventions are lacking.

An important approach to the possible development of a specific therapy for exacerbations of AIP is represented by the work of Watson *et al.* (1973, 1974, 1977, 1978) on the use of intravenous hematin infusions in the treatment of this disease. Hematin infusion in acute porphyria was initiated on the assumption that since hepatic ALA synthase is induced during acute attacks of hereditary hepatic porphyria the induction process should be subject to end-product repression by hemin, as has been shown experimentally (Granick, 1966; Waxman *et al.*, 1966). An original trial of such therapy was made by Bonkowsky *et al.* (1971) in one patient with AIP and the patient responded to the hematin infusion with marked diminution of serum ALA and PBG. This finding has been confirmed by a number of subsequent studies. Watson *et al.* (1977) have summarized their experience with hematin therapy which included 143 hematin infusions in 22 patients with acute porphyria. They observed prompt and often dramatic recovery after hematin infusion on 25 occasions in 31 acute attacks in 20 patients (Watson *et al.*, 1977, 1978). Two patients, however, died; one apparently having developed respiratory paralysis prior to the hematin infusion (Watson *et al.*, 1978).

Watson (1975) postulated that deficient synthesis of heme in hepatic porphyria might result in a corresponding deficiency of cytochrome P-450, which, in turn, could in some way provoke nervous system injury through decreased capacity of the cytochrome-P-450-dependent mixed-function oxidase system. Whether this mechanism occurs in nervous system tissues is not known. It should be noted, however, that nervous tissues, namely avian dorsal root ganglia maintained in tissue culture, has been shown to possess the enzymatic capacity to form porphyrins from the precursor ALA (Whetsell *et al.*, 1978). Such capacity is mainly localized in supporting cells, e.g., Schwann cells, rather than in neuronal cells (Whetsell *et al.*, 1978). The porphyrin–heme pathway in these cells can be readily impaired by environmental chemicals, e.g., lead, and such impairment of the heme pathway interferes with the normal development of myelin in this tissue type (Sassa *et al.*, 1979c). Thus it is not only likely that neuronal cells have an intrinsically low rate of heme formation, but that one of the metabolic supporting functions of nonneuronal cells may be to provide heme, or a heme precursor, to nerve cells for use in cytochrome formation. This function may be impaired by a genetic deficiency of heme formation, as may exist in AIP or by acquired defects in heme synthesis resulting from the biological impacts of environmental chemicals.

Watson *et al.* (1977) have also called attention to the necessity to consider variability in the role of heme destruction and the induction of hepatic heme oxygenase (Tenhunen *et al.*, 1969; Maines and Kappas, 1974) as factors bearing on the relation between acute attacks of porphyria and cellular heme content and synthesis in liver. Agents which stimulate the *de novo* synthesis of heme oxygenase provide a mechanism through which cellular heme depletion could be rapidly produced by an environmental chemical. Metals, for example, are very potent inducers of heme oxygenase, and can produce marked concurrent perturbations of ALA synthase activity and profound declines in cellular content of heme as well as of specific hemoproteins, such as cytochrome P-450 (Maines and Kappas, 1974, 1976, 1977; Drummond and Kappas, 1979). Other substances, including endotoxin and certain organic chemicals, also induce heme oxygenase and accelerate cellular degradation of heme (Bissell and Hammaker, 1976*a*; DeMatteis, 1978). These actions would all be expected to be highly detrimental to individuals carrying those gene lesions of human porphyria which may substantially impair heme content. The role of factors which enhance cellular heme degradation in precipitating exacerbations of the hereditary hepatic porphyrias has been little studied, but is a potentially important area of clinical research.

Hereditary Coproporphyria (HCP)

Clinical Findings

HCP is similar to AIP and variegate porphyria in that its clinical symptomatology is essentially neurovisceral in nature. Unlike patients with AIP, however, those with HCP may also display cutaneous photosensitivity. Acute attacks of HCP are also known to be provoked by chemicals that precipitate acute attacks of AIP. HCP occurs in both sexes and at the very least more than one-half of HCP subjects are asymptomatic or exhibit only mild symptoms (Jaeger *et al.*, 1975). The disorder is clearly inherited in an autosomal dominant fashion but clinical expression and severity are quite variable and symptomatically heterogeneous. For example, in a review of 111 cases of HCP (Brodie *et al.*, 1977*b*), 35 clinically symptomatic patients and 76 latent carriers were recognized (Langhof *et al.*, 1965; Smart *et al.*, 1965; With, 1965; Barnes and Whittaker, 1965; Birchfield and Cowger, 1966; Goldberg *et al.*, 1967; Haeger-Aronsen *et al.*, 1968; Connon and Turkington, 1968; Dean *et al.*, 1969; Gajdos *et al.*, 1969; Lomholt and With, 1969; Sasaki *et al.*, 1969; Kaufman and Marver,

1970; McIntyre *et al.*, 1971; Jaeger *et al.*, 1975). Signs and symptoms which have been frequently recorded are abdominal pain (80%), vomiting (34%), photosensitivity (29%), neurological and psychiatric symptoms (23%), and severe constipation (20%) (Brodie *et al.*, 1977*b*). Apparently, the disorder is less severe than AIP, since only two fatalities have been reported, both due to respiratory failure (Brodie *et al.*, 1977*b*). Like AIP, there is a female preponderance of cases in acute attacks (5:2). Drugs and chemicals were implicated as precipitating factors in 54% of acute attacks and barbiturates were responsible in 34% of cases (Brodie *et al.*, 1977*b*).

Biochemical Findings

The predominant biochemical finding in clinically expressed HCP is a marked elevation of urinary COPRO III excretion. In some HCP patients, greater than 95% of urinary porphyrins are COPRO, while in others both URO and COPRO are increased (URO ≪ COPRO) together with smaller increases of hepta-, hexa-, and pentacarboxylic porphyrins. Feces also contain large amounts of COPRO with increased amounts of more carboxylated porphyrins. Approximately 15% of fecal COPRO from a patient with HCP was shown to be chelated with copper, probably by a nonenzymatic incorporation of the metal into the porphyrin (Carlson *et al.*, 1978).

Fecal PROTO may also be increased, but always less than COPRO. During acute attacks, urinary ALA, PBG, and COPRO concentrations are markedly elevated. "X porphyrin", i.e., a hydrophilic porphyrin–peptide complex, is not usually elevated in HCP (Goldberg *et al.*, 1967; Haeger-Aronsen *et al.*, 1968), while the "X porphyrin" is elevated in variegate porphyria and porphyria cutanea tarda. HCP patients have also been found to excrete excessive amounts of 17-oxosteroids in the urine during acute attacks (McIntyre *et al.*, 1971; Paxton *et al.*, 1974), but not during remission (McIntyre *et al.*, 1971).

Gene Defect

Hepatic ALA synthase activity in patients with expressed HCP has been found to be elevated (Sasaki *et al.*, 1969; Kaufman and Marver, 1970; McIntyre *et al.*, 1971), but the enzyme was reported to be normal in activity in a patient during remission, at a time when urinary ALA and

PBG levels were also normal (McIntyre *et al.*, 1971). The enzyme activity was also shown to be at a normal level of activity in the bone marrow, in particular, during relapse (Sasaki *et al.*, 1969). As in the case of AIP, the pattern of porphyrin excretion suggests that elevated hepatic ALA synthase activity may be a secondary phenomenon associated with an enzymatic defect in HCP patients which primarily affects the step between COPROgen and PROTOgen, i.e., COPROgen oxidase. Direct measurement of COPROgen oxidase in tissues from HCP patients has confirmed a deficiency of this enzyme in HCP. Since COPROgen oxidase is a mitochondrial enzyme, the assay has to be performed in tissues other than circulating erythrocytes. Elder *et al.* (1976*a*) determined COPROgen oxidase activity in cultured skin fibroblasts from three patients with HCP in two family lineages and found that the enzyme activity in HCP individuals was approximately one-half that of fibroblasts from normal subjects or patients with other types of porphyria. These three patients are presumably heterozygotes for COPROgen oxidase deficiency and the mode of inheritance is compatible with an autosomal dominant transmission. Nordmann and his associates (Nordmann *et al.*, 1977; Grandchamp and Nordmann, 1977) also reported a COPROgen oxidase deficiency in lymphocytes from 17 individuals with HCP (six patients and eleven carriers without clinical symptoms but with high levels of fecal COPRO). HCP individuals, both patients (211 $\pm$ 45 pmol PROTO/mg protein/hr, 37°C) and carriers (235 $\pm$ 66), displayed approximately one-half the enzyme activity of control subjects (473 $\pm$ 76). These data support the idea that COPROgen oxidase deficiency is the primary gene defect in HCP. COPROgen oxidase deficiency has also been demonstrated in the buffy coat, probably consisting mainly of leukocytes, of blood from patients with HCP (Brodie *et al.*, 1977*b*).

    According to Elder *et al.* (1976*a*), a 50% COPROgen oxidase deficiency in HCP is still far in excess of (~30 times) of normal UROgen I synthase activity in liver (Strand *et al.*, 1970). The normally low activity of UROgen I synthase activity may explain the excessive excretion of ALA and PBG during acute attacks of HCP; thus the biochemical findings during acute attacks of HCP can be considered, as in AIP, to reflect the rate-limiting state of UROgen I synthase when ALA synthase activity is induced. This condition is probably less readily brought about in HCP than in AIP, because UROgen I synthase activity is normal in HCP, while it is decreased by 50% in AIP. This fact probably also explains the greater prevalence of latent carriers of the gene defect in HCP as compared with

AIP in several family studies (Goldberg *et al.*, 1967; Lomholt and With, 1969; Strand *et al.*, 1972; Brodie *et al.*, 1977*b*).

Recently, a remarkable case of homozygous HCP was described by Grandchamp *et al.* (1977). The patient was 20 years old when she became acutely ill. This event was apparently related to her pregnancy at the time. She was found to excrete very large amounts of ALA (34 mg/g creatinine), PBG (39 mg/g creatinine) and COPRO (8.47 μg/g creatinine) in urine. Fecal COPRO was 250 times the normal level. The patient's lymphocyte COPROgen oxidase activity was only 2% of the control level rather than the expected level of 50% for a heterozygous carrier of this enzyme defect. The proband's parents were first cousins and both displayed COPROgen oxidase activity approximately 50% of that of normal controls. In addition, the proband's stepsister was found to be a latent HCP individual displaying a 50% COPROgen oxidase activity. Thus it appears that this case represents a unique example of a homozygous gene defect in the dominantly transmitted porphyrias. Grandchamp *et al.* (1977) also suspected that a patient reported by Berger and Goldberg (1955) might be another homozygous HCP. The latter case was also the son of first cousins who were latent cases of HCP. The stature of the patient was small and the patient excreted very large amounts of COPRO in urine and feces.

Apparently, homozygous deficiency of COPROgen oxidase leading to a decline in enzyme activity exceeding 95% does not affect heme formation in erythroid cells, since both cases of HCP (Berger and Goldberg, 1955; Grandchamp *et al.*, 1977) did not show signs of anemia except during the late crisis associated with pregnancy in the first patient. These findings also suggest the possibility that COPROgen oxidase activity, even when the enzyme deficiency is extreme (95%), may not be rate-limiting for heme formation in other nonhepatic tissues. The possibility that COPROgen oxidase deficiency does not express itself in erythroid cells is not likely, since the enzyme deficiency can be identified not only in the liver cells of patients (Elder *et al.*, 1976*b*), but also in cultured fibroblasts (Elder *et al.*, 1976*b*) and lymphocytes (Nordmann *et al.*, 1977; Grandchamp and Nordmann, 1977; Grandchamp *et al.*, 1977). In a recent study from this laboratory utilizing cultured fibroblasts and cultured lymphocytes from heterozygous HCP patients, we have obtained no evidence that COPROgen oxidase becomes rate-limiting in the conversion of ALA to PROTO, thus supporting the idea that even profound deficiencies of this enzyme activity may not limit porphyrin–heme formation. Activity of the disease in HCP gene carriers may, as in AIP, therefore be dependent largely on associated metabolic events, i.e., chemical and hormonal influences which induce ALA synthase, and the gene carrier state may simply represent a heritable susceptibility to the deleterious consequences of such metabolic factors.

Variegate Porphyria (VP)

VP is another autosomal dominant form of inherited hepatic porphyria characterized by symptoms very similar to those seen in AIP and HCP. VP, also called South African porphyria or "protocoproporphyria hereditaria," is quite common among South African whites. In South Africa, Dean (1971) has been able to trace 236 cases of VP in 13 families to a Dutch couple who settled at the Cape of Good Hope in 1688. An average of 50% of the children of a porphyric parent in South Africa inherit VP and the distribution of the disorder between the sexes is approximately equal. This form of porphyria was called "variegate porphyria," because the disease could be present in various forms, i.e., an acute attack having neurovisceral symptoms, as a cutaneous type of porphyria, or as a completely latent disorder (Dean, 1971). Dean (1963, 1971) has estimated that there are approximately 8000 VP patients, a prevalence of three per 1000 white inhabitants, in South Africa. Since the discovery of VP in South Africa, it has been demonstrated that the disease also occurs in Europe, the USA, and other countries.

Clinical Findings

VP displays either cutaneous or neurological symptoms or both. These symptoms include abdominal pain (90%), tachycardia (83%), vomiting (80%), constipation (80%), hypertension (55%), motor neuropathy (53%), pain in the limbs (51%) or back (50%), fever (38%), urinary frequency (30%), dysuria (28%), and confusional state (32%), or other abnormal behavior (23%) (Eales, 1963); in a study by Eales (1963) over 33% of VP patients displayed skin photosensitivity alone, while 15% had never displayed skin symptoms.

Biochemical Findings

Since none of the clinical symptoms of VP distinguishes this disease from other hepatic porphyrias, the diagnosis of VP must be made on the biochemical findings in excreta, particularly in feces. Most VP patients excrete large amounts of PROTO with lesser increases of COPRO in feces (Dean and Barnes, 1959; Eales *et al.*, 1963, 1966*a,b*; Sweeney, 1963; Herbert, 1966; Mustajoki, 1978; Husquinet *et al.*, 1978), a pattern not observed in AIP, HCP, or porphyria cutanea tarda (PCT). As in AIP or HCP, urinary PBG excretion during acute attacks is markedly elevated, but it is generally normal during remission of this disease; in contrast, normal levels of urinary PBG rarely occur in AIP if the disorder has once become clinically expressed (Dean, 1963, 1971). In contrast to PCT or AIP, urinary COPRO is greatly increased between acute attacks of VP and may be accompanied by a large increase of URO as well. Urinary porphyrin patterns during acute attacks thus may be confusing in terms of diagnosis.

In 1968, Rimington *et al.* described ether-insoluble, hydrophilic porphyrins in the feces from patients with VP (Rimington *et al.*, 1968). The authors termed these compounds collectively as "X-porphyrin" which normally is present in trace amounts but increases markedly in VP (Rimington *et al.*, 1968; Elder *et al.*, 1974). More recent studies have demonstrated that "X-porphyrin" is also found in excess in PCT and occasionally in AIP and HCP, thus proving to be of less significance in the differential diagnosis of VP and PCT (Eales and Grosser, 1971; Moore *et al.*, 1972). Elder *et al.* (1974) suggested that the "X-porphyrin" in VP is different from that in PCT.

Gene Defect

The gene defect of VP remains as yet unclear. The porphyrin excretion pattern, i.e., predominantly PROTO in feces, suggests that the biochemical defect of VP may be at the terminal step of the heme biosynthetic pathway. Pimstone *et al.* (1973) determined ferrochelatase activity in muscle mitochondria from patients with VP but did not find an abnormality in this enzyme activity. In contrast, Becker *et al.* (1977) reported that ferrochelatase activity in bone marrow normoblasts from VP patients is approximately 50% less than that of normal controls. They also reported that ferrochelatase activity in the lysates of EPP normoblasts is 20% of normal but that the enzyme activity was normal when measured in intact normoblasts. Based on these findings, the latter authors concluded that VP is due to a gene mutation causing an inactive ferrochelatase while in EPP the mutation results in an unstable ferrochelatase. These findings still leave several questions unanswered: (1) why does the putative unstable ferrochelatase in EPP not cause a major disturbance of heme formation in the liver?; (2) why does the same ferrochelatase defect (though one enzyme protein is presumably more labile than the other) express itself in two distinct forms of porphyria?; (3) why is ferrochelatase in VP decreased specifically in erythroid cells (Becker *et al.*, 1977), but not in muscle cells (Pimstone *et al.*, 1973)?

These questions need to be settled since another defect in heme synthesis, i.e., PROTOgen oxidase deficiency, in VP has recently been postulated, and some evidence has been obtained in support of the existence of the latter enzymatic abnormality. Smith *et al.* (1976) originally proposed that the biochemical defect of VP may be a deficiency in PRO-TOgen oxidase activity. Recently Brenner and Bloomer (1979, 1980) reported that PROTOgen oxidase activity is decreased in cultured fibroblasts from patients with VP (43% of normal) while ferrochelatase activity is normal. These findings are in apparent conflict with those by Viljoen *et al.* (1979) who found a reduced ferrochelatase activity in fibroblasts from patients with VP.

An interesting combination of a case of VP and a case of PCT occurring in the same family has been described (Watson *et al.*, 1975, 1976). A woman, 54 years old, was diagnosed as having VP because of an acute neurologic relapse with biochemical characteristics of VP. Her brother, 59 years old, was found to have the biochemical abnormalities of PCT (Watson *et al.*, 1975). The occurrence of PCT in the proband's sibling may be interpreted as indicating that the brother's PCT was acquired. However, although this possibility cannot be wholly excluded, a later study which demonstrated that a niece of both of these individuals displayed a biochemically well-defined PCT (Watson *et al.*, 1976) suggests another interesting possibility—namely that the proband (VP) may represent a double heterozygote for VP and PCT while the two PCT cases may be heterozygotes for PCT.

Porphyria Cutanea Tarda (PCT)

PCT is characterized clinically by skin photosensitivity and biochemically by excessive urinary excretion of porphyrins. In contrast to the hereditary hepatic porphyrias, acute attacks of neurovisceral symptoms do not occur. Drugs which are known to precipitate acute attacks of hereditary hepatic porphyrias, for example, have no effect on PCT and urinary excretion of PBG is always within normal limits although occasional mild elevations of ALA excretion may occur. PCT is probably the most frequent form of porphyria and it has been diagnosed in all parts of the world (Schmid *et al.*, 1954; Elder *et al.*, 1972; Pimstone, 1975; Meyer and Schmid, 1978; Tschudy and Lamon, 1980); however, the Bantu population in South Africa is known to have a particularly high prevalence of the disease (Lamont *et al.*, 1961; Eales *et al.*, 1975; Pimstone, 1975). This disorder occurs in association with a variety of other disorders (Elder *et al.*, 1972; Tschudy and Lamon, 1980; Pimstone, 1975; Meyer and Schmid, 1978). In striking constrast to the well-defined hereditary forms of porphyria, a PCT syndrome may clearly occur in normal subjects. For example, in Turkey an outbreak of PCT occurred between 1956 and 1961 in several thousand normal individuals following ingestion of seed wheat inadvertently contaminated with hexachlorobenzene (Schmid, 1960; Cam and Nigogosyan, 1963), and PCT has also occurred as the result of occupational exposure to TCDD (Bleiberg *et al.*, 1964; Jirasek *et al.*, 1976; Strik, 1978*a,b*), methyl chloride (Chalmers *et al.*, 1940) and vinyl chloride (Lange *et al.*, 1976). Photosensitivity remarkably similar to that observed in PCT may also occur as the result of high dosage therapy with furosemide (Burry and Lawrence, 1976) or nalidixic acid (Birkett *et al.*, 1969) in patients with chronic renal failure or long-term tetracycline usage in subjects without renal complications (Epstein *et al.*, 1976). Thus, the majority of PCT cases are considered as acquired disorders rather than as genetically determined ones. However, the disease may also occur as a hereditary disorder presenting symptoms exactly similar to those of the acquired one, as noted below.

Clinical Findings

Skin lesions usually begin with erythema, and then progress to vesicular and bullous lesions which may eventually ulcerate. Chronic skin lesions include pigmentation, depigmentation, scars, hypertrichosis, and, occasionally, sclerodermic changes. Lupus erythematosus is known to occur in association with PCT more often than in the normal population (Harris *et al.*, 1966; Hetherington *et al.*, 1970; Cram *et al.*, 1973) although the reason for this association is not known. Unlike CEP or EPP, rapid skin responses to light exposure are usually not seen in PCT.

Several characteristic features of PCT merit special comment: (1) the association of liver disease, (2) the involvement of iron metabolism, (3) the effects of alcohol and estrogens, and (4) the peculiar sensitivity to chloroquine (Bickers *et al.*, 1979). Variable degrees of liver dysfunction are common among PCT patients, particularly in association with excessive alcoholic intake. This may occur with or without cirrhosis (Eales and Linder, 1962; Uys and Eales, 1963; Taddeini and Watson, 1968; Elder *et al.*, 1972; Tschudy and Lamon, 1980; Harber and Bickers, 1975; Pimstone, 1975; Elder, 1977; Felsher and Kushner, 1977; Meyer and Schmid, 1978). The histopathology of PCT liver reveals frequent siderosis and characteristic changes associated with alcoholism (Taddeini and Watson, 1968; Turnbull, 1971; Tschudy and Lamon, 1980; Meyer and Schmid, 1978). Patients with PCT occurring in association with estrogen usage alone usually display less hepatic injury than those with alcoholism (Haberman *et al.*, 1975; Harber and Bickers, 1975; Bickers *et al.*, 1979). Although the accompanying hepatic injury is clearly important in the expression of the PCT syndrome, it is not clear to what extent such liver cell injury is essential to the clinical presentation of PCT. For example, PCT is rare in patients with typical Laennec's cirrhosis (Hällen and Krook, 1963; Taddeini and Watson, 1968; Meyer and Schmid, 1978). For these reasons Taddeini and Watson (1968) suggested that there may be an underlying constitutional abnormality which might enhance the liability of the liver to the development of PCT and the morphological injury.

Abnormal iron metabolism appears to be another contributing factor in the clinical development of PCT. Iron overload in PCT liver has been well documented by a number of studies and iron removal by venesection is known to be highly effective in the treatment of PCT (Ippen, 1977). In a study of 30 patients with PCT, Lundvall *et al.* (1970) reported a 2.5-fold increase of the mean iron content in the liver in PCT patients compared with control subjects. The total iron-binding capacity is usually normal but with an increase of the percentage of iron saturation of transferrin (>60%) in more than one-half of PCT patients. Total serum iron and the percentage of transferrin saturation, however, do not reach the levels observed in hemochromatosis. Increased serum iron may be, in part, due to excess alcoholic intake (Sullivan and Herbert, 1964) since it often decreases on hospitalization of patients, presumably because of obligatory cessation of alcohol usage (Hourihane and Weir, 1970; Turnbull *et al.*, 1973). A decline in serum ferritin concentration appears to correlate better with the decreasing urinary porphyrin excretion after repeated phlebotomy than do changes in total iron binding capacity or in serum iron concentration (Sweeney and Jones, 1979). Mobilizable iron pools as assessed by the amount of iron mobilized after venesection or chelation therapy are also expanded in PCT—to an average 1800 mg/day in PCT (Lundvall *et al.*, 1970) vs. 500 mg/day in normal females (Pritchard and Mason, 1964), or 750 mg/day in normal males (Haskins *et al.*, 1952, Balcerzak *et al.*, 1968, Turnbull, 1971). The elevated serum iron in PCT cannot be due to hemolysis because (1) red cell survival is usually normal (Berman *et al.*, 1963), (2) hemoglobin levels are usually at the upper limit of normal (Lamont *et al.*, 1961) and some patients may even show a tendency to polycythemia (Epstein and Pinski, 1965), (3) hematological profiles are normocytic–normochromic (Ippen, 1977), and (4) ferrokinetic studies indicate iron overload in the liver in some, if not most, cases (Kramer, 1963; Lundvall *et al.*, 1970; Turnbull, 1971). Curiously in the Bantu population in which a high coincidence of hemochromatosis and PCT is known, patients with porphyria may show less stainable iron in the liver than patients without porphyria (Lamont *et al.*, 1961). Like the accompanying hepatic disorder, iron overload alone does not precipitate PCT; thus patients with hemochromatosis rarely develop the PCT syndrome. These findings suggest that hepatic injury or iron overload may accentuate PCT when it is present in the latent form, but that these factors may not be sufficient in themselves to cause PCT in normal subjects.

PCT may sometimes appear to be related to the use of estrogens in men for the treatment of prostatic carcinoma or in women for the treatment of menopausal symptoms (Watson, 1960; Warin, 1963; Becker, 1965; Copeman *et al.*, 1966; Zimmerman *et al.*, 1966; Levere, 1966; Taddeini and Watson, 1968; Roenigk and Gottlob, 1970). The widespread use of estrogens by women has apparently changed the patient profile of PCT. Although earlier studies reported a preponderance of the disease in male subjects (Brunsting *et al.*, 1951), a recent study based on the observation of 40 cases of PCT has noted an equal incidence of PCT in men and women (Grossman *et al.*, 1979). The onset of the PCT syndrome after estrogen therapy is greatly variable ranging from 1 month to 5 years or more. It should be noted, however, that a vastly greater population of patients receives estrogens for cancer or other treatment, than develops PCT or significant abnormalities in porphyrin metabolism, implying that there are individual determinants in the development of this disorder (Theologides *et al.*, 1964; Taddeini and Watson, 1968).

Chloroquine treatment in PCT is usually associated with a febrile reaction, general malaise, nausea, and vomiting. Following this generalized reaction, a large amount of URO is excreted in the urine. The initial generalized reaction as well as the subsequent porphyrin excretion are not seen in nonporphyric individuals, including those with cirrhosis (Taddeini and Watson, 1968). The apparent explanation is that chloroquine becomes concentrated in lysosomes and in mitochondria where porphyrins also are accumulated; in these organelles the drug forms a unique molecular complex with porphyrins which is then released from hepatocytes and excreted in urine (Scholnick and Marver, 1968; Scholnick *et al.*, 1973). A transitory liver cell injury occurs with the release of the porphyrin–chloroquine complex and proteolytic enzymes from lysosomes.

Biochemical Findings

The urine from patients with PCT contains large amounts of porphyrins and usually appears pinkish to brownish. When exposed to longwavelength ultraviolet light, the urine emits the intense red fluorescence characteristic of porphyrins. This red fluorescence is invariably present even when the urine appearance is normal to the eye. Fluorescence examination should be carried out after acidifying the urine sufficiently since porphyrin fluorescence in an acidic solution is more stable than in a neutral solution. The main urinary porphyrins are URO and heptacarboxylic porphyrin with lesser increases of hexa- or pentacarboxylic porphyrins, and COPRO (Lamont *et al.*, 1961; Sweeney, 1963; Nacht *et al.*, 1970; Dowdle *et al.*, 1970; Doss *et al.*, 1971; Eales *et al.*, 1975; Elder, 1977). URO is approximately 70% type I isomer, hepta- and hexacarboxylic porphyrins are about 90% type III, and pentacarboxylic porphyrin and COPRO are about equally types I and III (Chu and Chu, 1967; Dowdle *et al.*, 1970; Nacht *et al.*, 1970). PCT liver also contains large amounts of porphyrins which can be visualized as an intense red fluorescence by appropriate illumination of biopsied specimens with ultraviolet light. The immediate appearance of red fluorescence in the liver upon exposure to ultraviolet light indicates that some of the porphyrins in liver are present in the oxidized rather than in the reduced form, i.e., the porphyrinogens. The liver areas with iron or porphyrin accumulation do not necessarily overlap, thus there is no direct correlation between iron and porphyrin distribution in liver cells (Petryka *et al.*, 1978). In a few patients studied, the isomer distribution in the hepatic and fecal porphyrins was very similar to that of the urinary porphyrins (Chu and Chu, 1967; Dowdle *et al.*, 1970). Porphyrin concentrations of erythrocytes and bone marrow are normal (Schmid *et al.*, 1954). Urinary excretion of ALA and PBG is usually within normal limits (Lamont *et al.*, 1961; Dowdle *et al.*, 1970; Doss *et al.*, 1971; Eales *et al.*, 1975).

Ether-soluble fecal porphyrins are increased and consist partly of COPRO (<50%) (Elder, 1977), and there is as well a marked increase of ether-insoluble porphyrins (Taddeini and Watson, 1968; Moore *et al.*, 1972; Grosser and Eales, 1973). The increase in the ether-soluble fraction is due to the presence of isocopro-, diethylisocopro-, and hydroxyisocoproporphyrin rather than to COPRO itself (Taddeini and Watson, 1968; Moore *et al.*, 1972; Grosser and Eales, 1973). These porphyrins are thought to be derived from dehydroisocoproporphyrin (Elder, 1972). The fecal PROTO concentration is normal in about 60% of patients. It is probably also within normal limits in other patients with an apparent increase of PROTO if specific measurements of dicarboxylic porphyrins in feces are performed (Elder, 1977). The main porphyrins in the ether-insoluble fraction in PCT feces are heptacarboxylic porphyrins (Sweeney, 1963; Herbert, 1966; Taddeini and Watson, 1968), with lesser increase of URO (Elder, 1977), and are present as hydrophilic porphyrin–peptide conjugates, i.e., "X-porphyrin" (Elder *et al.*, 1974).

The laboratory diagnosis of PCT is possible by means of porphyrin analysis based on solvent extraction, but such diagnosis requires an analysis of porphyrins in both urine and feces; fecal porphyrin analysis is the more crucial, however. According to more recent data by Elder (1975), it is possible to diagnose PCT unequivocally by measuring the ratio of "isocoproporphyrin" to COPRO in feces. This can be done by either thin-layer or high-pressure liquid chromatographic methods which are highly specific and quantitative for the individual porphyrins. The ratio of isocoproporphyrin to COPRO was >0.1 for patients with PCT, while it was <0.01 for patients with VP or normal controls (Elder, 1975, 1977). It appears that this ratio may be the most specific laboratory finding which characterizes PCT.

Enzymatic Abnormality

Well-documented cases of familial PCT are very uncommon although a number of suggestions have been made for the existence of a genetic predisposition or constitutional abnormality in PCT (Watson, 1960; Waldenström and Haeger-Aronsen, 1963, 1967; Taddeini and Watson, 1968). These suggestions are based on the following findings: (1) although most

PCT is associated with an alcoholic history and overt liver disease, the majority of alcoholics do not develop PCT nor is PCT common among alcoholics; (2) large numbers of individuals treated with estrogens do not develop PCT although the involvement of estrogens in the onset of PCT is clearly documented in certain patients with this form of porphyria; finally, (3) there are a few reports of PCT cases with mildly elevated porphyrin excretion in family lineages (Waldenström and Haeger-Aronsen, 1963; Taddeini and Watson, 1968). Nevertheless, the existence of a genetic predisposition to PCT has remained hypothetical until recently. Enzymic studies directed towards identifying a biochemical defect in PCT have focused on UROgen decarboxylase because of the porphyrin excretion patterns in this disease. Blekkenhorst et al. (1976b) reported normal UROgen decarboxylase activity in erythrocytes from two patients with PCT and their immediate kin utilizing an assay similar to that of Kushner et al. (1975). In contrast, Kushner et al. (1975, 1976) found abnormally decreased UROgen decarboxylase activity both in liver (25% of controls values) and in erythrocytes (60% of control values) of selected PCT patients. The findings by the latter authors are supported by the results of more recent studies by Tiepermann and Doss (1978) and Felsher et al. (1978).

The decreased UROgen decarboxylase activity reported in the livers of patients with PCT (Kushner and Barbuto, 1975; Kushner et al., 1976; Elder et al., 1978) has also been observed in experimental animals made porphyric by hexachlorobenzene (Taljaard et al., 1971; Elder et al., 1976a; San Martin de Viale et al., 1976) and by TCDD (Goldstein et al., 1973, 1976, 1977). In rats fed with hexachlorobenzene for 11 weeks, the activity of UROgen decarboxylase in the liver was only 18% of control values (Elder et al., 1976a). Hexachlorobenzene had no effect on erythrocyte UROgen decarboxylase activity (San Martin de Viale, 1977). Decreased erythrocyte UROgen decarboxylase activity in PCT (Kushner and Barbuto, 1975; Kushner et al., 1976; Benedetto et al., 1978; Felsher et al., 1978; Tiepermann and Doss, 1978), however, has not been confirmed by some workers (Blekkenhorst et al., 1976a,b; Elder and Tovey, 1977; Elder et al., 1978; deVerneuil et al., 1978; Sweeney and Jones, 1979), nor has this enzyme deficiency been found in fibroblasts from PCT patients (Elder et al., 1978).

The apparent contradictions among investigators studying UROgen decarboxylase may be, in part, due to differences in methods utilized for the assay of the enzyme activity. Kushner et al. (1975, 1976) and Felsher *et al.* (1978) used an indirect assay for UROgen decarboxylase activity which is based on the enzymatic generation of the substrate, i.e., tritium-labeled UROgen I, and thin-layer chromatographic separation and liquid scintillation counting of the reaction products. A constant amount of hemoglobin and UROgen was added to each reaction mixture. Enzymatic activity was expressed as the sum of generation of hepta-, hexa-, penta-, and tetracarboxylic porphyrins per hour at 37°C. Elder and Tovey (1977), Elder *et al.* (1976a, 1978) and deVerneuil *et al.* (1978), on the other hand, utilized an assay more specific to the decarboxylation reaction by utilizing pentacarboxylic porphyrinogen which was prepared by chemical reduction from the corresponding porphyrin. This assay determines only the last of the four sequential decarboxylations catalyzed by UROgen decarboxylase and is less complicated in the interpretation of its results. Thus, there is a possibility that the two assay techniques for UROgen decarboxylase may not necessarily determine the same enzymatic activity. It is known, however, that a single enzyme catalyzes the entire series of these decarboxylations from UROgen to COPROgen and that the decarboxylase activities toward UROgen and pentacarboxylic porphyrinogen copurify from human erythrocytes (Elder and Tovey, 1977). The fact that the hepatic enzyme defect can be demonstrated with the pentacarboxylic porphyrinogen as substrate, but not in erythrocytes or in cultured fibroblasts (Elder *et al.*, 1978) indicates that, at least by utilizing the assay of Elder *et al.* (1976a, 1978), it is possible to recognize differences in UROgen decarboxylase deficiency between erythrocytes and liver cells. Another reason, although unlikely, which might explain the apparent discrepancy in the findings concerning UROgen decarboxylase activity in PCT erythrocytes is that there may be differences in the populations of PCT individuals in the USA (Kushner *et al.*, 1975, 1976; Felsher *et al.*, 1978) and in Europe (Elder *et al.*, 1978; deVerneuil *et al.*, 1978) or South Africa (Blekkenhorst *et al.*, 1976a,b). A recent study by deVerneuil *et al.* (1978) utilizing pentacarboxylic porphyrinogen as the substrate demonstrated erythrocyte UROgen decarboxylase activity in sporadic cases of PCT to be normal in confirmation of the results of Elder *et al.* (1978). On the other hand, erythrocyte UROgen decarboxylase activity in three families in which familial PCT was identified was approximately 50% of controls—not only in the proband PCT patient, but also in one parent and in siblings and relatives in different generations (deVerneuil *et al.*, 1978). These UROgen-decarboxylase-deficient subjects consisted of individuals of both sexes and without signs of hepatic disease. DeVerneuil *et al.* (1978) also compared the rate of the decarboxylation reaction utilizing UROgen, hepta-, and hexacarboxylic porphyrinogen as the substrate with that using pentacarboxylic porphyrinogen (Elder and Tovey, 1977; Elder *et al.*, 1976a, 1978), and found the results comparable, suggesting that the same enzyme catalyzes the successive decarboxylations from UROgen to COPROgen (deVerneuil *et al.*, 1978). Their data confirm an earlier finding obtained with UROgen decarboxylase in avian erythrocytes (Tomio *et al.*, 1970). The findings of deVerneuil *et al.* (1978) suggest that there are at least two distinct types of PCT: (1) the sporadic type which is usually associated with alcoholic hepatic injury and siderosis and (2) the familial type which can occur in children and women in the same family without overt hepatic disease. The former, sporadic type probably represents the majority of PCT patients who are characterized by decreased UROgen decarboxylase activity in liver cells (Elder and Tovey, 1977; Elder *et al.*, 1978) but not in erythrocytes. The fact that hexachlorobenzene treatment in rats decreases UROgen decarboxylase activity in the liver but not in erythrocytes (Elder *et al.*, 1976a; San Martin de Viale *et al.*, 1977) suggests that sporadic PCT patients may be the human equivalent of hexachlorobenzene-induced experimental porphyria in animals. The most striking human parallel to this animal model of porphyria is of course the epidemic of hexachlorobenzene-induced porphyria which took place in Turkey in the 1960s. Thus sporadic PCT with its presumably acquired deficiency of hepatic UROgen decarboxylase activity is important in that it may reflect the biochemical response of the liver in the normal population to environmental toxins.

A familial type of PCT, though rare, clearly exists as a distinct entity from sporadic PCT with the major difference between the two types being that the deficiency of UROgen decarboxylase activity probably characterizes all cells in the familial type, while the enzyme deficiency appears to be confined to the liver in the sporadic type. It should also be noted that the mode of inheritance of erythrocyte UROgen decarboxylase deficiency in familial PCT is consistent with an autosomal dominant trait and that familial PCT can occur without an accompanying hepatic disease. In the study by Benedetto *et al.* (1978), a familial form of PCT including eight members in three generations was recognized. Four individuals had cutaneous photosensitivity and six subjects, including the four clinically manifest patients, had excessive urinary URO excretion. All of these six subjects were characterized by low UROgen decarboxylase activity in erythrocytes using enzymatically generated UROgen as the substrate, as described by Kushner *et al.* (1975, 1976). Based on these findings, the authors suggested that PCT exists in overt, subclinical, or latent forms. Since examination of fecal porphyrins in this family was not carried out, Dean (1978) has raised a question concerning the diagnosis of PCT in this study. Benedetto *et al.* (1978) contended that the urinary porphyrin patterns in their study were compatible with the diagnosis of PCT and that this was confirmed by the decreased erythrocyte UROgen decarboxylase activity—a characteristic finding only of PCT (Felsher *et al.*, 1978); it would clearly have been useful, nevertheless, if fecal porphyrin analysis had been performed. As noted previously, the isocoproporphyrin/COPRO ratio in feces appears to make an unequivocal distinction between PCT and VP (Elder, 1975). It is of great importance to make the distinction between PCT and VP since the latter disorder is characterized by similar cutaneous symptoms but also neurovisceral symptoms often precipitated by drugs.

Acute Hepatic Porphyria Resulting from ALA Dehydratase Deficiency

A new type of hepatic porphyria with marked decrease in ALA dehydratase activity has recently been reported by Doss *et al.* (1977). Two males of unrelated origin, both 15 years of age, displayed neurological characteristics of acute hepatic porphyria with repeated intermittent acute manifestations having occurred since 6 years of age. The predominant biochemical feature in both cases was a highly elevated level of urinary excretion of ALA and porphyrins (mostly COPRO III with some pentacarboxylic porphyrin). Lead poisoning was excluded by normal blood lead levels. Both patients had erythrocyte ALA dehydratase activities below 1% of normal. ALA dehydratase in relatives of the patients ranged from one-fourth to one-third of the normal level. In contrast, UROgen I synthase activity in the patients, as well as in their relatives, was normal. The acute hepatic porphyric syndrome in these two patients is probably caused by the profound deficiency of ALA dehydratase.

Porphyria in Association with Other Disorders

Tyrosinemia and AIP

Tyrosinemia is an inborn error of metabolism characterized by hepatic cirrhosis in early childhood and multiple renal tubular defects with hypophosphatemic rickets (Gentz et al., 1965; LaDu and Gjessing, 1978). The derangement in tyrosine metabolism is thought to result from a low activity of 4-hydroxyphenylpyruvate dioxygenase as well as that of fumarylacetoacetase. Gentz et al. (1967) reported a patient with tyrosinemia who presented symptoms similar to those of AIP, e.g., severe colicky abdominal pain, nausea, vomiting, fever, hypertension, tachycardia, severe neuritic pain, and inability to walk. Subsequent studies demonstrated that urinary ALA excretion was elevated in three other tyrosinemia patients and even in those without AIP-like symptoms (Gentz et al., 1969a,b; Strife et al., 1977). An increased activity of ALA synthase was reported in an accompanying hepatoma (Gentz et al., 1969a,b) or in liver tissue (Kang and Gerald, 1970) removed from patients with tyrosinemia. A recent study by Lindblad et al. (1977) demonstrated that urine from patients with hereditary tyrosinemia contains succinylacetone which inhibits the activity of ALA dehydratase. Consistent with this finding, ALA dehydratase of three patients with tyrosinemia was less than 5% of that in erythrocytes and less than 1% of that in liver cells of control subjects. Strife et al. (1977) also reported that erythrocyte ALA dehydratase activity was approximately 15%–25% of controls during remission, but decreased to an undetectable level during relapse. The presence in tyrosinemia of a potent inhibitor of ALA dehydratase, i.e., succinylacetone, explains the elevated urinary ALA excretion and suggests that the primary enzyme defect in hereditary tyrosinemia is decreased activity of fumarylacetoacetase (Lindblad et al., 1977). These findings indicate that symptoms indistinguishable from those of AIP in tyrosinemia do not represent the coincidence of AIP and tyrosinemia, but rather are the consequence of a secondary effect of the primary metabolic-genetic derangement on the activity of a heme pathway enzyme. Although the clinical symptoms of patients with tyrosinemia and elevated urinary excretion of ALA are remarkably similar to those observed in AIP patients in acute attacks, the early onset of a porphyria-like syndrome (before puberty) and normal levels of urinary PBG in hereditary tyrosinemia (Gentz et al., 1969a,b) are entirely different from the related findings in AIP.

The normal levels of PBG in association with acute AIP-like symptoms in tyrosinemia also suggest that the neurological symptoms in tyrosinemia or in AIP may not be the result of accumulation of PBG or its oxidized derivative, porphobilin, in nerve cells (Feldman et al., 1971). Rather, such symptoms may possibly be caused by accumulated ALA (Müller and Snyder, 1977) or, alternatively, by a possible heme deficiency in nervous tissues as a result of the nearly complete inhibition of ALA dehydratase in tyrosinemia and by the 50% deficiency of UROgen I synthase in AIP.

PCT Accompanying Other Disorders

PCT may occur together with other diseases and this is not surprising since PCT may be largely considered an acquired disease resulting from exposures to environmental toxins in, perhaps, genetically susceptible individuals. There have been several reports of a cutaneous blistering eruption occurring in patients with chronic renal failure undergoing hemodialysis (Gilchrest et al., 1975; Korting, 1975; Keczkes and Farr, 1976; Thivolet et al., 1977; Griffon-Euvrard et al., 1977; Poh-Fitzpatrick et al., 1978, 1980). Most of these cases have been termed "pseudoporphyria" or "chronic bullous dermatosis of hemodialysis" because of the absence of elevated porphyrins in plasma, urine, or feces (Gilchrest et al., 1975; Keczkes and Farr, 1976; Griffon-Euvrard et al., 1977; Thivolet et al., 1977; Poh-Fitzpatrick et al., 1980), but three cases have been associated with significant increases of porphyrins in urine (Korting, 1975), or feces, plasma, and blister fluid (Poh-Fitzpatrick et al., 1978). Thin-layer chromatographic analysis of plasma porphyrins in one patient revealed URO and heptacarboxylic porphyrin in relative concentrations of 92:8 (Poh-Fitzpatrick et al., 1978). Fecal porphyrin analysis showing a marked excess of isocoproporphyrin in this case was also compatible with a diagnosis of PCT. The basis for the PCT syndrome induced by hemodialysis is not clear. Photosensitizing chemicals derived from the dialysis system have been suggested as responsible for the development of the PCT-like cutaneous symptoms (Thivolet et al., 1977). For example, aluminum has been suspected to be a causative agent of the PCT-like symptoms since porphyria has been described in rats fed aluminum hydroxide gel and in patients using the gel for therapeutic purposes. Aluminum accumulation has been observed in patients who have undergone hemodialysis treatment (Brivet et al., 1978). However, an attempt to demonstrate aluminum in plasma of patients with hemodialysis history was without success (Brivet et al., 1978). It would be useful to know whether chemicals leached out of the dialysis system are capable of inhibiting hepatic UROgen decarboxylase activity; however, it should be noted that URO, which is normally readily excreted in urine, undergoes little removal from plasma by hemodialysis. Thus the porphyrin would accumulate in plasma, reach the skin, and could then result in photosensitive lesions (Poh-Fitzpatrick *et al.*, 1978).

PCT has been reported to occur in association with systemic lupus erythematosus (Hetherington *et al.*, 1970; Cram *et al.*, 1973) and Felty's syndrome (Eales *et al.*, 1972; Rimington *et al.*, 1972). AIP has also been reported in association with systemic lupus erythematosus (Harris *et al.*, 1966).

Tio *et al.* (1957) reported a case of PCT associated with a benign porphyrin-producing hepatic adenoma. The patient displayed cutaneous photosensitivity and increased amounts of porphyrins in urine, largely URO, and COPRO and PROTO in feces. Cutaneous photosensitivity and excessive porphyrin excretion were normalized after the removal of the tumor which contained large amounts of porphyrins. Porphyrin producing malignant primary hepatomas have also been reported (Thompson *et al.*, 1970; Waddington, 1972; Rimbaud *et al.*, 1973); the association between these two conditions is rare.

Hereditary Porphyrias in Animals

There are a few congenital erythropoietic porphyrias which have been identified in animals, although no hereditary hepatic porphyrias have so far been described. Except for CEP in cattle, which is, in many respects, the equivalent of the human form of CEP, the other forms of porphyria in animals are somewhat different from their human counterparts.

Congenital Erythropoietic Porphyria (CEP) in Cattle

Bovine CEP is an erythropoietic disease which is essentially identical to human CEP in its symptomatology. Bovine as well as human CEP is characterized by skin photosensitivity and overproduction of type I URO and COPRO in some developing erythroblasts (Watson *et al.*, 1958, 1959). The bovine disease was originally reported in cattle in Swaziland in South Africa (Fourie, 1936) and later in Denmark (Jørgensen and With, 1955), England (Amoroso *et al.*, 1957), and in the United States (Watson *et al.*, 1958). The disease is inherited in an autosomal recessive fashion and heterozygous cattle are symptom-free.

Photoinjury occurs in areas of skin exposed to sunlight in human CEP patients, while it occurs only in white areas of the skin in black and white Holstein porphyric animals (Watson *et al.,* 1958). These animals have hemolytic anemia as indicated by the presence of anemia, reticulocytosis, and splenomegaly. Erythrocyte survival studies in two CEP cows showed a shortened median survival time (26 and 47 days) as compared to a normal of 160 days (Kaneko, 1963). In contrast to the well-known beneficial effect of splenectomy in human CEP, splenectomy performed in one porphyric cow did not affect the disease (Watson *et al.,* 1958). Excessive amounts of porphyrins in bovine CEP have been found, as in human CEP patients, in teeth, bone marrow, and spleen, but not in the liver.

One distinct difference between the human and the bovine form of CEP is the striking preponderance of PROTO in erythrocytes in bovine CEP, while URO and COPRO are the major porphyrins in erythrocytes in the human disease (Watson *et al.,* 1958; Chu and Chu, 1962). Urinary porphyrins of the porphyric cattle were mainly COPRO and URO with some pentacarboxylic porphyrin. All of the porphyrins of cattle were mainly type I isomers (Chu and Chu, 1962). This finding suggests that bovine CEP disease is associated with the overproduction of both type I and type III porphyrin isomers. It is interesting to note, however, that erythrocyte URO increased considerably and PROTO diminished when a CEP cow became pregnant (Watson *et al.,* 1958).

A recent study with bone marrow from a CEP bull showed a striking increase in ALA synthase activity with normal to moderate increases in ALA dehydratase and succinyl CoA synthase (Batlle *et al.,* 1979). When incubated with PBG, URO formation was seven times greater than normal in the porphyric bone marrow at 45°C (at which temperature UROgen III cosynthase is known to be inactivated), but only three times greater at 37°C. It was suggested that in this porphyric animal there was an increased activity of UROgen I synthase and ALA synthase with a diminished activity of UROgen III cosynthase (Batlle *et al.,* 1979). A similar conclusion was also made with respect to human CEP subjects (Moore *et al.,* 1978); and direct proof of diminished activity of UROgen III cosynthase had been provided earlier by Levin (1968) using blood hemolysates from porphyric cattle.

Erythropoietic Protoporphyria (EPP) in Cattle

Ruth *et al.* (1977) described the occurrence in cattle of a disease resembling human EPP. Four calves and one adult cow were found to have marked cutaneous photosensitivity without the discoloration of urine or teeth which characterizes CEP in humans (Taddeini and Watson, 1968; Tschudy and Lamon, 1980; Meyer and Schmid, 1978) and cattle (Watson *et al.,* 1959). Erythrocyte PROTO in affected calves was markedly elevated—35,000–46,000 µg PROTO/100 ml of packed red cells vs. 100–133 for controls—(Ruth *et al.,* 1977). The levels of erythrocyte PROTO in EPP calves were at least seven times higher than those in human EPP. Fecal PROTO and urobilinogen were also markedly increased. Unlike human EPP, which is a dominant disease, bovine EPP is a recessive disorder. The single adult EPP cow is known to have lost cutaneous photosensitivity (Ruth *et al.,* 1977), although all affected calves display this symptom. Spectral analysis has shown that erythrocyte and plasma PROTO in affected cows consists essentially of free PROTO as in human EPP, while controls and carriers have had about one-half free PROTO and one-half zinc-chelated PROTO. Ferrochelatase activity in homozygous animals as determined by the rate of mesoporphyrin utilization was approximately 10% of controls in liver, kidney, heart, spleen, and lung, and that for heterozygotes was about 40% of controls in liver. ALA synthase activity was increased only in the liver (3 times normal) while ALA dehydratase activity was normal in homozygous animals. Ferrochelatase deficiency in this disorder can also be demonstrated in cultured fibroblasts (Table V). As shown in Table V, PROTO accumulation in cells from homozygotes incubated with ALA was markedly elevated compared with that found in control cells. Cells from heterozygotes displayed intermediate levels of PROTO accumulation from ALA. These results clearly indicate that cultured fibroblasts from these EPP cows carry the gene defect of the disease, as do liver and bone marrow cells, and that the extent of functional ferrochelatase deficiency is dependent on gene dosage in these cells.

TABLE V. PROTO Formation from ALA in Cultured Bovine Skin Fibroblasts[a,b]

	Number of animals	PROTO formation (pmol/mg protein/24 hr)
Normal	6	42.27 ± 10.18
EPP heterozygotes	3	108.34 ± 4.67
EPP homozygotes	5	290.2 ± 40.78

[a] From Sassa, S., Schwartz, S., and Ruth, G., submitted for publication, 1981.

[b] PROTO formation was studied in cultured skin fibroblasts which were incubated in the presence of 0.6 mM ALA for 24 hr.

Swine Porphyria

Congenital erythropoietic porphyria (CEP) has also been observed in pigs (Clare and Stephens, 1944; Jørgensen and With, 1955, 1963). Skin lesions due to photosensitivity are mild and manifest only during the summer; they are limited to skin areas covered with white hair (Jørgensen and With, 1963). The diagnosis of porphyria therefore depends mainly on the detection of red fluorescence in porphyrin-laden bones and teeth or on biochemical analysis of urine or erythrocytes. The porphyric newborn pig also displays brown or black discoloration of the teeth. Porphyric pigs excrete large amounts of URO, hepta- and hexacarboxylic porphyrins in urine, although striking variation in porphyrin excretion has been observed both in quantity and quality from time to time in the same animal (Jørgensen and With, 1963). Porphyrins are excreted largely in the reduced form as porphyrinogens and become oxidized to porphyrins after acidification or upon standing in air. The porphyrins excreted are mainly type I. Excessive PBG excretion in urine is not found in this disease.

In contrast to the human and bovine forms of CEP, the swine disease is inherited as an autosomal dominant trait. Porphyric pigs show normal fertility although this appears not to be the case in porphyric cattle (Jørgensen and With, 1963).

Feline Porphyria

Congenital erythropoietic porphyria has been reported in a family of domestic short-haired cats (Tobias, 1964; Glenn et al., 1968), and the disease has also been found in a family of Siamese cats (Giddens et al., 1975). The inheritance pattern of the porphyric trait in the black domestic cat pedigree was compatible with autosomal dominant transmission, as in swine CEP; the nature of the inheritance in the Siamese cat family is not known. Porphyric kittens were readily recognized by the brownish discoloration and intense red fluorescence of their teeth under ultraviolet light illumination (more intense in the deciduous teeth than in the permanent teeth) (Glenn et al., 1968). The principal porphyrin excreted in urine was URO I and excess PBG was also found. The Siamese porphyric cats were distinct in that CEP in them was associated with severe anemia, hepatosplenomegaly, and a peculiar renal disease characterized by mesangial hypertrophy and tubular degeneration. They also excreted large amounts of URO I (Giddens et al., 1975). An apparent difference in the symptoms of CEP in the two porphyric lineages in cats emphasizes the possibility that CEP in other species, i.e., man, may not necessarily be a single disorder.

Porphyria in Squirrels

The fox squirrel (*Sciurus niger*) has a physiological porphyria in the sense that all members of this species excrete large amounts of type I URO and COPRO. Turner (1937) examined a large number of bones from various mammalian, avian, and amphibian sources and found red-stained bones only in Sciuridae, i.e., the fox squirrel. All fox squirrels show, as in human CEP, pink to reddish discoloration in teeth, bones, and tissues due to accumulation of URO I, and markedly elevated concentrations of porphyrins in spleen, blood, and urine. The concentration of URO I in the spleen is much higher than in liver, kidney, or bone marrow probably due to sequestration of porphyrin-laden red cells (Levin and Flyger, 1973).

UROgen III cosynthase activity in blood hemolysates and bone-marrow cells of the fox squirrel is greatly diminished (1/10) as compared with this enzyme activity in the bone marrow of the grey squirrel (Levin and Flyger, 1973). The fox squirrel enzyme is also much more unstable than the grey squirrel enzyme, which may account for the extremely low activity found in fox squirrel tissues and the marked overproduction of URO I in this species. In contrast to human CEP which is characterized by extensive photoinduced dermatosis and hemolytic anemia, fox squirrels do not display skin photosensitivity (Flyger and Levin, 1977). Although hemolytic anemia has not been demonstrated in fox squirrels, the URO concentration of red cells and spleen of these animals can be greatly augmented by bleeding (Levin and Flyger, 1973), suggesting that the principal site of porphyrinogenesis is in erythroid tissue.

Other Possible Animal Models of Porphyria

These include *Pteria vulgaris,* a mussel from the Persian Gulf containing URO I in its shell (Fischer and Haarer, 1932); birds of the genus *Turacus* from Africa containing the copper complex of URO I in their wing feathers (Fischer and Hilger, 1923); and *Asio flammeus,* a bird of subarctic habitat which contains a porphyrin (probably URO) in its feathers (Derrien and Turchini, 1926). The blood and bones of certain mammalian fetuses have also been shown to contain URO and COPRO (Derrien, 1924; Borst and Konigsdorfer, 1929; Fikentscher, 1935). In the annelids and the asteroid echinoderms, exposure to sunlight is known often to result in cutaneous photosensitivity, marked edema and, eventually, death (Mangum, 1978). The occurrence of PROTO is also well known in brown eggshells and in uteri of Rhode Island Reds, Japanese quail, and other hens (Schwartz and Ruth, 1978). It is known that in molluscs porphyrin (URO I) accumulates in the shell when they do not produce hemoglobin, and, interestingly, the porphyrin is lacking in shells of molluscs that do produce hemoglobin (Kennedy, 1975). Harderian glands (bilobed tubuloalveolar glands located deep within the orbit of the eye) are known to contain markedly elevated levels of ALA synthase activity (Margolis, 1971; Eida *et al.*, 1975) and PROTO (Derrien and Turchini, 1924). ALA synthase activity in Harderian glands is not affected by hemin (Margolis, 1971), but may be regulated by female sex hormones since the enzyme activity is higher in female rats than in males (Margolis, 1971) and the activity in females shows cyclic oscillations in relation to the estrous cycle (Moore *et al.*, 1977; Payne *et al.*, 1979).

The occurrence of porphyrins in a wide variety of invertebrate species has recently been summarized in a comprehensive review by Kennedy (1975). In general, free porphyrins are found only sporadically in nature and, when found, only in trace amounts, indicating that the heme biosynthetic pathway, unless impaired by genetic lesions or the impacts of environmental chemicals, normally operates with a very high degree of efficiency (Lemberg, 1954).

ACKNOWLEDGMENTS. This work was supported in part by USPHS grant ES-01055, National Foundation grant I-350, and Clinical Research Center grant number MO1 RR 00102. The authors are deeply indebted to Dr. Karl E. Anderson for his helpful comments and to Mrs. Heidemarie Robinson for her preparation of this manuscript.

REFERENCES

Aldrich, R. A., Hawkinson, V., Grinstein, M., and Watson, C. J., 1951, Photosensitive or congenital porphyria with hemolytic anemia. I. Clinical and fundamental studies before and after splenectomy, *Blood* 6:685–698.

Alvares, A. P., Leigh, S., Cohn, J., and Kappas, A., 1972, Lead and methyl mercury: Effects of acute exposure on cytochrome P450 and the mixed function oxidase system in the liver, *J. Exp. Med.* 135:1406–1409.

Amoroso, E. C., Loosmore, R. M., Rimington, C., and Tooth, B. E., 1957, Congenital porphyria in bovines: First living cases in Britain, *Nature* 180:230–231.

Anderson, K. E., Sassa, S., Peterson, C. M., and Kappas, A., 1977, Increased erythrocyte uroporphyrinogen-I-synthetase, δ-aminolevulinic acid dehydratase and protoporphyrin in hemolytic anemias, *Am. J. Med.* 63:359–364.

Anderson, K. E., Bradlow, H. L., Sassa, S., and Kappas, A., 1979, Studies in porphyria VIII. Relationship of the 5α-reductase metabolism of steroid hormones to clinical expression of the genetic defect in acute intermittent porphyria, *Am. J. Med.* 66:644–650.

Anderson, P. M., 1979, Studies of human heme biosynthetic enzymes and acute intermittent porphyria. A dissertation submitted to the Graduate Faculty in Biomedical Sciences in partial fulfillment of the requirement for the degree of Doctor of Philosophy, The City University of New York.

Anderson, P. M., and Desnick, R. J., 1979, Purification and properties of δ-aminolevulinic acid dehydratase from human erythrocytes, *J. Biol. Chem.* 254:6924–6930.

Anderson, P. M., and Desnick, R. J., 1980, Purification and properties of uroporphyrinogen I synthase from human erythrocytes, Identification of stable enzyme-substrate intermediates, *J. Biol. Chem.* 255:1993–1999.

Aoki, Y., 1978, Crystallization and characterization of a new protease in mitochondria of bone marrow cells, *J. Biol. Chem.* 253:2026–2032.

Aoki, Y., Wada, O., Urata, G., Takaku, F., and Nakao, K., 1971, Purification and some properties of δ-aminolevulinate (ALA) synthetase in rabbit reticulocytes, *Biochem. Biophys. Res. Comm.* 42:568–575.

Aoki, Y., Urata, G., Takaku, F., and Katsunuma, N., 1975, A new protease inactivating δ-aminolevulinic acid synthetase in mitochondria of human bone marrow, *Biochem. Biophys. Res. Comm.* 65:567–574.

Astrup, E. G., 1978, Family studies on the activity of uroporphyrinogen I synthase in diagnosis of acute intermittent porphyria, *Clin. Sci. Mol. Med.* 54:251–256.

Balcerzak, S. P., Westerman, M. P., Heinle, E. W., and Taylor, F. H., 1968, Measurement of iron stores using deferoxamine, *Ann. Intern. Med.* 68:518–525.

Barnes, H. D., and Whittaker, N., 1965, Hereditary coproporphyria with acute intermittent manifestations, *Br. Med. J.* ii:1102–1104.

Barnes, R., Jones, M. S., Jones, O. T. G., and Porra, R. T., 1971, Ferrochelatase and δ-aminolaevulinate synthetase in brain, heart, kidney and liver of normal and porphyric rats. The induction of δ-aminolaevulinate synthetase in kidney cytosol and mitochondria by allylisopropylacetamide, *Biochem. J.* 124:633–637.

Battersby, A. R., Fookes, C. J. R., Matcham, G. W., and McDonald, E., 1979, Order of assembly of the four pyrrole rings during biosynthesis of the natural porphyrins, *J. C. S. Chem. Comm.* 539–541.

Batlle, A. M. del C., Benson, A., and Rimington, C., 1965, Purification and properties of coproporphyrinogenase, *Biochem. J.* 97:731–740.

Batlle, A. M. del C., deXifra, E. A. W., Stella, A. M., Bustos, N., and With, T. K., 1979, Studies on porphyrin biosynthesis and the enzymes involved in bovine congenital erythropoietic porphyria, *Clin. Sci.* 57:63–70.

Beale, S. I., and Castelfranco, P. A., 1974, The biosynthesis of δ-aminolevulinic acid in higher plants. II. Formation of ^{14}C-δ-aminolevulinic acid from labeled precursors in greening plant tissues, *Plant Physiol.* 53:297–303.

Beale, S. I., Gough, S. P., and Granick, S., 1975, Biosynthesis of δ-aminolevulinic acid from the intact carbon skeleton of glutamic acid in greening barley, *Proc. Natl. Acad. Sci. USA* 72:2719–2723.

Beauvais, P., Klein, M. L., Denave, L., and Martel, C., 1976, Porphyrie aigue intermittente, *Arch. Franç. Ped.* **33**:987–992.

Becker, D. M., Viljoen, J. D., Katz, J., and Kramer, S., 1977, Reduced ferrochelatase activity: A defect common to porphyria variegata and protoporphyria, *Br. J. Haemat.* **36**:171–179.

Becker, F. T., 1965, Porphyria cutanea tarda induced by estrogens, *Arch. Dermatol.* **92**:252–256.

Benedetto, A. V., Kushner, J. P., and Taylor, J. S., 1978, Porphyria cutanea tarda in three generations of a single family, *N. Engl. J. Med.* **298**:358–362.

Berger, H., and Goldberg, A., 1955, Hereditary coproporphyria, *Brit. Med. J.* **ii**:85–88.

Berman, J., Friedmann, B., and Brousil, J., 1963, Survival of erythrocytes in porphyria cutanea tarda (in Czech), *Vnitř. Lĕk.* **9**:336 [cited from Turnbull (1971)].

Bevan, D. R., Bodlaender, P., and Shemin, D., 1980, Mechanism of porphobilinogen synthase. Requirement of Zn^{2+} for enzyme activity, *J. Biol. Chem.* **255**:2030–2035.

Bickers, D. R., Pathak, M. A., and Magnus, I. A., 1979, The porphyrias, in: *Dermatology in General Medicine* (T. B. Fitzpatrick, A. Z. Eisen, K. Wolff, I. M. Friedberg, and K. F. Austen, eds.), Second Ed., pp. 1072–1105, McGraw-Hill, New York.

Birchfield, R. I., and Cowger, M. L., 1966, Acute intermittent porphyria with seizures. Anticonvulsant medication-induced metabolic changes, *Am. J. Dis. Child.* **112**:561–565.

Bird, T. D., Hamernyik, P., Nutter, J. Y., and Labbe, R. F., 1979, Inherited deficiency of delta-aminolevulinic acid dehydratase, *Am. J. Hum. Genet.* **31**:662–668.

Birkett, D. A., Garratts, M., and Stevenson, D. J., 1969, Phototoxic bullous eruptions due to nalidixic acid, *Br. J. Dermatol.* **81**:342–344.

Bishop, D. F., 1976, Differentiation between isozymes of guinea pig δ-aminolevulinate synthetase by affinity chromatography, *Fed. Proc.* **35**:1658.

Bissell, D. M., and Hammaker, L. E., 1976a, Cytochrome P-450 heme and the regulation of hepatic heme oxygenase activity, *Arch. Biochem. Biophys.* **176**:91–102.

Bissell, D. M., and Hammaker, L. E., 1976b, Cytochrome P-450 heme and the regulation of δ-aminolevulinic acid synthetase in the liver, *Arch. Biochem. Biophys.* **176**:103–112.

Bleiberg, J., Wallen, M., Brodkin, R., and Applebaum, I. L., 1964, Industrially acquired porphyria, *Arch. Dermatol.* **89**:793–797.

Blekkenhorst, G., Pimstone, N. R., and Eales, L., 1976a, Porphyria cutanea tarda in South Africa: The metabolic basis of disordered haem biosynthesis, in: *Porphyrins in Human Diseases* (M. Doss, ed.), pp. 299–311, Karger, Basel.

Blekkenhorst, G. H., Pimstone, N. R., Webber, B. L., and Eales, L., 1976b, Hepatic haem metabolism in porphyria cutanea tarda (PCT): Enzymatic studies and their relation to liver ultrastructure, *Ann. Clin. Res.* **8**(Suppl. 17):108–121.

Bloomer, J. R., 1980, Characterization of deficient heme synthase activity in protoporphyria with cultured skin fibroblasts, *J. Clin. Invest.* **65**:321–328.

Bloomer, J. R., Brenner, D. A., and Mahoney, M. J., 1977, Study of factors causing excess protoporphyrin accumulation in cultured skin fibroblasts from patients with protoporphyria, *J. Clin. Invest.* **60**:1354–1361.

Bogorad, L., 1958a, The enzymatic synthesis of porphyrins from porphobilinogen. I. Uroporphyrin I, *J. Biol. Chem.* **233**:501–509.

Bogorad, L., 1958b, The enzymatic synthesis of porphyrins from porphobilinogen. II. Uroporphyrin III, *J. Biol. Chem.* **233**:510–515.

Bonkowsky, H. L., Tschudy, D. P., Collins, A., Doherty, J., Bossenmaier, I., Cardinal, R., and Watson, C. J., 1971, Repression of the overproduction of porphyrin precursors in acute intermittent porphyria by intravenous infusions of hematin, *Proc. Natl. Acad. Sci. USA* **68**:2725–2729.

Bonkowsky, H. L., Bloomer, J. R., Ebert, P. S., and Mahoney, M. J., 1975a, Heme synthase deficiency in human protoporphyria. Demonstration of the defect in liver and cultured skin fibroblasts, *J. Clin. Invest.* **56**:1139–1148.

Bonkowsky, H. L., Tschudy, D. P., Weinbach, E. C., Ebert, P. S., and Doherty, J. M., 1975b, Porphyrin synthesis and mitochondrial respiration in acute intermittent porphyria: Studies using cultured human fibroblasts, *J. Lab. Clin. Med.* **85**:93–102.

Bornstein, J. C., Pickett, J. B., and Diamond, I., 1979, Inhibition of the evoked release of acetylcholine by the porphyrin precursor δ-aminolevulinic acid, *Ann. Neurol.* **5**:94–96.

Borst, M. and Königsdorfer, H., Jr., 1929, *Untersuchungen über Porphyrie mit besonderer Berücksichtigung der Porphyria congenita*, Hirzel, Leipzig.

Bottomley, S. S., Tanaka, M., and Everett, M. A., 1975, Diminished erythroid ferrochelatase activity in protoporphyria, *J. Lab. Clin. Med.* **86**:126–131.

Bradlow, H. L., Gillette, P. N., Gallagher, T. F., and Kappas, A., 1973, Studies in porphyria. II. Evidence for a deficiency of steroid Δ^4-5α reductase activity in acute intermittent porphyria, *J. Exp. Med.* **138**:754–763.

Brennan, M. J. W., and Cantrell, R. C., 1979, δ-Aminolaevulinic acid is a potent agonist for GABA autoreceptors, *Nature* **280**:514–515.

Brenner, D. A., and Bloomer, J. R., 1979, The enzymatic defect in variegate porphyria, *Clin. Res.* **27**:274a.

Brenner, D. A., and Bloomer, J. R., 1980, The enzymatic defect in variegate porphyria. Studies with human cultured skin fibroblasts, *N. Engl. J. Med.* **302**:765–769.

Briggs, D., Condie, L., Sedman, R., and Tephly, T., 1976, δ-Aminolevulinic acid synthetase in the heart, *J. Biol. Chem.* **251**:4996–5001.

Brivet, F., Drueke, T., Guillemette, J., 1978, Porphyria cutanea tarda-like syndrome in hemodialysed patients, *Nephron* **20**:258–266.

Brockelhurst, D., 1978, Low-cost uroporphyrinogen I synthase screening for acute intermittent porphyria, *Clin. Chem.* **24**:730–731.

Brodie, M. J., Moore, M. R., Thompson, G. G., and Goldberg, A., 1977a, The treatment of acute intermittent porphyria with laevulose, *Clin. Sci. Mol. Med.* **53**:365–371.

Brodie, M. J., Thompson, G. G., Moore, M. R., Beattie, A. D., and Goldberg, A., 1977b, Hereditary coproporphyria, *Quart. J. Med.* **46**:229–241.

Brodie, M. J., Moore, M. R., Thompson, G. G., Goldberg, A., and Holti, G., 1978, Haem biosynthesis in peripheral blood in erythropoietic protoporphyria, *Clin. Exp. Dermatol.* **2**:381–388.

Brooker, J. D., May, B. K., and Elliott, W. H., 1980, Synthesis of δ-aminolevulinate synthase *in vitro* using hepatic mRNA from chick embryos with induced porphyria, *Eur. J. Biochem.* **106**:17–24.

Brunsting, L. A., Mason, H. L., and Aldrich, R. A., 1951, Adult form of chronic porphyria with cutaneous manifestations: Report of 17 additional cases, *J. Am. Med. Assoc.* **146**:1207.

Bugany, H., Flothe, L., and Weser, U., 1971, Kinetics of metal chelatase of rat liver mitochondria, *FEBS Lett.* **13**:92–94.

Burry, J. N., and Lawrence, J. R., 1976, Phototoxic blisters from high furosemide dosage, *Br. J. Dermatol.* **99**:445–449.

Cam, C., and Nigogosyan, G., 1963, Acquired toxic porphyria cutanea tarda due to hexachlorobenzene. Report of 348 cases caused by this fungicide, *J. Am. Med. Assoc.* **183**:88–91.

Canivet, J., and Pelhard-Considère, M., 1958, Étude de l'hémolyse dans deux cas de porphyria congénitale, *Rev. Fr. Etud. Clin. Biol.* **3**:27–32.

Carlson, R. E., Dolphin, D., and Bernstein, M., 1978, Copper coproporphyrin excretion in familial coproporphyria, *Clin. Chem.* **24**:2009–2012.

Chalmers, J. N. M., Gillam, A. E., and Kench, J. E., 1940, Porphyrinuria. In a case of industrial methyl chloride poisoning, *Lancet* **ii**:806–807.

Chatterjea, J. B., 1964, Erythropoietic porphyria, *Blood* **24**:806–807.

Cheh, A., and Neilands, J. B., 1973, Zinc, an essential metal ion for beef liver δ-aminolevulinate dehydratase, *Biochem. Biophys. Res. Comm.* **55**:1060–1063.

Chu, T. C., and Chu, E. J.-H., 1962, Porphyrins from congenitally porphyric (pink-tooth) cattle, *Biochem. J.* **83**:318–325.

Chu, T. C., and Chu, E. J.-H., 1967, Porphyrin patterns in different types of porphyria, *Clin. Chem.* **13**:371–387.

Clare, N. T., and Stephens, E. H., 1944, Congenital porphyria in pigs, *Nature* **153**:252–253.

Clark, K. G. A., and Nicholson, D. C., 1971, Erythrocyte protoporphyrin and iron uptake in erythropoietic protoporphyria, *Clin. Sci.* **41**:363–370.

Coleman, D. L., 1966, Purification and properties of δ-aminolevulinate dehydratase from tissues of two strains of mice, *J. Biol. Chem.* **241**:5511–5517.

Condie, L. W., Tephly, T. R., and Baron, J., 1976a, Studies on heme synthesis in the rat adrenal, *Ann. Clin. Res.* **8**(suppl. 17):83–88.

Condie, L. W., Baron, J., and Tephly, T. R., 1976b, Studies on adrenal δ-aminolevulinic acid synthetase, *Arch. Biochem. Biophys.* **172**:123–129.

Connon, J. J., and Turkington, V., 1968, Hereditary coproporphyria, *Lancet* **ii**:263–264.

Cookson, G. H., and Rimington, C., 1954, Porphobilinogen, *Biochem. J.* **57**:476–484.

Copeman, P. W. M., Cripps, D. J., and Summerly, R., 1966, Cutaneous hepatic porphyria and oestrogens, *Br. Med. J.* **i**:461–463.

Cornford, P., 1964, Transformation of porphobilinogen into porphyrins by preparations from human erythrocytes, *Biochem. J.* **91**:64–73.

Cram, D. L., Epstein, J. H., and Tuffanelli, D. L., 1973, Lupus erythematosus and porphyria: Coexistence in seven patients, *Arch. Dermatol.* **108**:779–784.

Cripps, D. J., and MacEachern, W. N., 1971, Hepatic and erythropoietic porphyria, *Arch. Pathol.* **91**:497–505.

Cripps, D. J., Hawgood, R. S., and Magnus, I. A., 1966, Iodine tungsten fluorescence microscopy for porphyrin fluorescence. A study on erythropoietic protoporphyria, *Arch. Dermatol.* **93**:129–134.

Cutler, M. G., Dick, J. M., and Moore, M. R., 1978, Effect of delta-aminolevulinic acid on frog nerve-muscle function, *Life Sci.* **23**:2233–2238.

Danby, C. W. E., Koval, A., and Wyllie, J., 1966, Acquired porphyria cutanea tarda: Two cases treated by repeated phlebotomies, *Can. Med. Assoc. J.* **94**:1358–1359.

Dean, G., 1963, Prevalence of the porphyrias, *S. Afr. J. Lab. Clin. Med.* **9**:145–151.

Dean, G., 1971, *The Porphyrias: A Story of Inheritance and Environment,* Second edn., Pitman Medical, London.

Dean, G., 1978, Porphyria: Cutanea tarda or variegata?, *N. Engl. J. Med.* **298**:1317.

Dean, G., and Barnes, H. D., 1959, Porphyria in Sweden and South Africa, *S. Afr. Med. J.* **33**:246–253.

Dean, G., Kramer, S., and Lamb, P., 1969, Coproporphyria, *S. Afr. Med. J.* **43**:138–142.

De Goeij, A. F. P. M., Christianse, K., and van Steveninck, I., 1975, Decreased haem synthetase activity in blood cells of patients with erythropoietic protoporphyria, *Eur. J. Clin. Invest.* **5**:397–400.

DeLeo, V. A., Poh-Fitzpatrick, M., Matthews-Roth, M., and Harber, L. C., 1976, Erythropoietic protoporphyria. Ten years experience, *Am. J. Med.* **60**:8–22.

DeMatteis, F., 1973, Drug interactions in experimental hepatic porphyria. A model for the exacerbation by drug of human variegate porphryia, *Enzyme* **16**:266–275.

DeMatteis, F., 1978, Loss of microsomal components in drug induced liver damage, in cholestasis and after administration of chemicals which stimulate heme catabolism, *Pharmacol. Ther. A.* **2**:693–725.

DeMatteis, F., and Cantoni, L., 1979, Alteration of the porphyrin nucleus of cytochrome P-450 caused in the liver by treatment with allyl-containing drugs. Is the modified porphyrin N-substituted? *Biochem. J.* **183**:99–103.

DeMatteis, F., and Gibbs, A. H., 1980, Drug-induced conversion of liver haem into modified porphyrins. Evidence for two classes of products. *Biochem. J.* **187**:285–288.

DeMatteis, F., Abbritti, G., and Gibbs, A., 1973, Decreased liver activity of porphyrin-metal chelatase in hepatic porphyria caused by 3,5-diethoxycarbonyl-1,4-dihydrocollidine. Studies in rats and mice, *Biochem. J.* **134**:717–727.

DeMatteis, F., Gibbs, A. H., and Tephly, T. A., 1980, Inhibition of protohaem ferro-lyase in experimental porphyria. Isolation and partial characterization of a modified porphyrin inhibitor. *Biochem. J.* **188**:145–152.

Derrien, E., 1924, Note préliminaire sur quelques faits nouveaux pour l'histoire naturelle des porphyrines animales, *Compt. Rend. Soc. Biol.* **91**:634–636.

Derrien, E., and Turchini, J., 1924, Sur l'accumulation d'une porphyrine dans la glande de Harder des Rongeurs du genre mus et sur son mode d'excrétion. *C. R. Soc. Biol.* **91**:637–639.

Derrien, E., and Turchini, J., 1926, Sur la biologie des porphyrines naturelles, *Bull. Soc. Chim. Biol.* **8**:218–219.

De Verneuil, H., Aitken, G., and Nordmann, Y., 1978, Familial and sporadic porphyria cutanea. Two different diseases, *Hum. Genet.* **44**:145–151.

DeVerneuil, H., Grandchamp, B., and Nordmann, Y., 1980, Some kinetic properties of human red cell uroporphyrinogen decarboxylase, *Biochim. Biophys. Acta* **611**:174–186.

Doss, M., and Tiepermann, R., 1978, Uroporphyrinogen-Synthase in Erythrocyten bei akuter intermittierender Porphyrie: Neue pathobiochemische Aspekte, *J. Clin. Chem. Clin. Biochem.* **16**:111–118.

Doss, M., Meinhof, W., and Look, D., 1971, Porphyrins in liver and urine in acute intermittent and chronic hepatic porphyrias, *S. Afr. J. Lab. Clin. Med.* **17**:50–56.

Doss, M., von Tiepermann, R., Schneider, J., and Schmid, H., 1977, New type of hepatic porphyria with porphobilinogen synthase defect and intermittent acute clinical manifestation, *Klin. Wochensch.* **57**:1123–1127.

Dowdle, E. B., Mustard, P., and Eales, L., 1967, δ-Aminolaevulinic acid synthetase activity in normal and porphyric human livers, *S. Afr. Med. J.* **2**:1093–1096.

Dowdle, E., Goldswain, P., Spong, N., and Eales, L., 1970, The pattern of porphyrin isomer accumulation and excretion in symptomatic porphyria, *Clin. Sci.* **39**:147–158.

Doyle, D., and Schimke, R. T., 1969, The genetic and developmental regulation and hepatic δ-aminolevulinate dehydratase in mice, *J. Biol. Chem.* **244**:5449–5459.

Drummond, G. S., and Kappas, A., 1979, Manganese and zinc blockade of enzyme induction: Studies with microsomal heme oxygenase, *Proc. Natl. Acad. Sci. USA* **76**:5331–5335.

Druyan, R., DeBernard, B., and Rabinowitz, M., 1969, Turnover of cytochromes labeled with δ-aminolevulinic acid—^{3}H in rat liver, *J. Biol. Chem.* **244**:5874–5878.

Eales, L., 1963, Porphyria as seen in Cape Town: A survey of 250 patients and some recent studies, *S. Afr. J. Lab. Clin. Med.* **9**:151–161.

Eales, L., and Grosser, Y., 1971, The porphyrin peptides: Practical implications, *S. Afr. J. Lab. Clin. Med.* **17**:160–163.

Eales, L., and Linder, G. C., 1962, Porphyria—the acute attack. An analysis of 80 cases, *S. Afr. Med. J.* **36**:284–292.

Eales, L., Dowdle, E. B., Saunders, S. J., and Sweeney, G. D., 1963, The diagnostic importance of faecal porphyrins in the differentiation of the porphyrias. II. Values in the cutaneous porphyrias, *S. Afr. J. Lab. Clin. Med.* **9**:126–137.

Eales, L., Dowdle, E. B., Levey, M. J., and Sweeney, G. D., 1966a, The diagnostic importance of faecal porphyrins in the differentiation of the porphyrias. III. South African genetic porphyria (variegate porphyria)—the acute attack, *S. Afr. Med. J.* **40**:380–382.

Eales, L., Levey, M. J., and Sweeney, G. D., 1966b, The place of screening tests and quantitative investigations in the diagnosis of the porphyrias, with particular reference to variegate and symptomatic porphyria, *S. Afr. Med. J.* **40**:63–71.

Eales, L., Sears, W. G., King, K. B., Levey, M. J., and Rimington, C., 1972, Symptomatic porphyria in a case of Felty's syndrome. I. Clinical and routine biochemical studies, *Clin. Chem.* **18**:459–461.

Eales, L., Grosser, T., and Sears, W. G., 1975, The clinical biochemistry of the human hepatocutaneous porphyrias in the light of recent studies of newly identified intermediates and porphyrin derivatives, *Ann. N.Y. Acad. Sci.* **244**:441–471.

Ebert, P. S., Hess, P. A., Frykholm, B. C., and Tschudy, D. P., 1979, Succinylacetone, a potent inhibitor of heme biosynthesis: Effect on cell growth, heme content and δ-aminolevulinic acid dehydratase activity of malignant murine erythroleukemia cells, *Biochem. Biophys. Res.Comm.* **88**:1382–1390.

Eida, K., Kubota, N., Nishigaki, T., and Kikutani, M., 1975, Harderian gland. V. Effect of dietary pantothenic acid deficiency on porphyrin biosynthesis in Harderian gland of rats, *Chem. Pharm. Bull.* **23**:1–4.

Eisen, H., Keppel-Ballivet, F., Georgopoulos, C. P., Sassa, S., Granick, J., Pragnell, I., and Ostertag, W., 1978, Biochemical and genetic analysis of erythroid differentiation in Friend-virus-transformed murine erythroleukemia cells, in: *Differentiation of Normal and Neoplastic Hematopoietic Cells* (B. Clarkson, P. A. Marks, and J. E. Till, eds.), pp. 277–294, Cold Spring Harbor Conference on Cell Proliferation, Vol. 5, Cold Spring Harbor Lab.

Elder, G. H., 1972, Identification of a group of tetracarboxylate porphyrins containing one acetate and three propionate beta-substituents, in faeces from patients with symptomatic cutaneous hepatic porphyria, and from rats with porphyria due to hexachlorbenzene, *Biochem. J.* **126**:877–891.

Elder, G. H., 1975, Differentiation of porphyria cutanea tarda symptomatica from other types of porphyria by measurement of isocoproporphyrin in faeces, *J. Clin. Pathol.* **28**:601–607.

Elder, G. H., 1977, Porphyrin metabolism in porphyria cutanea tarda, *Sem. Hematol.* **14**:227–242.

Elder, G. H., and Tovey, J. A., 1977, Uroporphyrinogen decarboxylase activity of human tissues, *Biochem. Soc. Trans.* **5**:1470–1472.

Elder, G. H., Gray, C. H., and Nicholson, D. C., 1972, The porphyrias: A review, *J. Clin. Pathol.* **25**:1013–1033.

Elder, G. H., Magnus, I. A., Handa, F., and Doyle, M., 1974, Faecal "X porphyrin" in the hepatic porphyrias, *Enzyme* **17**:29–38.

Elder, G. H., Evans, J. O., and Matlin, S. A., 1976a, The effect of the porphyrinogenic compound, hexachlorobenzene, on the activity of hepatic uroporphyrinogen decarboxylase in the rat, *Clin. Sci. Mol. Med.* **51**:71–80.

Elder, G. H., Evans, J. O., Thomas, N., Cox, R., Brodie, M. J., Moore, M. R., Goldberg, A., and Nicholson, D. C., 1976*b*, The primary enzyme defect in hereditary coproporphyria, *Lancet* **ii**:1217–1219.

Elder, G. H., Lee, G. B., and Tovey, J. A., 1978, Decreased activity of hepatic uroporphyrinogen decarboxylase in sporadic porphyria cutanea tarda, *N. Engl. J. Med.* **299**:274–278.

Epstein, J. H., and Pinski, J. B., 1965, Porphyria cutanea tarda. Association with abnormal iron metabolism, *Arch. Dermatol.* **92**:362–366.

Epstein, J. H., and Redeker, A. G., 1965, Porphyria cutanea tarda symptomatica (PCT-S). A study of the effect of phlebotomy therapy, *Arch. Dermatol.* **92**:286–290.

Epstein, J. H., Tuffanelli, D. L., Seibert, J. S., and Epstein, W. L., 1976, Porphyria-like cutaneous changes induced by tetracycline hydrochloride photosensitization, *Arch. Dermatol.* **112**:661–666.

Eriksen, L., and Eriksen, N., 1974, Porphyrin distribution and porphyrin excretion in human congenital erythropoietic porphyria, *Scand. J. Clin. Lab. Invest.* **33**:323–332.

Eriksen, L., and Seip, M., 1973, Congenital erythropoietic porphyria. A family study, *Clin. Genet.* **4**:166–172.

Estabrook, R. W., Martinez-Zedillo, G., Young, S., Peterson, J. A., and McCarthy, J., 1975, The interaction of steroids with liver microsomal cytochrome P-450—a general hypothesis, *J. Steroid Biochem.* **6**:419–425.

Falk, J. E., 1964, *Porphyrins and Metalloporphyrins. Their General, Physical and Coordination Chemistry, and Laboratory Methods,* Elsevier Publishing, Amsterdam.

Feldman, D. S., Levere, R. D., Lieberman, J. S., Cardinal, R. A., and Watson, C. J., 1971, Presynaptic neuromuscular inhibition by porphobilinogen and porphobilin, *Proc. Natl. Acad. Sic. USA* **68**:383–386.

Felsher, B. F., and Kushner, J. P., 1977, Hepatic siderosis and porphyria cutanea tarda: Relation of iron excess to the metabolic defect, *Semin. Hemat.* **14**:243–251.

Felsher, B. F., Norris, M. E., and Shih, J. C., 1978, Red-cell uroporphyrinogen decarboxylase activity in porphyria cutanea tarda and in other forms of porphyria, *New Engl. J. Med.* **299**:1095–1098.

Fikentscher, R., 1935, Über Porphyrinbefunde im Serum von Feten und Neugeborenen, *Klin. Wschr.* **14**:569–571.

Finelli, V. N., Murthy, L. Peirano, W. B., and Petering, H. G., 1974, δ-Aminolevulinate dehydratase, a zinc dependent enzyme, *Biochem. Biophys. Res. Comm.* **60**:1418–1424.

Finelli, V. N., Klauder, D. S., Karaffa, M. A., and Petering, H. G., 1975, Interaction of zinc and lead on δ-aminolevulinate dehydratase, *Biochem. Biophys. Res. Commun.* **65**:303–311.

Fischer, H., and Haarer, E., 1932, Über Uroporphyrin aus Muschelschalen, *Z. physiol. Chem.* **204**:101–104.

Fischer, H., and Hilger, J., 1923, Zur Kenntnis der natürlichen Porphyrine. II. Über das Turacin, *Z. physiol. Chem.* **128**:167–174.

Flyger, V., and Levin, E. Y., 1977, Congenital erythropoietic porphyria, *Am. J. Pathol.* **87**:269–272.

Fourie, P. J. J., 1936, The occurrence of congenital porphyrinuria (pink tooth) in cattle in South Africa (Swaziland) Onderstepoort, *J. Vet. Sci. Animal Indust.* **2**:535–565.

Freshney, R. I., and Paul, J., 1971, The activities of three enzymes of heme synthesis during hepatic erythropoiesis in the mouse embryo, *J. Embryol. Exp. Morphol.* **26**:313–322.

Frydman, R. B., and Feinstein, G., 1974, Studies on porphobilinogen deaminase and uroporphyrinogen III cosynthase from human erythrocytes, *Biochim. Biophys. Acta* **350**:358–373.

Frydman, B., Frydman, R. B., and Tomaro, M. L., 1973a, Pyrrolooxygenases: A new type of oxygenases, *Mol. Cell. Biochem.* **2**:121–136.

Frydman, R. B., Tomaro, M. L., Wanschelbaum, A., Andersen, E. M., Awruch, J., and Frydman, B., 1973b, Porphobilinogen oxygenase from wheat germ: Isolation, properties and products formed, *Biochemistry* **12**:5253–5262.

Frydman, R. B., Tomaro, M. L., Frydman, B., and Wanschelbaum, A., 1975, Porphobilinogen excretion in chemical induced porphyria. Reversal by induction of porphobilinogen oxygenase, *FEBS Lett.* **51**:206–210.

Gajdos, A., Weil, J., Gajdos-Török, M., and Coupry, A., 1969, Une nouvelle variété de porphyrie, la cop.roporphyria héréditaire mixte ou variegata, *Rev. Franc. Études Clin. Biol.* **14**:279–282.

Gassman, M., Pluscec, J., and Bogorad, L., 1966, δ-Aminolevulinic acid transaminase from *Chlorella* and *Phaseolus*, *Plant Physiol.* **41**(Suppl.):xiv.

Gassman, M., Pluscec, J., and Bogorad, L., 1968, δ-Aminolevulinic acid transaminase in *Chlorella vulgaris*, *Plant Physiol.* **43**:1411–1414.

Gayathri, A. K., Rao, M. R. S., and Padmanaban, G., 1973, Studies on the induction of δ-aminolevulinic acid synthetase in mouse liver, *Arch. Biochem. Biophys.* **155**:299–306.

Gentz, J., Jagenburg, R., and Zetterström, R., 1965, An inborn error of tyrosine metabolism with cirrhosis of the liver and multiple renal tubular defects (de Toni–Fanconi syndrome), *J. Pediatr.* **66**:670–696.

Gentz, J., Lindblad, B., Lindstedt, S., Levy, L., Shasteen, W., and Zetterström, R., 1967, Dietary treatment in tyrosinemia (tyrosinosis), *Am. J. Dis. Child.* **113**:31–37.

Gentz, J., Heinrich, J., Lindblad, B., Lindstedt, S., and Zetterström, R., 1969a, Enzymatic studies in a case of hereditary tyrosinemia with hepatoma, *Acta Paed. Scand.* **58**:393–396.

Gentz, J., Johansson, S., Lindblad, B., Lindstedt, S., and Zetterström, R., 1969b, Excretion of δ-aminolevulinic acid in hereditary tyrosinemia, *Clin. Chim. Acta* **23**:257–263.

Giddens, W. E., Jr., Labbe, R. F., Swango, L. J., and Padgett, G. A., 1975, Feline congenital erythropoietic porphyria associated with severe anemia and renal disease. Clinical, morphologic and biochemical studies, *Am. J. Pathol.* **80**:367–380.

Gilchrest, B., Rowe, J. W., and Mihm, M. C., Jr., 1975, Bullous dermatosis of hemodialysis, *Ann. Intern. Med.* **83**:480–483.

Glenn, B. L., Glenn, H. G., and Omtvedt, I. T., 1968, Congenital porphyria in the domestic cat (*Felis catus*): Preliminary investigations on inheritance pattern, *Am. J. Vet. Res.* **29**:1653–1657.

Goldberg, A., 1959, Acute intermittent porphyria. A study of 50 cases, *Quart. J. Med.* **28**:183–209.

Goldberg, A., and Rimington, C., 1962, *Diseases of Porphyrin Metabolism*, Charles C. Thomas, Springfield, Ill.

Goldberg, A., Rimington, C., and Lochhead, A., 1967, Hereditary coproporphyria, *Lancet* **i**:632–636.

Goldberg, A., Moore, M. R., Beattie, A. D., Hall, P. E., McCallum, J., and Grant, J. K., 1969, Excessive urinary excretion of certain porphyrinogenic steroids in human acute intermittent porphyria, *Lancet* **i**:115–118.

Goldberg, L., and St. John, A. C., 1976, Intracellular protein degradation in mammalian and bacterial cells: Part 2, *Ann. Rev. Biochem.* **45**:747–803.

Goldstein, J. A., Hickman, P., Bergman, H., and Vos, J. G., 1973, Hepatic porphyria induced by 2,3,7,8-tetrachlorodibenzo-p-dioxin, *Biochem. Biophys. Res. Comm.* **6**:919–928.

Goldstein, J. A., McKinney, J. D., Lucier, G. W., Hickman, P., Bergman, H., and Moore,

J. A., 1976, Toxicological assessment of hexachlorobiphenyl isomers and 2,3,7,8-te-trachlorodibenzofuran in chicks. II. Effects on drug metabolism and porphyrin accu-mulation, *Toxicol. Appl. Pharm.* **36**:81–92.

Goldstein, J. A., Friesen, M., Linder, R. E., Hickman, P., Hass, J. R., and Bergman, H., 1977, Effects of pentachlorophenol on hepatic drug-metabolizing enzymes and porphyria related to contamination with chlorinated dibenzo-p-dioxins and dibenzofurans, *Biochem. Pharm.* **26**:1549–1557.

Grandchamp, B., and Nordmann, Y., 1977, Decreased lymphocyte coproporphyrinogen III oxidase activity in hereditary coproporphyria, *Biochem. Biophys. Res. Comm.* **74**:1089–1095.

Grandchamp, B., Phung, N., Grelier, M., and Nordmann, Y., 1976, The spectrophotometric determination of uroporphyrinogen I synthase activity, *Clin. Chim. Acta* **70**:113–118.

Grandchamp, B., Phung, N., and Nordmann, Y., 1977, Homozygous case of hereditary coproporphyria, *Lancet* **ii**:1348–1349.

Granick, J. L., and Sassa, S., 1978, Hemin control of heme biosynthesis in mouse Friend-virus transformed erythroleukemia cells in culture, *J. Biol. Chem.* **253**:5402–5406.

Granick, J. L., Sassa, S., Granick, S., Levere, R. D., and Kappas, A., 1973, Studies in lead poisoning. II. Correlations between the ratio of activated and inactivated δ-ami-nolevulinic acid dehydratase of whole blood and the blood lead level, *Biochem. Med.* **8**:149–159.

Granick, S., 1950, The structural and functional relationships between heme and chlorophyll, *Harvey Lect.* **44**:220–245.

Granick, S., 1966, The induction *in vitro* of the synthesis of δ-aminolevulinic acid synthetase in chemical porphyria. A response to certain drugs, sex hormones and foreign chemicals, *J. Biol. Chem.* **241**:1359–1375.

Granick, S., and Kappas, A., 1967, Steroid induction of porphyrin synthesis in liver cell culture. I. Structural basis and possible physiological role in the control of heme for-mation, *J. Biol. Chem.* **242**:4587–4593.

Granick, S., and Mauzerall, D., 1958a, Enzymes of porphyrin synthesis in red blood cells, *Ann. N.Y. Acad. Sci.* **75**:115–121.

Granick, S., and Mauzerall, D., 1958b, Porphyrin biosynthesis in erythrocytes. II. Enzymes converting δ-aminolevulinic acid to coproporphyrinogen, *J. Biol. Chem.* **232**:1119–1140.

Granick, S., and Mauzerall, D., 1961, The metabolism of heme and chlorophyll, in: *Met-abolic Pathways*, Vol. 2 (D. M. Greenberg, ed.), pp. 525–616, Academic Press, New York.

Granick, S., and Sassa, S., 1971, δ-Aminolevulinic acid synthetase and the control of heme and chlorophyll synthesis, in: *Metabolic Regulation*, Vol. 5 (H. J. Vogel, ed.), pp. 77–141, *Metabolic Pathways*. Academic Press, New York.

Granick, S., and Urata, G., 1963, Increase in activity of δ-aminolevulinic acid synthetase in liver mitochondria induced by feeding of 3,5-dicarbethoxy-1,4-dihydrocollidine, *J. Biol. Chem.* **238**:821–827.

Granick, S., Sinclair, P., Sassa, S., and Grieninger, G., 1975, Effects by heme, insulin, and serum albumin on heme and protein synthesis in chick embryo liver cells cultured in a chemically defined medium, and a spectrofluorometric assay for porphyrin compo-sition, *J. Biol. Chem.* **250**:9215–9225.

Gray, C. H., 1952, Isotope studies in porphyria, *Br. Med. Bull.* **8**:229–233.

Gray, C. H., and Neuberger, A., 1952, Effect of splenectomy in a case of congenital porphyria, *Lancet* **i**:851–854.

Gray, C. H., Muir, H., and Neuberger, A., 1950a, Studies in congenital porphyria. 3. The incorporation of ^{15}N into the haem and glycine of haemoglobin, *Biochem. J.* **47**:542–548.

Gray, C. H., Neuberger, A., and Sneath, P. H. A., 1950*b*, Studies in congenital porphyria. 2. Incorporation of ^{15}N in the stercobilin in the normal and in the porphyric, *Biochem. J.* **47**:87–92.

Gray, C. H., Kulczycka, A., Nicholson, D. C., Magnus, I. A., and Rimington, C., 1964, Isotope studies on a case of erythropoietic protoporphyria, *Clin. Sci.* **26**:7–15.

Griffon-Euvrard, S., Thivolet, J., and Laurent, G., Calemard, E., Gaillemin, J., Perrot, H., and Ortonne, J. P., 1977, Recherche de la pseudo-porphyrie cutanée tardive chez 100 hémodialyses, *Dermatologica* **155**:193–199.

Grinstein, M., Aldrich, R. A., Hawkinson, V., and Watson, C. J., 1949, An isotopic study of porphyrin and hemoglobin metabolism in a case of porphyria, *J. Biol. Chem.* **179**:983–984.

Gross, S., 1964, Hematologic studies on erythropoietic porphyria: A new case with severe hemolysis, chronic thrombocytopenia and folic acid deficiency, *Blood* **23**:762–775.

Grosser, Y., and Eales, L., 1973, Patterns of fecal porphyrin excretion in the hepatocutaneous porphyrias, *S. Afr. Med. J.* **27**:2162–2168.

Grossman, M. E., Bickers, D. R., Poh-Fitzpatrick, M. B., Deleo, V. A., and Harber, L. C., 1979, Porphyria cutanea tarda. Clinical features and laboratory findings in 40 patients, *Am. J. Med.* **67**:277–285.

Haberman, H. F., Rosenberg, F., and Menon, I. A., 1975, Porphyria cutanea tarda: Comparison of cases precipitated by alcohol and estrogens, *Can. Med. Assoc. J.* **113**:653–655.

Haeger-Aronson, B., 1963, Erythropoietic porphyria: A new type of inborn error of metabolism, *Am. J. Med.* **35**:450–454.

Haeger-Aronson, B., and Krook, G., 1966, Erythropoietic protoporphyria. A study of known cases in Sweden, *Acta Med. Scand.* **245**(Suppl.):48–55.

Haeger-Aronson, B., Stathers, G., and Swahn, G., 1968, Hereditary coproporphyria: Study of a Swedish family, *Ann. Intern. Med.* **69**:221–227.

Haining, R. C., Cowger, M. L., Shurtleff, D. B., and Labbe, R. F., 1968, Congenital erythropoietic porphyria. I. Case report, special studies and therapy, *Am. J. Med.* **45**:624–637.

Hällén, J., and Krook, H., 1963, Follow-up studies on an unselected ten-year material of 360 patients with liver cirrhosis in one community, *Acta Med. Scand.* **173**:479–493.

Harber, L. C., and Bickers, D. R., 1975, The porphyrias: Basic science aspects, clinical diagnosis and management, in: *The Year Book of Dermatology* (F. D. Malkinson and R. W. Pearson, eds.), pp. 9–47, Year Book, Chicago.

Harris, M. Y., Mills, G. C., and Levin, W. C., 1966, Coexistent systemic lupus erythematosus and porphyria, *Arch. Intern. Med.* **117**:425–428.

Hasegawa, E., Smith, C., and Tephly, T. R., 1970, Induction of hepatic mitochondrial ferrochelatase by phenobarbital, *Biochem. Biophys. Res. Comm.* **40**:517–523.

Haskins, D., Stevens, A. R., Jr., Finch, S., and Finch, C. A., 1952, Iron metabolism. Iron stores in man as measured by phlebotomy, *J. Clin. Invest.* **31**:543–547.

Hayashi, N., Yoda, B., and Kikuchi, G., 1969, Mechanism of allylisopropylacetamide-induced increase of δ-aminolevulinate synthetase in liver mitochondria. IV. Accumulation of the enzyme in the soluble fraction of rat liver, *Arch. Biochem. Biophys.* **131**:83–91.

Heilmeyer, L., and Clotten, R., 1964, The erythropoietic porphyria, *Acta Haematol.* **31**:137–149.

Heilmeyer, L., and Clotten, R., 1965, Die Störung der Porphyrinsynthese bei der Güntherschen Porphyria congenita, *Acta Haematol.* **34**:65–71.

Heilmeyer, L. E., Clotten, R., Kerp, I., Merker, H., Parra, C. A., and Wetzel, H. P., 1963, Porphyria erythropoietica congenita Günther: Bericht über zwei Familien mit Erfassung der Merkmalsträger, *Dtsch. Med. Wschr.* **88**:2449–2456.

Heilmeyer, L., Clotten, R., and Heilmeyer, L., Jr., 1966, *Disturbances in Heme Synthesis* (translated by M. Steiner), Charles C. Thomas, Springfield, Ill.

Herbert, F. K., 1966, Porphyrins excreted in various types of porphyria, *Clin. Chim. Acta* **13**:19–30.

Hetherinton, G. W., Jetton, R. L., and Knox, J. M., 1970, The association of lupus erythematosus and porphyria, *Br. J. Dermatol.* **82**:118–124.

Hickman, R., Saunders, S. J., and Eales, L., 1967, Treatment of symptomatic porphyria by venesection, *S. Afr. Med. J.* **41**:456–460.

Higuchi, M., and Bogorad, L., 1975, The purification and properties of uroporphyrinogen I synthase and uroporphyrinogen III cosynthase. Interactions between the enzymes, *Ann. N.Y. Acad. Sci.* **244**:401–418.

Hoare, D. S., and Heath, H., 1958, Intermediates in the biosynthesis of porphyrins from porphobilinogen by *Rhodopseudomonas spheroides, Nature* **181**:1592–1593.

Hofstad, F., Seip, M., and Eriksen, L., 1973, Congenital erythropoietic porphyria with a hitherto undescribed porphyrin pattern, *Acta Paed. Scand.* **62**:380–384.

Hourihane, D. O'B., and Weir, D. G., 1970, Suppression of erythropoiesis by alcohol, *Br. Med. J.* **i**:86–89.

Husquinet, H., Noirfalise, A., and Parent, M.-Th., 1978, Porphyria variegata: Etude d'une grande famille, *J. Genet. Hum.* **26**:367–383.

Hutton, J. J., and Gross, S. R., 1970, Chemical induction of hepatic porphyria in inbred strains of mice, *Arch. Biochem. Biophys.* **141**:284–292.

Ibrahim, N. G., Gruenspecht, N. R., and Freedman, M. L., 1978, Hemin feedback inhibition of reticulocyte δ-aminolevulinic acid synthetase and δ-aminolevulinic acid dehydratase, *Biochem. Biophys. Res. Comm.* **80**:722–728.

Ibrahim, N. G., Lutton, J. D., and Levere, R. D., 1979, *In vitro* development of heme enzymes in erythroid colonies (EC): The role of heme oxygenase (HO), *Blood* **54**:140a.

Ippen, H., 1961, Allgemein Symptome: der späten Hautporphyrie (Porphyria cutanea tarda) als Hinweise für deren Behandlung, *Dtsch. Med. Wschr.* **86**:127–133.

Ippen, H., 1977, Treatment of porphyria cutanea tarda by phlebotomy, *Sem. Hematol.* **14**:253–259.

Irving, R. A., Mainwaring, W. I. P., and Spooner, P. M., 1976, The regulation of haemoglobin synthesis in cultured chick blastoderms by steroid related to 5β-androstane, *Biochem. J.* **154**:81–93.

Jackson, A. H., Sancovich, H. A., Ferramola, A. M., Evans, N., Games, D. E., Matlin, S. A., Elder, G. H., and Smith, S. G., 1975, Macrocyclic intermediates in the biosynthesis of porphyrins, *Phil. Trans. Royal Soc. London B.* **273**:119–134.

Jaeger, A., Tempe, J. D., Geisler, F., Nordmann, Y., and Mantz, J. M., 1975, La coproporphyrie héréditaire, *Nouv. Presse Med.* **4**:2783–2787.

Jirásek, L., Kalenský, J., Kubec, K., Pazderova, J., and Lukás, E., 1976, Chlorakne, Porphyria cutanea tarda und andere Intoxikationen durch Herbizide, *Hautarzt* **27**:328–333.

Jøergensen, S. K., and With, T. K., 1955, Congenital porphyria in swine and cattle in Denmark, *Nature* **176**: 156–158.

Jørgensen, S. K., and With, T. K., 1963, Porphyria in domestic animals. Danish observations in pigs and cattle and comparison with human porphyria, *Ann. N.Y. Acad. Sci.* **104**:701–709.

Johnson, A., and Jones, O. T. G., 1964, Enzymic formation of hemes and other metalloporphyrins, *Biochim. Biophys. Acta* **93**:171–173.

Jones, K. G., and Sweeney, G. D., 1977, Association between induction of aryl hydrocarbon hydroxylase and depression of uroporphyrinogen decarboxylase activity, *Res. Comm. Chem. Pathol. Pharmacol.* **17**:631–637.

Jones, M. S., and Jones, O. T. G., 1969, The structural organization of heme synthesis in rat liver mitochondria, *Biochem. J.* **113:**507–514.

Jordan, P. M., and Seehra, J. S., 1980, Mechanism of action of 5-aminolevulinic acid dehydratase: Stepwise order of addition of the two molecules of 5-aminolevulinic acid in the enzymic synthesis of porphobilinogen, *J. C. S. Chem. Comm.* 240–242.

Kaneko, J. J., 1963, Erythrokinetics and iron metabolism in bovine porphyria erythropoietica, *Ann. N.Y. Acad. Sci.* **104:**689–700.

Kang, E. S., and Gerald, P. S., 1970, Hereditary tyrosinemia and abnormal pyrrole metabolism, *J. Pediatr.* **77:**397–406.

Kaplan, B. H., 1971, δ-Aminolevulinic acid synthetase from the particulate fraction of liver of porphyric rats, *Biochim. Biophys. Acta* **235:**381–388.

Kaplowitz, N., Javitt, N., and Harber, L. C., 1968, Isolation of erythrocytes with normal protoporphyrin levels in erythropoietic protoporphyria, *N. Engl. J. Med.* **278:**1077–1081.

Kappas, A., and Granick, S., 1968, Steroid induction of porphyrin synthesis in liver cell culture II. The effects of heme, uridine diphosphate glucuronic acid, and inhibitors of nucleic acid and protein synthesis on the induction process, *J. Biol. Chem.* **243:**346–351.

Kappas, A., Song, C. S., Levere, R. D., Sachson, R. A., and Granick, S., 1968, The induction of δ-aminolevulinic acid synthetase *in vivo* in chick embryo liver by natural steroids, *Proc. Natl. Acad. Sci. USA* **61:**509–513.

Kappas, A., Song, C. S., Sassa, S., Levere, R. D., and Granick, S., 1969, The occurrence of substances in human plasma capable of inducing the enzyme δ-aminolevulinate synthetase in liver cells, *Proc. Natl. Acad. Sci. USA* **64:**557–564.

Kappas, A., Bradlow, H. L., Gillette, P. N., and Gallagher, T. F., 1972, Studies in porphyria. I. A defect in the reductive transformation of natural steroid hormones in the hereditary liver disease, acute intermittent porphyria, *J. Exp. Med.* **136:**1043–1053.

Kappas, A., Sassa, S., Granick, S., and Bradlow, H. L., 1974, Endocrine-gene interaction in the pathogenesis of acute intermittent porphyria, in: *Brain Dysfunction in Metabolic Disorders* (F. Plum, ed.) (*Res. Publ. Assoc. Nerv. Ment. Dis.* **53:**225–237).

Kappas, A., Bradlow, H. L., Bickers, D. R., and Alvares, A. P., 1977, Induction of a deficiency of steroid Δ^4-5α-reductase activity in liver by a porphyrinogenic drug, *J. Clin. Invest.* **59:**159–164.

Karibian, D., and London, I. M., 1965, Control of heme synthesis by feedback inhibition, *Biochem. Biophys. Res. Comm.* **18:**243–249.

Kato, R., 1967, Effect of administration of 3-aminotriazole on the activity of microsomal drug metabolizing enzyme systems of rat liver, *Jpn. J. Pharmacol.* **17:**56–64.

Kaufman, K., and Marver, H. S., 1970, The biochemical defects in two types of human hepatic porphyria, *N. Engl. J. Med.* **283:**954–958.

Keczkes, K., and Farr, M., 1976, Bullous dermatosis of chronic renal failure, *Br. J. Dermatol.* **95:**541–546.

Kench, J. D., Langley, F. A., and Wilkinson, J. F., 1953, Biochemical and pathological studies of congenital porphyria, *Quart. J. Med.* n.s. **22:**285–294.

Kennedy, G. Y., 1975, Porphyrins in invertebrates, *Ann. N.Y. Acad. Sci.* **244:**662–673.

Knudson, K. B., Sparberg, M., and Lecocq, F., 1977, Porphyria precipitated by fasting, *N. Engl. J. Med.* **277:**350–351.

Korting, G. W., 1975, Über Porphyria-cutanea-tarda-artige Hautveränderungen bei Langzeithämodialysepatienten, *Dermatologica* **150:**58–61.

Kramer, S., 1963, Iron metabolism in the porphyrias, *S. Afr. J. Lab. Clin. Med.* **9:**283–287.

Kramer, S., Viljoen, E., Mayer, A. M., and Metz, J., 1965, The anemia of erythropoietic porphyria with the first description of the disease in an elderly patient, *Br. J. Haematol.* **11:**666–675.

Kushner, J. P., and Barbuto, A. J., 1975, An inherited defect in porphyria cutanea tarda (PCT): Decreased uroporphyrinogen decarboxylase activity (urodecarb), *Clin. Res.* 23:403A.

Kushner, J. P., Steinmuller, D. P., and Lee, G. R., 1975, The role of iron in the pathogenesis of porphyria cutanea tarda. II. Inhibition of uroporphyrinogen decarboxylase, *J. Clin. Invest.* 56:661–667.

Kushner, J. P., Barbuto, A. J., and Lee, G. R., 1976, An inherited enzymatic defect in porphyria cutanea tarda. Decreased uroporphyrinogen decarboxylase activity, *J. Clin. Invest.* 58:1089–1097.

Labbe, R. F., and Hubbard, N., 1961, Metal specificity of the iron-protoporphyrin chelating enzyme from rat liver, *Biochim. Biophys. Acta* 52:130–135.

La Du, B. N., and Gjessing, L. R., 1978, Tyrosinosis and tyrosinemia, in: *The Metabolic Basis of Inherited Disease* (J. B. Stanbury, J. B. Wyngaarden, and D. S. Frederickson, eds.), pp. 256–267, McGraw-Hill, New York.

Lamont, N. M., Hathorn, M., and Joubert, S. M., 1961, Porphyria in the African. A study of 100 cases, *Quart. J. Med.* n.s. 30:373–392.

Lange, C. E., Block, H., Veltman, G., and Doss, M., 1976, Urinary porphyrins among PVC workers, in: *Porphyrins in Human Diseases* (M. Doss, ed.), pp. 352–355, Basel, Kager.

Langelaan, D. E., Losowsky, M. S., and Toothill, C., 1970, Heme synthetase activity in human blood cells, *Clin. Chim. Acta* 27:453–459.

Langhof, M., Müller, H., and Rietschel, L., 1961, Untersuchungen zur familiären proto-porphyrinämischen Lichturticaria, *Arch. Klin. Exp. Dermatol.* 212:506–518.

Langhof, H., Heilmeyer, R. C., and Rietschel, L., 1964, Die erythropoetische Protopor-phyrie: Protoporphyrinämische Lichturticaria, *Dtsch. Med. Wschr.* 27:1289–1293.

Langhof, H., Franken, E., and Kluge, K., 1965, Kombinierte hereditäre Porphyria Heptica (Hereditäre Koproporphyrie), *Hautarzt* 16:101.

Larriza, P., 1962, The problem of erythropoietic porphyria in the light of the latest advances in biochemistry and morphology, *Panminerva Med.* 4:315–323.

Lemberg, M. R., 1954, Porphyrins in nature, *Fortschr. Chem. Org. Naturst.* 11:300–349.

Levere, R. D., 1966, Stilbesterol-induced porphyria: Increase in hepatic δ-aminolevulinic acid synthetase, *Blood* 28:569–572.

Levere, R. D., and Granick, S., 1965, Control of hemoglobin synthesis in the cultured chick blastoderm by δ-aminolevulinic acid synthetase: Increase in the rate of hemoglobin formation with δ-aminolevulinic acid, *Proc. Natl. Acad. Sci. USA* 54:134–137.

Levere, R. D., and Granick, S., 1967, Control of hemoglobin synthesis in the cultured chick blastoderm, *J. Biol. Chem.* 242:1903–1911.

Levere, R. D., Kappas, A., and Granick, S., 1967, Stimulation of hemoglobin synthesis in chick blastoderms by certain 5β-androstane and 5β-pregnane steroids, *Proc. Natl. Acad. Sci. USA.* 58:985–990.

Levin, E. Y., 1968, Uroporphyrinogen III cosynthetase in bovine erythropoietic porphyria, *Science* 161:907–908.

Levin, E. Y., 1971, Enzymatic properties of uroporphyrinogen III cosynthetase, *Biochem-istry* 10:4669–4675.

Levin, E. Y., and Flyger, V., 1973, Erythropoietic porphyria of the fox squirrel *Sciurus niger, J. Clin. Invest.* 52:96–105.

Levin, W., Jacobson, M., and Kuntzman, R., 1972, Incorporation of radioactive δ-ami-nolevulinic acid into microsomal cytochrome P-450. Selective breakdown of the hem-oprotein by allyl-isopropylacetamide and carbon tetrachloride, *Arch. Biochem. Biophys.* 148:262–269.

Lindblad, B., Lindstedt, S., and Steen, G., 1977, On the enzymic defects in hereditary tyrosinemia, *Proc. Natl. Acad. Sci. USA.* **74**:4641–4645.

Llambias, E. B. C., and Batlle, A. M. del C., 1971, Porphyrin biosynthesis VIII. Avian erythrocyte porphobilinogen deaminase—uroporphyrinogen III cosynthetase, its purification, properties, and the separation of its compounds, *Biochim. Biophys. Acta* **227**:180–191.

Lomholt, J. C., and With, T. K., 1969, Hereditary coproporphyria. A family with unusually few and mild symptoms, *Acta Med. Scand.* **186**:83–85.

Louw, M., Neethling, A. C., Percy, V. A., Carstens, M., and Shanley, B. C., 1977, Effects of hexachlorobenzene feeding and iron overload on enzymes of haem biosynthesis and cytochrome P450 in rat liver, *Clin. Sci. Mol. Med.* **53**:111–115.

Lundvall, O., Weinfeld, A. D., and Lundin, P., 1970, Iron storage in porphyria cutanea tarda, *Acta Med. Scand.* **188**:37–53.

Lynch, P. J., and Miedler, L. J., 1965, Erythropoietic protoporphyria. A report of a family and a clinical review, *Arch. Dermatol.* **92**:351–356.

Magnus, I. A., Jarrett, A., Prankerd, T. A. J., and Rimington, C., 1961, Erythropoietic protoporphyria: A new porphyria syndrome with solar urticaria due to protoporphyrinaemia, *Lancet* **ii**:448–451.

Magnussen, C. R., Levine, J. B., Doherty, J. M., Cheesman, J. O., and Tschudy, D. P., 1974, A red cell enzyme method for the diagnosis of acute intermittent porphyria, *Blood* **44**:857–868.

Maines, M. D., and Kappas, K., 1974, Cobalt induction of hepatic heme oxygenase; with evidence that cytochrome P-450 is not essential for this enzymatic activity, *Proc. Natl. Acad. Sci. USA* **71**:4293–4297.

Maines, M. D., and Kappas, A., 1976, Studies on the mechanism of induction of haem oxygenase by cobalt and other metal ions, *Biochem. J.* **154**:125–131.

Maines, M. D., and Kappas, A., 1977, Metals as regulators of heme catabolism, *Science* **198**:1215–1221.

Mangum, C. P., 1978, Nonhuman models of hereditary porphyrias, *Science* **201**:1043.

Margolis, F. L., 1971, Regulation of porphyrin biosynthesis in the Harderian gland of inbred mouse strains, *Arch. Biochem. Biophys.* **145**:373–381.

Marver, H. S., Collins, A., Tschudy, D. P., and Rechcigl, M., Jr., 1966, δ-Aminolevulinic acid synthetase II. Induction in rat liver, *J. Biol. Chem.* **241**:4323–4329.

Masuya, T., 1969, Pathophysiological observations on porphyrias, *Acta Haematol. Jpn.* **32**:25–74.

Mauzerall, D., and Granick, S., 1956, Occurrence and determination of δ-aminolevulinic acid and porphobilinogen in urine, *J. Biol. Chem.* **219**:435–446.

Mauzerall, D., and Granick, S., 1958, Porphyrin biosynthesis in erythrocytes. III. Uroporphyrinogen and its decarboxylase, *J. Biol. Chem.* **232**:1141–1162.

McIntyre, N., Pearson, A. J. G., Allan, D. J., Craske, S., West, G. M. L., Moore, M., Beattie, A. D., Paxton, J., and Goldberg, A., 1971, Hepatic δ-aminolevulinic acid synthetase in acute attack of hereditary coproporphyria and during remission, *Lancet* **i**:560–564.

McKay, R., Druyan, R., Getz, G. S., and Rabinowitz, M., 1969, Intramitochondrial localisation of δ-aminolaevulate synthetase and ferrochelatase in rat liver, *Biochem. J.* **114**:455–461.

Meyer, U. A., 1973, Intermittent acute porphyria. Clinical and biochemical studies of disordered heme biosynthesis, *Enzyme* **16**:336–342.

Meyer, U. A., and Schmid, R., 1978, The porphyrias, in: *The Metabolic Basis of Inherited*

Diseases (J. B. Stanbury, J. B. Wyngaarden, and D. S. Frederickson, eds.), 5th Edn., pp. 1166–1220, McGraw-Hill, New York.

Meyer, U. A., Strand, L. J., Doss, M., Rees, A. C., and Marver, H. S., 1972, Intermittent acute porphyria—demonstration of a genetic defect in porphobilinogen metabolism, *N. Engl. J. Med.* **286**:1277–1282.

Miyagi, K., Cardinal, R., Bassenmaier, I., and Watson, C. J., 1971, The serum porphobilinogen and hepatic porphobilinogen deaminase in normal and porphyric individuals, *J. Lab. Clin. Med.* **78**:683–695.

Miyagi, K., Petryka, Z. J., Bassenmaier, I., Cardinal, R., and Watson, C. J., 1976, The activities of uroporphyrinogen synthetase and cosynthetase in congenital erythropoietic porphyria (CEP), *Am. J. Hematol.* **1**:3–21.

Miyagi, K., Kaneshima, M., Kawakami, J., Nakada, F., Petryka, Z. J., and Watson, C. J., 1979, Uroporphyrinogen I synthase from human erythrocytes: Separation, purification, and properties of isoenzymes, *Proc. Natl. Acad. Sci. USA* **76**:6172–6176.

Moore, M. R., and Meredith, P. A., 1976, The association of delta-aminolaevulinic acid with the neurological and behavioral effects of lead exposure, in: *Trace Substances in Environmental Health. X. A Symposium* (D. D. Hemphil, ed.), pp. 363–371, University of Missouri, Columbia.

Moore, M. R., Thompson, G. G., and Goldberg, A., 1972, Amounts of faecal porphyrin-peptide conjugates in porphyrias, *Clin. Sic.* **43**:299–302.

Moore, M. R., Thompson, G. G., Payne, A. P., and McGadey, J., 1977, Cyclical oscillations in the activity of δ-aminolevulinate synthase and porphyrin synthesis in the Harderian gland during the oestrous cycle of the golden hamster (*Mesocricetus auratus*), *Biochem. Soc. Trans.* **5**:1475–1478.

Moore, M. R., Thompson, G. G., Goldberg, A., Ippen, H., Seubert, A., and Seubert, S., 1978, The biosynthesis of haem in congenital (erythropoietic) porphyria, *Int. J. Biochem.* **9**:933–938.

Müller, W. E., and Snyder, S. H., 1977, δ-Aminolevulinic acid: Influences on synaptic GABA receptor binding may explain CNS symptoms of porphyria, *Ann. Neurol.* **2**:340–342.

Mustajoki, P., 1976, Red cell uroporphyrinogen I synthetase in acute intermittent porphyria, *Ann. Clin. Res.* **8**(suppl. 17):133–138.

Mustajoki, P., 1978, Variegate porphyria, *Ann. Intern. Med.* **89**:238–244.

Mustajoki, P., and Koskelo, P., 1976, Hereditary hepatic porphyrias in Finland, *Acta Med. Scand.* **200**:171–178.

Nacht, S., San Martín de Viale, L. C., and Grinstein, M., 1970, Human porphyria cutanea tarda. Isolation and properties of the urinary porphyrins, *Clin. Chim. Acta* **27**:445–452.

Nakakuki, M., Yamauchi, K., Hayashi, N., and Kikuchi, G., 1980, Purification and some properties of δ-aminolevulinate synthase from the rat liver cytosol fraction and immunochemical identity of the cytosolic enzyme and the mitochondrial enzyme, *J. Biol. Chem.* **255**:1738–1745.

Nakao, K., Wada, O., Kitamura, T., Uono, K., and Urata, G., 1966, Activity of aminolaevulinic acid synthetase in normal and porphyric human livers, *Nature* **210**:838–839.

Nandi, D. L., and Shemin, D., 1968, δ-Aminolevulinic acid dehydratase of *Rhodopseudomonas spheroides*. III. Mechanism of porphobilinogen synthesis, *J. Biol. Chem.* **234**:1236–1242.

Neuberger, A., and Turner, J. M., 1963, γ,δ-Dioxovalerate aminotransferase activity in *Rhodopseudomonas spheroides*, *Biochim. Biophys. Acta* **67**:342–345.

Neuwirt, J., and Ponka, P., 1977, *Regulation of Haemoglobin Synthesis*, Martinus Nijhoff/ Medical Division. The Hague, The Netherlands.

Nicoll, R. A., 1976, The interaction of porphyrin precursors with GABA receptors in the isolated frog spinal cord, *Life Sci.* **19**:521–526.

Nitowsky, H. M., Sassa, S., Nakagawa, M., and Jagani, N., 1978, Prenatal diagnosis of congenital erythropoietic porphyria (abs.), *Pediatr. Res.* **12**:455.

Nordmann, Y., Grandchamp, B., Grelier, M., Phung, N., and deVerneuil, H., 1976, Detection of intermittent acute porphyria trait in children, *Lancet* **ii**:201–202.

Nordman, Y., Grandchamp, B., Phung, N., deVerneuil, H., Grelier, M., and Noiré, J., 1977, Coproporphyrinogen-oxidase deficiency in hereditary coproporphyria, *Lancet* **i**:140.

Ohashi, A., and Kikuchi, G., 1972, Mechanism of allylisopropylacetamide-induced increase of δ-aminolevulinate synthetase in liver mitochondria. VI. Multiple molecular forms of δ-aminolevulinate synthetase in the cytosol and mitochondria of induced cock liver, *Arch. Biochem. Biophys.* **153**:34–46.

Ohashi, A., and Kikuchi, G., 1979, Purification and some properties of two forms of δ-aminolevulinate synthase from rat liver cytosol, *J. Biochem. (Tokyo)* **85**:239–247.

Onisawa, J., and Labbe, R. F., 1963, Effects of diethyl-1,4-dihydro-2,4,6-trimethylpyridine-3,5-dicarboxylate on the metabolism of porphyrins and iron, *J. Biol. Chem.* **238**:724–727.

Padmanaban, G., Rao, M. R. S., and Malathi, K., 1973, A model for the regulation of δ-aminolaevulinate synthetase induction in rat liver, *Biochem. J.* **134**:847–857.

Pain, R. W., Welch, F. W., Woodroffe, A. J., Handley, D. A., and Lockwood, W. H., 1975, Erythropoietic uroporphyria of Günther first presenting at 58 years with positive family studies, *Br. Med. J.* **3**:621–623.

Paterniti, J., Simone, J., and Beattie, D., 1978, Detection and regulation of δ-aminolevulinic acid synthetase activity in the rat brain, *Arch. Biochem. Biophys.* **189**:86–91.

Paterniti, J. R., Jr. and Beattie, D. S., 1979, δ-Aminolevulinic acid synthetase from rat liver mitochondria, Purification and properties, *J. Biol. Chem.* **253**:6112–6118.

Patton, G. M., and Beattie, D. S., 1973, Studies on hepatic δ-aminolevulinic acid synthetase, *J. Biol. Chem.* **248**:4467–4474.

Paul, S., Bickers, D. R., Levere, R. D., and Kappas, A., 1974, Inhibited induction of hepatic δ-aminolevulinate synthetase in pregnancy, *FEBS Lett.* **41**:192–194.

Paxton, J. W., Moore, M. R., Beattie, A. D., and Goldberg, A., 1974, 17-Oxosteroid conjugates in plasma and urine of patients with acute intermittent porphyria, *Clin. Sci. Mol. Med.* **46**:207–222.

Payne, A. P., McGadey, J., Moore, M. R., and Thompson, G. G., 1979, Changes in Harderian gland activity in the female golden hamster during the oestrous cycle, pregnancy and lactation, *Biochem. J.* **178**:597–604.

Percy, V. A., and Shanley, B. C., 1977, Porphyrin precursors in blood, urine and cerebrospinal fluid in acute porphyria, *S. Afr. Med. J.* **52**:214–222.

Peterson, L. R., Hamernyik, P., Bird, T. D., and Labbe, R. F., 1976, Erythrocyte uroporphyrinogen I synthase activity in diagnosis of acute intermittent porphyria, *Clin. Chem.* **22**:1835–1840.

Petryka, Z. J., Kostich, N. D., and Dhar, G. J., 1978, Location of iron and porphyrin in the liver in porphyria cutanea tarda, *Acta Histochem.* **63**:168–176.

Piepkorn, M. W., Hamernyik, P., and Labbe, R. F., 1978, Modified erythrocyte uroporphyrinogen I synthase assay, and its clinical interpretation, *Clin. Chem.* **24**:1751–1754.

Pierach, C. A., and Edwards, P. S., 1978, Neurotoxicity of δ-aminolevulinic acid and porphobilinogen, *Exp. Neurol.* **62**:810–814.

Pimstone, N. R., 1975, The hepatic aspects of the porphyrias, in: *Modern Trends in Gastroenterology* (A. E. Read, ed.), pp. 373–417, Butterworths, London.

Pimstone, N. R., Blekkenhorst, G., and Eales, L., 1973, Enzymatic defects in hepatic porphyria. Preliminary observations in patients with porphyria cutanea tarda, and variegate porphyria, *Enzyme* **16**:354–366.

Piñol Aguadé, J., Castells, A., Indacochea, A., and Rodés, J., 1969, A case of biochemically unclassifiable hepatic porphyria, *Br. J. Dermatol.* **81**:270–275.

Piñol Aguadé, J., Herrero, C., Almeida, J., Castells Mas, A., Ferrando, J., De Asprer, J., Palou, A., and Giménez, A., 1975, Porphyrie hépato-érythrocytaire. Une nouvelle forme de porphyrie, *Ann. Derm. Syphiligr.* **102**:125–136.

Piomelli, S., Lamola, A. A., Poh-Fitzpatrick, M. B., Seaman, C., and Harber, L. C., 1975, Erythropoietic protoporphyria and lead intoxication: The molecular basis for difference in cutaneous photosensitivity. I. Different rates of disappearance of protoporphyrin from the erythrocytes, both *in vivo* and *in vitro*, *J. Clin. Invest.* **56**:1519–1527.

Poh-Fitzpatrick, M. B. Bellet, N., DeLeo, V. A., Grossman, M. E., and Bickers, D. R., 1978, Porphyria cutanea tarda in two patients treated with hemodialysis for chronic renal failure, *N. Engl. J. Med.* **299**:292–294.

Porra, R. J., and Jones, O. T. G., 1963, Studies on ferrochelatase. 1. Assay and properties of ferrochelatase from a pig liver mitochondrial extract, *Biochem. J.* **87**:181–185.

Porra, R. J., and Falk, J. E., 1964, The enzymic conversion of coproporphyrinogen III into protoporphyrin IX, *Biochem. J.* **90**:69–75.

Porra, R. J., and Grimme, L. H., 1978, Tetrapyrrole biosynthesis in algae and higher plants: A discussion of the importance of the 5-aminolaevulinate synthase and the dioxovalerate transaminase pathways in the biosynthesis of chlorophyll, *Int. J. Biochem.* **9**:883–886.

Porter, P. N., Meints, R. H., and Mesner, K., 1979, Enhancement of erythroid colony growth in culture by hemin, *Exp. Hematol.* **7**:11–16.

Porter, S., and Lowe, B. A., 1963, Congenital erythropoietic protoporphyria: I. Case reports, clinical studies and porphyrin analyses in two brothers, *Blood* **22**:521–531.

Poulson, R., 1976, The enzymic conversion of protoporphyrinogen IX to protoporphyrin IX in mammalian mitochrondria, *J. Biol. Chem.* **251**:3730–3733.

Poulson, R., and Polgase, W. J., 1974a, Aerobic and anaerobic coproporphyrinogenase activities in extracts from *Saccharomyces cerevisiae:* Purification and characterization, *J. Biol. Chem.* **249**:6367–6371.

Poulson, R., and Polgase, W. J., 1974b, Site of glucose repression of heme biosynthesis, *FEBS Lett.* **40**:258–260.

Poulson, R., and Polgasc, W. J., 1975, The enzymic conversion of protoporphyrinogen IX to protoporphyrin IX. Protoporphyrinogen oxidase activity in mitochondrial extracts of *Saccharomyces cerevisiae, J. Biol. Chem.* **250**:1269–1274.

Pritchard, J. A., and Mason, R. A., 1964, Iron stores of normal adults and replenishment with oral iron therapy, *J. Am. Med. Assoc.* **190**:897–901.

Rasmussen, G. L., and Kushner, J. P., 1979, The enzymatic decarboxylation of the naturally occurring isomers of uroporphyrinogens by human erythrocytes, *J. Lab. Clin. Med.* **93**:54–59.

Redeker, A. G., and Bronow, R. S., 1964, Erythropoietic protoporphyria presenting as hydroa aestivale, *Arch. Dermatol.* **89**:104–109.

Redeker, A. G., and Bryan, H. G., 1964, Erythropoietic protoporphyria, *Lancet* i:1449–1450.

Redeker, A. G., and Sterling, R. E., 1968, The "glucose effect" in erythropoietic protoporphyria, *Arch. Intern. Med.* **121**:446–448.

Redeker, A. G., Bronow, R. S., and Sterling, R. E., 1963, Erythropoietic protoporphyria, *S. Afr. J. Lab. Clin. Med.* **9**:235–238.

Reed, W. B., Wuepper, K. D., Epstein, J. H., Redeker, A., Simonson, R. J., and McKusick, V. A., 1970, Erythropoietic protoporphyria. A clinical and genetic study, *J. Am. Med. Assoc.* **214**:1060–1066.

Rifkind, A. B., 1976, Drug-induced exacerbations of porphyria, *Primary Care* **3**:665–685.

Rifkind, A. B., 1979, Maintenance of microsomal hemoprotein concentrations following inhibition of ferrochelatase activity of 3,5-diethoxycarbonyl-1,4-dihydrocollidine in chick embryo liver, *J. Biol. Chem.* **254**:4636–4644.

Rifkind, A. B., Sassa, S., Merkatz, I. R., Winchester, R., Harber, L., and Kappas, A., 1974, Stimulators and inhibitors of hepatic porphyrin formation in human sera, *J. Clin. Invest.* **53**:1167–1177.

Rimbaud, P., Meynadier, J., and Guilhou, J. J., 1973, La porphyrie cutanée tardive. A propos de deux observations associées à un cancer hépatique, *Sem. Hôp.* **49**:719–725.

Rimington, C., and Miles, P. A., 1951, A study of the porphyrins excreted in the urine by a case of congenital porphyria, *Biochem. J.* **50**:202–206.

Rimington, C., and With, T. K., 1973, Porphyrin studies in erythropoietic porphyria, *Danish Med. Bull.* **20**:5–12.

Rimington, C., Lockwood, W. H., and Belcher, R. V., 1968, The excretion of porphyrin-peptide conjugates in porphyria variegata, *Clin. Sci.* **35**:211–247.

Rimington, C., Sears, W. G., and Eales, L., 1972, Symptomatic porphyria in a case of Felty's syndrome. II. Biochemical investigations, *Clin. Chem.* **18**:462–470.

Robinson, S. M., Tsong, M., Brown, B., and Schmid, R., 1966, The sources of bile pigment in the rat: Studies of the "early labeled" fraction, *J. Clin. Invest.* **45**:1569–1586.

Roenigk, H. H., and Gottlob, M. E., 1970, Estrogen-induced porphyria cutanea tarda, *Arch. Dermatol.* **102**:260–266.

Romeo, G., and Levin, E. Y., 1969, Uroporphyrinogen III cosynthetase in human congenital erythropoietic porphyria, *Proc. Natl. Acad. Sci. USA* **63**:856–863.

Romeo, G., and Levin, E. Y., 1971, Uroporphyrinogen decarboxylase from mouse spleen, *Biochim. Biophys. Acta* **230**:330–341.

Rose, J. A., Hellman, E. S., and Tschudy, D. P., 1961, Effect of diet on the induction of experimental porphyria, *Metabolism* **10**:514–521.

Rosenthal, I. M., Lipton, E. L., and Asrow, G., 1955, Effect of splenectomy on porphyria erythropoietica, *Pediatrics* **15**:663–675.

Ross, J., and Sautner, D., 1976, Induction of globin mRNA accumulation by hemin in cultured erythroleukemic cells, *Cell* **8**:513–520.

Ross, J., Ikawa, Y., and Leder, P., 1972, Globin messenger RNA induction during erythroid differentiation of cultured leukemia cells, *Proc. Natl. Acad. Sci. USA* **69**:3620–3623.

Ruth, G. R., Schwartz, S., and Stephenson, B., 1977, Bovine protoporphyria: The first nonhuman model of this hereditary photosensitizing defect, *Science* **198**:199–201.

Rutherford, T., Thompson, G. G., and Moore, M. R., 1979, Heme biosynthesis in Friend erythroleukemia cells: Control by ferrochelatase, *Proc. Natl. Acad. Sci. USA* **76**:833–836.

San Martín de Viale, L. C., Rios de Molina, M. del C., Wainstok de Calmanovici, R., and Tomio, J. M., 1976, Experimental porphyria produced in rats by hexachlorobenzene. IV. Studies on the stepwise decarboxylation of uroporphyrinogen and phyriaporphyrinogen "in vivo" and "in vitro" in several tissues, in: *Porphyrins in Human Diseases* (M. Doss, ed.), pp. 445–452, Karger, Basel.

San Martín de Viale, L. C., Rios de Molina, M. del C., Wainstok de Calmanovici, R., and Tomio, J. M., 1977, Porphyrins and porphyrinogen carboxy-lyase in hexachlorobenzene induced porphyria, *Biochem. J.* **168**:393–400.

Sano, S., and Granick, S., 1961, Mitochondrial coproporphyrinogen oxidase and protoporphyrin formation, *J. Biol. Chem.* **236**:1173–1180.

Sancovich, H. A., Batlle, A. M. C., and Grinstein, M., 1969, Porphyrin biosynthesis. VI. Separation and purification of porphobilinogen deaminase and uroporphyrinogen isomerase from cow liver, *Biochim. Biophys. Acta* **191**:130–143.

Sasaki, H., Kaneko, K., and Tsuneyama, H., 1969, Activities of δ-aminolevulinic acid synthetase in the liver and bone marrow of hepatic coproporphyria (hereditary copro-porphyria), *Acta Med. Biol. (Niigata)* **17**:97–99.

Sassa, S., 1976, Sequential induction of heme pathway enzymes during erythroid differ-entiation of mouse Friend leukemia virus-infected cells, *J. Exp. Med.* **143**:305–315.

Sassa, S., 1980, Control of heme biosynthesis in erythroid cells, in: *Erythropoiesis and Differentiation in Friend Leukemia Cells, EMBO Workshop on CNR*, Urbino, Italy, pp. 219–228, Elsevier/North-Holland Biochem. Press, Amsterdam.

Sassa, S., and Bernstein, S. E., 1977, Levels of δ-aminolevulinate dehydratase, uropor-phyrinogen-I synthase, and protoporphyrin IX in erythrocytes from anemic mutant mice, *Proc. Natl. Acad. Sci. USA* **74**:1181–1184.

Sassa, S., and Granick, S., 1970, Induction of δ-aminolevulinic acid synthetase in chick embryo liver cells in culture, *Proc. Natl. Acad. Sci. USA* **67**:517–522.

Sassa, S., and Kappas. A., 1977, Induction of δ-aminolevulinate synthase and porphyrins in cultured liver cells maintained in chemically defined medium. Permissive effects of hormones on the induction process, *J. Biol. Chem.* **252**:2428–2436.

Sassa, S., and Urabe, A., 1979, Uroporphyrinogen I synthase induction in normal human bone marrow cultures: An early and quantitative response of erythroid differentiation, *Proc. Natl. Acad. Sci. USA* **76**:5321–5325.

Sassa, S., Granick, S., Bickers, D. S., Levere, R. D., and Kappas, A., 1973, Studies on the inheritance of human erythrocyte δ-aminolevulinate dehydratase and uroporphyr-inogen synthetase, *Enzyme* **16**:326–333.

Sassa, S., Granick, S., Bickers, D. R., Bradlow, H. L., and Kappas, A., 1974a, A microassay for uroporphyrinogen I synthase, one of three abnormal enzyme activities in acute intermittent porphyria, and its application to the study of the genetics of this disease, *Proc. Natl. Acad. Sci. USA* **71**:732–736.

Sassa, S., Levere, R. D., Solish, G., and Kappas, A., 1974b, Studies on the porphyrin-heme biosynthetic pathway in cultured human amniotic cells, *J. Clin. Invest.* **53**:70a–71a.

Sassa, S., Granick, S., and Kappas, A., 1975a, Effect of lead and genetic factors on heme biosynthesis in the human red cell, *Ann. N.Y. Acad. Sci.* **244**:419–440.

Sassa, S., Solish, G., Levere, R. D., and Kappas, A., 1975b, Studies in porphyria. IV. Expression of the gene defect of acute intermittent porphyria in cultured human skin fibroblasts and amniotic cells: Prenatal diagnosis of the porphyric trait, *J. Exp. Med.* **142**:722–731.

Sassa, S., Granick, J. L., Eisen, H., and Ostertag, W., 1978a, Regulation of heme biosyn-thesis in mouse Friend virus-transformed cells in culture, in: *In Vitro Aspects of Eryth-ropoiesis* (M. J. Murphy, ed.), pp. 135–142, Springer-Verlag, New York.

Sassa, S., Zalar, G. L., and Kappas, A., 1978b, Studies in porphyria. VII. Induction of uroporphyrinogen-I synthase and expression of the gene defect of acute intermittent porphyria in mitogen-stimulated human lymphocytes, *J. Clin. Invest.* **61**:499–508.

Sassa, S., Bradlow, H. L., and Kappas, A., 1979a, Steroid induction of δ-aminolevulinic acid synthase and porphyrins in liver. Structure-activity studies and the permissive effects of hormones on the induction process, *J. Biol. Chem.* **254**:10011–10020.

Sassa, S., Zalar, G. L., Poh-Fitzpatrick, M. B., and Kappas, A., 1979b, Studies in porphyria. IX: Detection of the gene defect of erythropoietic protoporphyria in mitogen-stimulated human lymphocytes, *Trans. Assoc. Am. Physicians* **92**:268–276.

Sassa, S., Whetsell, W. O., Jr., and Kappas, A., 1979c, Studies on porphyrin-heme bio-synthesis in organotypic cultures of chick dorsal root ganglia. II. The effect of lead, *Env. Res.* **19**:415–426.

Sato, A., and Takahasi, N., 1926, A new form of congenital hematoporphyria: Oligochromemia, porphyrinuria (megalosplenica congenita), *Am. J. Dis. Child.* **32**:325–333.

Saunders, S. J., 1963, Iron metabolism in symptomatic porphyria, *S. Afr. J. Lab. Clin. Med.* **9**:277–283.

Sawada, H., Takeshita, M., Sugita, Y., and Yoneyama, Y., 1969, Effect of lipid on protoheme ferrolyase, *Biochim. Biophys. Acta* **178**:145–155.

Schmid, R., 1960, Cutaneous porphyria in Turkey, *N. Engl. J. Med.* **263**:393–398.

Schmid, R., Schwartz, A., and Watson, C. J., 1954, Porphyrin content of bone marrow and liver in various forms of porphyria, *Arch. Intern. Med.* **93**:167–190.

Schmid, R., Schwartz, S., and Sundberg, D., 1955, Erythropoietic (congenital) porphyria: A rare abnormality of the normoblasts, *Blood* **10**:416–428.

Scholnick, P., and Marver, H. S., 1968, The molecular basis of chloroquine responsiveness in porphyria cutanea tarda, *Clin. Res.* **16**:258.

Scholnick, P., Marver, H. S., and Schmid, R., 1971, Erythropoietic protoporphyria: Evidence for multiple sites of excess protoporphyrin formation, *J. Clin. Invest.* **50**:203–207.

Scholnick, P. L., Hammaker, E., and Marver, H. S., 1972, Soluble δ-aminolevulinic acid synthetase of rat liver. II. Studies related to the mechanism of enzyme action and heme inhibition, *J. Biol. Chem.* **247**:4132–4137.

Scholnick, P. L., Epstein, J., and Marver, H. S., 1973, The molecular basis of the action of chloroquine in porphyria cutanea tarda, *J. Invest. Dermatol.* **61**:226–232.

Schumaker, H. M., Tishler, P. V., and Knighton, D. J., 1976, A spot test for uroporphyrinogen I synthase, the enzyme that is deficient in intermittent acute porphyria, *Clin. Chem.* **22**:1991–1994.

Schwartz, S., and Ruth, G. R., 1978, Nonhuman models of hereditary porphyrias, *Science* **201**:1043.

Schwartz, S., Johnson, J. A., Stephenson, B. D., Anderson, A. S., Edmondson, P. R., and Fusaro, R. M., 1971, Erythropoietic defects in protoporphyria: A study of factors involved in labelling of porphyrins and bile pigments from ALA-^{3}H and glycine-^{14}C, *J. Lab. Clin. Med.* **78**:411–434.

Sedman, R. M., and Tephly, T. R., 1980, Cardiac δ-aminolevulinic acid synthetase activity. Effects of fasting, cobaltous chloride and hemin, *Biochem. Pharmacol.* **29**:795–800.

Shemin, D., 1955, The succinate-glycine cycle: The role of δ-aminolaevulinic acid in porphyrin synthesis, *Porphyrin Biosynth. Metab. Ciba Found. Symp.* 4, pp. 4–22, Churchill, London.

Simard, H., Barry, A., Villeneuve, B., Petitclerc, C., Garneau, R., and Delâge, J.-M., 1972, Porphyrie erythropoiétique congénitale, *Can. Med. Assoc. J.* **106**:1002–1004.

Sinclair, P., and Granick, S., 1975, Heme control of the synthesis of delta-aminolevulinic acid synthetase in cultured chick embryo liver cells, *Ann. N.Y. Acad. Sci.* **244**:509–520.

Singer, J. W., Samuels, A. I., and Adamson, J. W., 1976, Steroids and hematopoiesis. I. The effect of steroids on *in vitro* erythroid colony growth structure/activity relationships, *J. Cell Physiol.* **88**:127–134.

Smart, G. A., Herbert, F. K., Whittaker, N., and Barnes, H. D., 1965, Contraindications of biological oxidation inhibitors in treatment of porphyria, *Lancet* **i**:318.

Smith, A. G., and Francis, J. E., 1979, Decarboxylation of porphyrinogens by rat liver uroporphyrinogen decarboxylase, *Biochem. J.* **183**:455–458.

Smith, A. G., Cabral, J. R. P., and DeMatteis, F., 1979, A difference between two strains of rats in their liver non-haem iron content and in their response to the porphyrogenic effect of hexachlorobenzene, *Chem. Biol. Interact.* **27**:353–363.

Smith, S. G., 1960, The porphobilinogen in fresh postmortem tissues from a case of acute intermittent porphyria, *Arch. Pathol.* **70**:361–362.

Smith, S. G., Jackson, A. H., and Jackson, T. R., 1976, Incubation of double labelled coproporphyrinogen with chicken red haemolysates. Chemical and TLC fractionation of extracts, *Ann. Clin. Res.* **8**(Suppl. 17):53–55.

Song, C. S., Moses, H. L., Rosenthal, A. S., Gelb, N. A., and Kappas, A., 1971, The influence of postnatal development on drug-induced hepatic porphyria and the synthesis of cytochrome P-450, *J. Exp. Med.* **134**:1349–1371.

Soslau, G., Stotz, E. H., and Lockshin, R. A., 1971, Stimulation of cytochrome c synthesis in the developing *polyphemus* moth by δ-aminolevulinic acid, *Biochemistry* **10**:3296–3249.

Soyka, L. F., and Deckert, F. W., 1974, Further studies on the inhibition of drug metabolism by pregnanolone and related steroids, *Biochem. Pharmacol.* **23**:1629–1639.

Soyka, L. F., and Long, R., 1972, *In vitro* inhibition of drug metabolism by metabolites of progesterone, *J. Pharm. Exp. Ther.* **182**:320–327.

Stich, von W., 1958, Die kongenitale Porphyrie, eine erythropathische hämolytische Anämie (Porphyrocytose), *Schw. Klin. Wschr.* **41**:1012–1014.

Strand, L. J., and Marver, H., 1970, Determination of δ-aminolevulinic acid synthetase (ALA-S) in cell culture: Naturally occurring inducers in normal human plasma, *Clin. Res.* **18**:345.

Strand, L. J., Felsher, B., Redeker, A. G., and Marver, H. S., 1970, Enzymatic abnormalities in heme biosynthesis in intermittent acute porphyria: Decreased hepatic conversion of porphobilinogen to porphyrins and increased δ-aminolaevulinic acid synthetase activity, *Proc. Natl. Acad. Sci. USA* **67**:1315–1320.

Strand, L. J., Meyer, U. A., Felsher, B. F., Redeker, A. G., and Marver, H. S., 1972, Decreased red cell uroporphyrinogen I synthetase activity in intermittent acute porphyria, *J. Clin. Invest.* **51**:2530–2536.

Strife, C. F., Zuroweste, E. L., Emmett, E. A., Finelli, V. N., Petering, H. G., and Berry, H. K., 1977, Tyrosinemia with acute intermittent porphyria: Aminolevulinic acid dehydratase deficiency related to elevated urinary aminolevulinic acid levels, *J. Pediat.* **90**:400–404.

Strik, J. J. T. W. A., 1978a, Porphyrinogenic action of polyhalogenated aromatic compounds, with special reference to porphyria and environmental impact, in: *Diagnosis and Therapy of Porphyrias and Lead Intoxication* (M. Doss, ed.), pp. 151–164, Springer-Verlag, Berlin.

Strik, J. J. T. W. A., 1978b, Toxicity of PBB's with special reference to porphyrinogenic action and spectral interaction with hepatic cytochrome P-450, *Env. Health Persp.* **23**:167–175.

Sullivan, L. W., and Herbert, V., 1964, Suppression of hematopoiesis by ethanol, *J. Clin. Invest.* **43**:2048–2062.

Sveinsson, S. L., Rimington, C., and Barnes, H. D., 1949, Complete porphyrin analysis of pathological urines, *Scand. J. Clin. Lab. Invest.* **1**:2–11.

Sweeney, G. D., 1963, Patterns of porphyrin excretion in South African porphyric patients, *S. Afr. J. Clin. Lab. Med.* **9**:182–189.

Sweeney, G. D., and Jones, K. G., 1979, Porphyria cutanea tarda: Clinical and laboratory features, *Can. Med. Assoc. J.* **120**:803–807.

Sweeney, G. D., Jones, K. G., Cole, F. M., Basford, D., and Kretynski, F., 1979, Iron deficiency prevents liver toxicity of 2,3,7,8-tetrachlorodibenzo-p-dioxin, *Science* **204**:332–335.

Sweeney, V. P., Pathak, M. A., and Asbury, A. K., 1970, Acute intermittent porphyria. Increased ALA-synthetase activity during an acute attack, *Brain* **93**:369–380.

Taddeini, L., and Watson, C. J., 1968, The clinical porphyrias, *Sem. Hematol.* **5**:335–369.

Takaku, F., Yano, Y., Aoki, Y., Nakao, K., and Wada, O., 1972, δ-Aminolevulinic acid synthetase activity of human bone marrow erythroid cells in various hematological disorders, *Tohoku J. Exp. Med.* **107**:217–228.

Taljaard, J. J. F., Shanley, B. C., and Joubert, S. M., 1971, Decreased uroporphyrinogen decarboxylase activity in "experimental symptomatic porphyria," *Life Sci.* **10**(Pt. 2):887–893.

Taljaard, J. J. F., Shanley, B. C., Deppe, W. M., and Joubert, S. M., 1972, Porphyrin metabolism in experimental hepatic siderosis in the rat. III. Effect of iron overload and hexachlorobenzene on liver haem biosynthesis, *Br. J. Haematol.* **23**:587–593.

Tenhunen, R., Marver, H. S., and Schmid, R., 1969, Microsomal heme oxygenase. Characterization of the enzyme, *J. Biol. Chem.* **244**:6388–6394.

Tephly, T. R., 1978, Inhibition of liver hemoprotein synthesis, in: *Heme and Hemoproteins— Handbook of Experimental Pharmacology* (F. DeMatteis and W. M. Aldridge, ed.), Vol. 44, pp. 81–94, Springer-Verlag, Berlin.

Tephly, T. R., Hasegawa, E., and Baron, J., 1971, Effect of drugs on heme synthesis in the liver, *Metabolism* **20**:200–214.

Tephly, T. R., Condie, L. W., and Piper, W. N., 1973, The role of substrates for glycine acyltransferase and the role of sulfanilamide and acetate in the reversal of chemically-induced porphyria in the rat, *Enzyme* **16**:187–195.

Theologides, A., Kennedy, B. J., and Watson, C. J., 1964, A study of urinary porphyrins and porphyrin precursors in patients with malignant disease receiving diethylstilbestrol, *Metabolism* **13**:391–395.

Thivolet, J., Euvrard, S., and Perrot, H., Moskovtchenko, J.-F., Claudy, A., and Ortonne, J.-P., 1977, La pseudo-porphyrie cutanée tardive des hémodialyses; aspects cliniques et histologiques à propos de 9 cas, *Ann. Dermatol. Vénérol.* **104**:12–17.

Thompson, R. P. H., Nicholson, D. C., Farnan, M. B., Whitemore, D. N., and Williams, R., 1970, Cutaneous porphyria due to a malignant primary hepatoma, *Gastroenterology* **59**:779–783.

Tiepermann, von R. v., and Doss, M., 1978, Uroporphyrinogen-Decarboxylase in Erythrocyten. Untersuchungen zum primären genetischer Enzymdefekt bei chronischer hepatischer Porphyrie, *J. Clin. Chem. Clin. Biochemie* **16**:513–517.

Tio, T. H., Leijnse, B., Jarrett, A., and Rimington, C., 1957, Acquired porphyria from a liver tumor, *Clin. Sci.* **16**:517–527.

Tobias, G., 1964, Congenital porphyria in a cat, *J. Am. Vet. Med. Ass.* **145**:462–463.

Tofilon, P. J., and Piper, W. N., 1980, Measurement and regulation of rat testicular δ-aminolevulinic acid synthetase activity, *Arch. Biochem. Biophys.* **201**:104–109.

Tomaro, M. L., Frydman, R. O., and Frydman, B., 1973, Porphobilinogen oxygenase from rat liver. Induction, isolation and properties, *Biochemistry* **12**:5263–5268.

Tomio, J. M., Garcia, R. C., San Martín de Viale, L. C., and Grinstein, M., 1970, Porphyrin biosynthesis. VII. Porphyrinogen carboxy-lyase from avian erythrocytes. Purification and properties, *Biochim. Biophys. Acta* **198**:353–363.

Tomita, Y., Ohashi, A., and Kikuchi, G., 1974, Induction of δ-aminolevulinate synthetase in organ culture of chick embryo liver by allylisopropylacetamide and 3,5-dicarbethoxy-1,4-dihydrocollidine, *J. Biochem.* **75**:1007–1015.

Tschudy, D. P., 1965, Biochemical lesions in porphyria, *J. Am. Med. Assoc.* **191**:718–730.

Tschudy, D. P., and Lamon, J. M., 1980, Porphyrin metabolism and the porphyrias, in: *Metabolic Control and Disease*, 8th Ed. (P. K. Bondy and L. E. Rosenberg, eds.), pp. 939–1007, W. B. Saunders, Philadelphia.

Tschudy, D. P., and Collins, A., 1957, Effect of 3-amino-1,2,4-triazole on δ-aminolevulinic acid dehydrase activity, *Science* **126**:168.

Tschudy, D. P., Welland, F. H., Collins, A., and Hunter, G., Jr., 1964, The effect of carbohydrate feeding on the induction of δ-aminolevulinic acid synthetase, *Metabolism* **13**:396–406.

Tschudy, D. P., Marver, H. S., and Collins, A., 1965a, A model for calculating messenger RNA half-life: Short-lived messenger RNA in the induction of mammalian δ-aminolevulinic acid synthetase, *Biochem. Biophys. Res. Comm.* **21**:480–487.

Tschudy, D. P., Perlroth, M. G., Marver, H. S., Collins, A., Hunter, G., Jr., and Rechcigl, M., Jr., 1965b, Acute intermittent porphyria: The first "over-production disease" localized to a specific enzyme, *Proc. Natl. Acad. Sci. USA* **53**:841–847.

Tschudy, D. P., Valsamis, M., and Magnussen, C. R., 1975, Acute intermittent porphyria: Clinical and selected research aspects, *Ann. Intern. Med.* **83**:851–864.

Tsukamoto, I., Yoshinaga, T., and Sano, S., 1979, The role of zinc with special reference to the essential thiol groups in δ-aminolevulinic acid dehydratase of bovine liver, *Biochim. Biophys. Acta.* **570**:167–178.

Turnbull, A., 1971, Iron metabolism in the porphyria, *Br. J. Dermatol.* **84**:380–383.

Turnbull, A., Baker, H., Vernon-Roberts, B., and Magnus, J. A., 1973, Iron metabolism in porphyria cutanea tarda and in erythropoietic protoporphyria, *Quart. J. Med.* **42**:341–355.

Turner, W. J., 1937, Studies on porphyria. I. Observations on the fox squirrel (*Sciurus niger*), *J. Biol. Chem.* **118**:519–530.

Urabe, A., Sassa, S., and Kappas, A., 1979, The influence of steroid hormone metabolites on the *in vitro* development of erythroid colonies derived from human bone marrow, *J. Exp. Med.* **149**:1314–1325.

Uys, C. J., and Eales, L., 1963, The histopathology of the liver in acquired porphyria, *S. Afr. J. Lab. Clin. Med.* **9**:190–197.

Varadi, S., 1958, Haematological aspects in a case of erythropoietic porphyria, *Br. J. Haematol.* **4**:270–280.

Varticovski, L., Kushner, J. P., and Burnham, B. F., 1980, Biosynthesis of porphyrin precursors. Purification and characterisation of mammalian L-alanine γ.δ-dioxovaleric acid aminotransferase, *J. Biol. Chem.* **355**:3742–3747.

Viljoen, D. J., Cayanis, E., Becker, D. M., Kramer, S., Dawson, B., and Bernstein, R., 1979, Reduced ferrochelatase activity in fibroblasts from patients with porphyria variegata, *Am. J. Hematol.* **6**:185–190.

Wada, O., Sassa, S., Takaku, F., Yano, Y., Urata, G., and Nakao, K., 1967, Different responses of the hepatic and erythropoietic δ-aminolevulinic acid synthetase of mice, *Biochim. Biophys. Acta* **148**:585–587.

Waddington, R. T., 1972, A case of primary liver tumour associated with porphyria, *Br. J. Surg.* **59**:653–654.

Wainwright, S. D., and Wainwright, L. K., 1966, Regulation of the initiation of hemoglobin synthesis in the blood island cells of chick embryos. I. Qualitative studies on the effects of actinomycin D and δ-aminolevulinic acid, *Can. J. Biochem.* **44**:1543–1560.

Wainwright, S. D., and Wainwright, L. K., 1967, Regulation of the initiation of hemoglobin synthesis in the blood island cells of chick embryos. II. Early onset and stimulation of hemoglobin formation induced by exogenous δ-aminolevulinic acid, *Can. J. Biochem.* **45**:344–347.

Waldenström, J., 1957, The porphyrias as inborn errors of metabolism, *Am. J. Med.* **22**:758–773.

Waldenström, J., and Haeger-Aronson, B., 1963, Different patterns of human porphyria, *Br. Med. J.* **2**:272–276.

Waldenström, J., and Haeger-Aronson, B., 1967, The porphyrias: A genetic problem, *Progr. Med. Genet.* **5**:58–101.

Warin, R. P., 1963, Porphyria cutanea tarda associated with estrogen therapy for prostatic carcinoma, *Br. J. Dermatol.* **75**:298–299.

Watson, C. J., 1954, Some studies of nature and clinical significance of porphobilinogen, *Arch. Intern. Med.* **93**:643–657.

Watson, C. J., 1960, The problem of porphyria—some facts and questions, *N. Engl. J. Med.* **263**:1205–1215.

Watson, C. J., 1966, Some recent advances in the problem of erythropoietic porphyria, *Acta Med. Scand.* **445**(suppl.):25–35.

Watson, C. J., 1975, Hematin and porphyria, *N. Engl. J. Med.* **293**:605–607.

Watson, C. J., and Schwartz, S., 1941, A simple test for urinary porphobilinogen, *Proc. Soc. Exp. Biol. Med.* **47**:393–394.

Watson, C. J., Perman, V., Spurrell, F. A., Hoyt, H. H., and Schwartz, S., 1958, Some studies of the comparative biology of human and bovine porphyria erythropoietica, *Trans. Assoc. Am. Physicians* **71**:196–206.

Watson, C. J., Perman, V., Spurrell, F. A., Hoyt, H. H., and Schwartz, S., 1959, Some studies of the comparative biology of human and bovine erythropoietic porphyria, *Arch. Intern. Med.* **103**:436–444.

Watson, C. J., Bossenmaier, I., and Cardinal, R., 1961, Acute intermittent porphyria—urinary porphobilinogen and other Ehrlich reactors in the diagnosis, *J. Am. Med. Assoc.* **175**:1087–1091.

Watson, C. J., Runge, W., Taddeini, L., Bossenmaier, I., and Cardinal, R., 1964a, A suggested control gene mechanism for the excessive production of types I and III porphyrins in congenital erythropoietic porphyria, *Proc. Natl. Acad. Sci. USA* **52**:478–485.

Watson, C. J., Taddeini, L., and Bossenmaier, I., 1964b, Present status of the Ehrlich aldehyde reaction for urinary porphobilinogen, *J. Am. Med. Assoc.* **190**:501–504.

Watson, C. J., Dhar, G. J., Bossenmaier, I., Cardinal, R., and Petryka, Z. J., 1973, Effect of hematin in acute porphyric relapse, *Ann. Int. Med.* **74**:80–83.

Watson, C. J., Bossenmaier, I., Cardinal, R., and Petryka, Z. J., 1974, Repression by hematin of porphyrin biosynthesis in erythrocyte precursors in congenital erythropoietic porphyria, *Proc. Natl. Acad. Sci. USA* **71**:278–282.

Watson, C. J., Cardinal, R. A., Bossenmaier, I., and Petryka, Z. J., 1975, Porphyria variegata and porphyria cutanea tarda in siblings: Chemical and genetic aspects, *Proc. Natl. Acad. Sci. USA* **72**:5126–5129.

Watson, C. J., Cardinal, R. A., Bossenmaier, I., and Petryka, Z. J., 1976, Porphyria variegata and porphyria cutanea tarda in siblings: Chemical and genetic aspects (Addendum), *Proc. Natl. Acad. Sci. USA* **73**:1323.

Watson, C. J., Pierach, C. A., Bossenmaier, I., and Cardinal, R., 1977, Postulated deficiency of hepatic heme and repair by hematin infusions in the "inducible" hepatic porphyrias, *Proc. Natl. Acad. Sci. USA* **74**:2118–2120.

Watson, C. J., Pierach, C. A., Bossenmaier, I., and Cardinal, R., 1978, Use of hematin in the acute attack of the "inducible" hepatic porphyrias, *Adv. Intern. Med.* **23**:265–286.

Waxman, A. D., Collins, A., and Tschudy, D. P., 1966, Oscillations of hepatic δ-aminolevulinic acid synthetase produced *in vivo* by heme, *Biochem. Biophys. Res. Comm.* **24**:675–683.

Weissberg, J. B., and Voytek, P. E., 1974, Liver and red cell porphobilinogen synthase in the adult and fetal guinea pig, *Biochim. Biophys. Acta* **364**:304–319.

Welland, F. H., Hellman, E. S., Gaddis, E. M., Collins, A., Hunter, G. W., Jr., and Tschudy, D. P., 1964, Factors affecting the excretion of porphyrin precursors by patients with acute intermittent porphyria. I. The effect of diet, *Metabolism* **13**:232–250.

Wetterberg, L., 1967, *A Neuropsychiatric and Genetical Investigation of Acute Intermittent Porphyria*, Scandinavian University Books, Svenska Bokforläget, Norstedts.

Wetterberg, L., 1975, Report on an international survey of safe and unsafe drugs in acute intermittent porphyria, in: *Supplement to the Proceedings of the First International Porphyria Meeting—Porphyrins in Human Diseases* (M. Doss, ed.), p. 191, Freiburg, Germany.

Whetsell, W. O., Jr., Sassa, S., Bickers, D., and Kappas, A., 1978, Studies on porphyrin-heme biosynthesis in organotypic cultures of chick dorsal root ganglion. I. Observation on neuronal and non-neuronal elements, *J. Neuropathol. Exp. Neurol.* **37**:497–507.

Whiting, M. J., 1976, Synthesis of δ-aminolaevulinate synthase by isolated liver polyribosomes, *Biochem. J.* **158**:391–400.

Whiting, M. J., and Elliott, W. H., 1972, Purification and properties of solubilized mitochondrial δ-aminolevulinic acid synthetase and comparison with the cytosol enzyme, *J. Biol. Chem.* **247**:6818–6826.

Whiting, M. J., and Granick, S., 1976a, δ-Aminolevulinic acid synthase from chick embryo liver mitochondria. I. Purification and some properties, *J. Biol. Chem.* **251**:1340–1346.

Whiting, M. J., and Granick, S., 1976b, δ-Aminolevulinic acid synthase from chick embryo liver mitochondria. II. Immunochemical correlation between synthesis and activity in induction and repression, *J. Biol. Chem.* **251**:1347–1353.

Wilson, E. L., Burger, P. E., and Dowdle, E. B., 1972, Beef-liver 5-aminolevulinic acid dehydratase. Purification and properties, *Eur. J. Biochem.* **29**:563–571.

Wilt, F. H., 1968, Heme and regulation of embryonic hemoglobin synthesis, *Biochem. Biophys. Res. Comm.* **33**:113–118.

With, T. K., 1965, Porphyria, *Lancet* **i**:916–917.

Woods, J. S., 1974, Studies on the role of heme in the regulation of δ-aminolevulinic acid synthetase during fetal hepatic development, *Mol. Pharmacol.* **10**:389–397.

Woods, J. S., and Dixon, R. L., 1970, Perinatal differences in delta-aminolevulinic acid synthetase activity, *Life Sci.* **9**(Pt. II):711–719.

Woods, J. S., and Dixon, R. L., 1972, Studies on the perinatal differences in the activity of hepatic δ-aminolevulinic acid synthetase, *Biochem. Pharmacol.* **21**:1735–1744.

Woods, J. S., and Murthy, V. V., 1975, δ-Aminolevulinic acid synthetase from fetal rat liver: Studies on the partially purified enzyme, *Mol. Pharmacol.* **11**:70–78.

Wu, W. H., Shemin, D., Richards, K. E., and Williams, R. C., 1974, The quaternary structure of δ-aminolevulinic acid dehydratase from bovine liver, *Proc. Natl. Acad. Sci. USA* **71**:1767–1770.

Yamauchi, K., Hayashi, N., and Kikuchi, G., 1980, Translocation of δ-aminolevulinate synthase from the cytosol to the mitochondria and its regulation by hemin in the rat liver, *J. Biol. Chem.* **255**:1746–1751.

Yannoni, C. A., and Robinson, S. H., 1975, Early-labelled haem in erythroid and hepatic cells, *Nature* **258**:330–331.

Yoda, B., Schacter, B. A., and Israels, L. G., 1974, Induction of δ-aminolevulinic acid synthetase in the kidney of chicks treated with porphyrinogenic drugs, *Biochim. Biophys. Acta* **372**:478–481.

Yoshikawa, H., and Yoneyama, Y., 1964, Incorporation of iron in the heme moiety of chromoproteins, in: *Iron Metabolism* (F. Gross, ed.), pp. 24–37, Springer-Verlag, Berlin.

Yoshinaga, T., and Sano, S., 1980, Coproporphyrinogen oxidase I. Purification, properties, reaction mechanism and activation by phospholipid, *J. Biol. Chem.* **255**:4722–4726.

Zail, S. S., Krawitz, P., Viljoen, E., and Kramer, S., 1967, The anemia of erythropoietic porphyria. II. Studies of some red cell intermediates, *Br. J. Haematol.* **13**:60–67.

Zimmerman, T. S., McMillin, J. M., and Watson, C. J., 1966, Onset of manifestations of hepatic porphyria in relation to the influence of female sex hormones, *Arch. Intern. Med.* **118:**229–240.

Zinkl, J., and Kaneko, J. J., 1973, Erythrocytic enzymes and glycolytic intermediates in the normal bovine and in bovine erythropoietic porphyria, *Comp. Biochem. Physiol.* **45A:**463–476.

Chapter 4

The Molecular Genetics of Thalassemia

Stuart H. Orkin and David G. Nathan

Division of Hematology
The Children's Hospital Medical Center
and Department of Pediatrics
Harvard Medical School
Boston, Massachusetts 02115

INTRODUCTION

In the past few years analysis of the genetics and disorders of human hemoglobins has undergone a revolution as new research tools have provided the means to obtain long-sought-after answers regarding gene structure and function. Study of human hemoglobins and their disorders, particularly the thalassemia syndromes, is at the forefront of the interaction of new genetic technology and medicine. It offers the promise of understanding an entire class of diseases at the molecular level in detail totally unforeseen just a short time ago. In this review we will attempt to summarize recent approaches and findings in this rapidly moving area and illustrate how these studies may provide important models for the analysis of other human diseases.

Our review will first address the immense technological developments that have led to the most significant new findings in the genetics of human hemoglobins. With this background, we will outline the structure and chromosomal organization of globin genes in vertebrates, and then summarize results obtained in the study of human hemoglobin disorders. We will attempt to identify those areas that seem fruitful for further studies and new developments.

THE NEW GENETICS

Study of the genetics of human hemoglobins necessitates examination
of the globin genes and their products. Initial progress in this area in the
early 1970s relied on techniques of molecular hybridization to assess
globin-specific nucleic acids (both RNA and DNA) (Weatherall and Clegg,
1979). Use of globin cDNA, a complement of globin mRNA synthesized
in vitro, first allowed quantitation of globin mRNA and globin genes in
total cellular DNA (Housman *et al.*, 1973; Kacian *et al.*, 1973; Taylor *et
al.*, 1974; Ottolenghi *et al.*, 1974). The specific findings of these approaches in the study of human disorders will be discussed below. The
development of recombinant DNA methods, however, provided a quantum leap in the degree of precision with which information regarding
globin gene structure could be acquired. Initial molecular cloning involved
amplifying copies of globin mRNAs in bacterial plasmids (Maniatis *et al.*,
1976; Little *et al.*, 1978; Wilson *et al.*, 1978). These globin cDNA clones
provided purified globin gene sequences that have been analyzed at the
nucleotide sequence level (Efstratiadis *et al.*, 1977; Wilson *et al.*, 1980;
Forget *et al.*, 1980; Little *et al.*, 1978) or that are employed as pure
molecular hybridization probes. These globin probes (either cDNA or
cloned cDNAs) have been utilized to great advantage in two major directions: (1) gene mapping and (2) gene cloning. As both bear heavily on
the results obtained in recent work they will be described here.

Gene mapping signifies the use of bacterial restriction endonucleases
in concert with gel electrophoresis to examine the organization of gene
sequences in cellular DNA. The approach most widely used is founded
on the pioneering technical approach of Southern (1975). Restriction endonucleases are highly specific enzymes that cleave double-stranded
DNAs at specific sites determined by DNA sequence (Roberts, 1980). To
date, more than one hundred enzymes with known recognition sequences
have been characterized. Digestion of cellular DNA with such enzymes
generates a reproducible collection of DNA fragments of widely varying
sizes. In the case of human DNA and a common endonuclease, Eco RI,
for example, nearly one million individual DNA fragments are generated
due to the vast complexity of the human genome. Southern developed
a methodology that permits location of specific DNA sequences among
this collection of DNA fragments. Restriction-endonuclease-digested
DNA is electrophoresed in agarose gel slabs that have considerable ca-

pacity for the resolution of DNA fragments. Southern's technique, commonly known as "blotting," involves transfer of DNA fragments from the gel slab to a nitrocellulose filter sheet after electrophoresis. The DNA fragments, retained on the nitrocellulose filter, are then incubated with highly radioactive nucleic acid probes (cDNA or plasmid cloned cDNA). The filters are washed and subjected to autoradiography to reveal the position of the DNA fragment(s) in the gel. Numerous technical improvements have been introduced to enhance both the sensitivity and reliability of this experimental approach. These include the treatment of the filter sheets to reduce nonspecific hybridization of the radioactive probes (Denhardt, 1966), development of filter sheets that can accept DNA (or RNA) fragments covalently to permit repeated use of the same material and prevent loss of nucleic acid during hybridization (Alwine et al., 1977), and the intensification of autoradiographic signals (Swanstrom and Shank, 1978). Information most commonly gained from gene mapping or gel blotting experiments involves the presence or absence of specific sequences in the DNA and the size of the DNA fragment(s) in which molecular hybridization occurs. As generally employed, the method is qualitative and not strictly quantitative. Using some knowledge about the location of restriction enzyme cleavage sites within the coding sequences of specific genes (determined by study of cDNA), one can generally use gene mapping to develop a physical map or restriction map of a region of the DNA (Jeffreys and Flavell, 1977a). These physical maps, as discussed below, can often provide insights into gene structure as well as the organization and number of structural gene copies in the cellular DNA (Jeffreys and Flavell, 1977b). A variation on this "southern blot" strategy is another approach, commonly referred to as "northern blotting." In this method electrophoresed RNA is bound to filter sheets (Alwine et al., 1977) and then probed for specific sequences using highly labeled DNAs. This technology allows detection of specific RNAs among many different types and identification of mRNA precursors.

A companion method of great power is molecular cloning of chromosomal genes using bacteriophage cloning vectors. Since single copy genes in mammalian DNAs represent less than one part per million of the total DNA, efficient methods are required for the isolation of a gene in pure form. Bacteriophage vectors (Leder et al., 1977; Blattner et al., 1977) developed by a number of laboratories fulfill this need, as genetic manipulations can be employed to select for recombinants with foreign DNAs and large numbers of independent phage can be handled and screened for specific gene sequences (Benton and Davis, 1977). Two strategies for chromosomal gene cloning have been most successful. The first, pioneered in the laboratory of Philip Leder, involves initial purification of a restriction fragment of DNA containing a gene sequence of interest (Tilghman *et al.*, 1977). This enrichment of a DNA fragment reduces the labor of screening phage for specific sequences. The enriched DNA fragments are ligated to bacteriophage cloning vector DNA, and viable phage are generated *in vitro* by a series of steps known as "packaging" in which the phage coat is assembled in the test tube (Sternberg *et al.*, 1977; Blattner *et al.*, 1978; Smithies *et al.*, 1978). The recombinant phage are grown as plaques on specific *E. coli* strains. At high density, these phage plaques can be screened for specific gene sequences using a rapid hybridization technique developed by Benton and Davis (1977). This technique is essentially a blot of the phage plaques in that a nitrocellulose filter is applied to a petri dish of phage plaques and the phage are absorbed *in situ*. Molecular hybridization and autoradiography reveal those phage containing gene sequences of interest. The method is sufficiently rapid that an entire mammalian genome cloned in phage can be analyzed confortably in a week's time. The prepurification strategy permits one to identify a restriction fragment of interest and attempt to isolate it quite directly. Disadvantages are the necessity for larger amounts of DNA for the purification steps and the inability to obtain a collection of phage clones containing different flanking regions (see below).

An alternate method developed largely by Maniatis *et al.* (1978) involves the construction of a "library" or entire genome bank within phage. In this approach, an essentially random array of DNA fragments is linked to phage vector DNA and phage representing the entire genome are generated. These phage are then screened by molecular hybridization for specific genes. This approach generally results in the acquisition of a number of phage clones spanning a region of interest. Its greatest advantage over the purification scheme is in its ability to establish linkage between genes in a cluster in the DNA (see below). Potential disadvantages are the necessity to screen large numbers of phage and, more significantly, the possibility that some specific DNA fragments may be lost or underrepresented in the "library" during its construction or amplification. In addition, the cloning of specific genomic DNA fragments may be limited at times by size constraints of the vectors commonly employed. In these instances, variations on typical cloning approaches may be utilized. In general, the use of one of these cloning strategies permits isolation of chromosomal DNA sequences identifiable by restriction mapping.

Complete structural analysis of genes includes knowledge of the nucleotide sequence. Molecular cloning provides the substrates for DNA sequencing. Two ingenious methods are commonly used for DNA sequencing: one a chemical degradation method of end-labeled DNA devised by Maxam and Gilbert (1977) and the other a chain-termination method of Sanger and Coulson (1975). The details of these methods are not reviewed here.

The new approach to genetics (Nathans, 1979), we believe, has additional approaches that will be used increasingly in the future. These include the modification of DNA sequences *in vitro* (chemical mutagenesis) (Hutchison *et al.*, 1978; Shortie and Nathans, 1978), the transcription of cloned genes *in vitro* (Manley *et al.*, 1980; Weil *et al.*, 1979), and the ability to introduce cloned (and even uncloned) genes into living cells in various ways (Mulligan *et al.*, 1979, Hamer *et al.*, 1979; Wigler *et al.*, 1979; Mantei *et al.*, 1979). In its broadest sense the "new genetics" now permits identification and isolation of specific genes, determination of their nucleotide sequences, and analysis of gene function *in vitro* and *in vivo*, both with and without modification of the DNA sequence *in vitro*. This is a powerful array of tools with which to study the genetics of human hemoglobin genes. To date, nearly all have been utilized, as we hope to elucidate below.

Globin Gene Organization in the Cellular DNA

Fundamental to an understanding of human hemoglobin genetics is consideration of globin gene structure and arrangement in the genome. We will first summarize some common features of globin gene structure found in vertebrates before proceeding to a discussion of human globin genes more specifically. The principles described here are derived from restriction mapping, molecular cloning, and DNA sequencing of genes from several species (mouse, rabbit, chicken, human) in a number of laboratories.

Normal hemoglobins are tetramers of two different types of globin, α-like and β-like chains. These globins are members of developmentally regulated families of proteins. The globins are expressed only in developing erythroid cells. Importantly, different members within each family of globins are expressed in a temporal sequence during development (Weatherall and Clegg, 1979). For example, within the human β-globin family, the β-globin gene is expressed at a high rate during adult life whereas the γ- or fetal human globin genes are expressed at a high rate during fetal life. In addition, embryonic globin genes are expressed predominantly during early fetal life. Somatic cell hybridization, chromosomal linkage studies, as well as family studies in the case of humans, have shown that the α-like and β-like globin gene families are unlinked in the genome; that is, they are present on different chromosomes (Deisseroth et al., 1977, 1978). However, within each family, the related genes are tightly clustered in the DNA. In the case of humans, the α-like globin genes reside on chromosome 16 whereas the β-like genes are on the short arm of chromosome 11 (Deisseroth et al., 1978; Lebo et al., 1979; Scott et al., 1979; Gusella et al., 1979; Sanders-Haigh et al., 1980).

Closely linked, related members of each globin family are often arranged in the chromosome in the temporal sequence of their expression (Dodgson et al., 1979; Hardison et al., 1979; Fritsch et al., 1980; Lauer et al., 1980). Thus, in humans, for example, the embryonic genes are located upstream or 5′ in the DNA, followed by the fetal globin genes, and then the adult globin genes in the β-like complex (Fritsch et al., 1980). Whether this arrangement is a reflection of the gene duplications giving rise to these globin genes or is of functional significance is unknown. However, the regularity of this arrangement among the globin genes of several (but not all) species suggests an important role.

One of the most surprising developments in recent years was the discovery of interrupted genes, first identified in mammalian DNA viruses but soon after found in rabbit and mouse globin genes (Jeffreys and Flavell, 1977b; Tilghman et al., 1978a). It is now apparent that all normal globin genes are mosaics, assembled as three coding segments interrupted by two blocks of DNA (intervening sequences) which are not represented in mature mRNA. The schematic representation of mammalian globin genes is shown in Fig. 1. Several features are striking. First, the interruption of the coding segments by the intervening sequences occurs in fixed positions in the globin genes in the homologous portions of the α- and β-like genes (Leder et al., 1978; Nishioka and Leder, 1979). Thus, the interruption of the coding sequences must be an ancient event, and gene duplication has not altered this pattern. Second, the length of the intervening sequences is not fixed among the species and between the

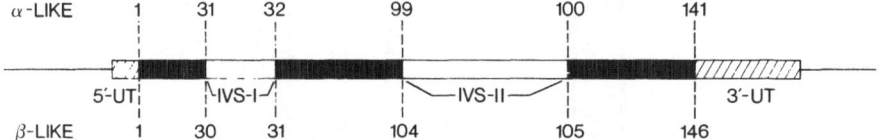

Fig. 1. Schematic representation of cellular globin genes. The globin gene is a mosaic formed from untranslated sequences, coding regions, and intervening sequences. All portions are represented in the initial mRNA precursors. The intervening sequences are removed in formation of final mRNA. Intervening sequences interrupt the α-like and β-like coding regions as indicated. The IVS-I is approximately 115–150 bp in the human α-, β-, and γ-globin genes and IVS-II about 150 bp in the α-genes and 850–950 bp in the β-like genes. Striped area: untranslated segments; black area: coding regions; white area: intervening sequences.

globin gene families. In general, the second intervening sequence of the β-like genes is considerably larger than that in α-like genes. Third, among related members of a gene family, the intervening sequences may have diverged extensively. For example, the human β- and δ-globin genes (see below) have intervening sequences (Flavell *et al.*, 1978; Mears *et al.*, 1978*b*) that are nonhomologous except at the borders of the intervening sequences and the coding regions (Lawn *et al.*, 1978). Fourth, electron-microscopic and hybridization studies, in which precursor mRNA is hybridized to cloned genes, and analysis of mRNA precursors have established that intervening sequences are transcribed into RNA within the nucleus and then are excised from the RNA precursors by a precise mechanism (RNA splicing) (Tilghman *et al.*, 1978*b*; Kinniburgh *et al.*, 1978). The junctions of the intervening and coding regions of the globin genes and other eukaryotic genes are similar and follow a consensus rule (the Chambon rule) in which intervening sequences generally begin with GT and end with AG (Breathnach *et al.*, 1978). Table I summarizes several intervening sequence junctions that have been characterized to date.

An unforeseen feature of globin gene clusters, revealed by gene cloning, is the presence of ancestral, nonfunctional structural genes amidst the complex of normal globin structural genes (Lauer *et al.*, 1980; Fritsch *et al.*, 1980; Hardison *et al.*, 1979; Lacy *et al.*, 1979; Nishioka *et al.*, 1980). It is conceivable during the duplication events that gave rise to multiple structural genes, expressed selectively during development, that errors led to the creation of structurally abnormal genes that have not been (or may not be) eliminated from the genome. In the human, as discussed below, a striking example of this is a gene identified by Lauer

TABLE I. Nucleotide Sequences at Intervening Sequence-Coding Region Junctions in Several Globin Genes

Globin gene	Nucleotide sequence [a]			Reference
Mouse β_major	IVS 1	GCAGGTTG	......TTAGGCTGCTG	Konkel et al. (1978)
	IVS 2	TCAGGGTG	......ACAGCTCCTG	
Mouse α	IVS 1	AGGTGAGAACA	......CTCCCAGGATG	Nishioka and Leder (1979)
	IVS 2	AAGTATGCGC	......CTCCGCAGCTC	
Rabbit β	IVS 1	GGCAGGTTGGTATCC	......TTTTCTCAGGCTGCTG	Hardison et al. (1979)
	IVS 2	TTCAGGGTGAGTTTGG	......TTTCCTACAGCTGCTG	
Human β	IVS 1	AGGTTGGTATCA	......CTTAGGCTGCTG	Maniatis (personal communication)
	IVS 2	AGGGTGAGTCTA	......CCCACAGCTCCTG	
Human α (1)	IVS 2	AAGGTGAGCGGGCC	......GCACAGCTCCTA	Michelson and Orkin (1980)
Protype		5'-end TCAGGTA......TXCAGG 3'-end		Breathnach et al. (1978)

[a] IVS, intervening sequence. IVS 1 interrupts codons 31 and 32 of the α sequence, and 30 and 31 of the β sequence (Fig. 1). IVS 2 interrupts codons 99 and 100 of the α sequence, and 104 and 105 of the β sequence. The underlined nucleotides indicate the GT/AG "Chambon" rule (Breathnach et al., 1978).

et al. (1980). Nonfunctional globin genes have also been identified amid the normal β-like genes in rabbits, humans (Hardison *et al.*, 1979; Fritsch *et al.*, 1980), and mice (Nishioka *et al.*, 1980) in other cloning studies.

DNA sequence analysis has revealed regions preceding globin genes that may serve as possible promoters, sites at which RNA polymerase binds in preparation for gene transcription (Konkel *et al.*, 1978, 1979; Hardison *et al.*, 1980; Lai *et al.*, 1979; Lomedico *et al.*, 1979; Benoist *et al.*, 1980; Van Ooyen *et al.*, 1979). The general sequence pattern resembles that of the Hogness box structure GTATAAATAG and generally occurs approximately 30 nucleotides before the cap site or presumed initiation point for transcription, the 5' end of the mRNA (Table II). Whether this general promoter-like sequence is truly a promoter or merely an element that fixes the transcription start position (a selector) has not yet been definitively demonstrated (Grosschedl and Birnstiel, 1980; Wasylyk *et al.*, 1980). In addition, specific DNA sequences near the end of the gene and just past the end of the gene are thought to be required for poly(A) addition to the 3' end of the globin mRNA and for termination of transcription (Konkel *et al.*, 1978).

Another general feature of regions containing globin sequences is the presence of interspersed repeated DNA sequences (Duncan *et al.*, 1979; Fritsch *et al.*, 1980; Shen and Maniatis, 1980). These repeated elements are present in several copies through the rabbit and human β-like regions and elsewhere. Repeated DNA sequences of the highly repeated human DNA family, now called "the Alu family" (Jelinek *et al.*, 1980), appear to be present among the β-like cluster in humans. The "Alu" family members are repeated roughly 500,000 times in the human genome. Several different repeated sequences are present in the human globin clusters. Whether the presence of these repeated units is merely a general reflection of the interspersion of repeated and single-copy genes throughout the genome or of more fundamental significance for globin gene expression is unknown.

Although the rationale for the development of intervening sequences is obscure, several theories have been advanced to explain their origin and benefit to the organism. It has become apparent that intervening sequences generally separate portions of final proteins that serve different functions (Gilbert, 1978; Doolittle, 1978). Most striking is the example of immunoglobulin genes where intervening sequences separate functional domains within the molecule (Sakano *et al.*, 1979; Gough *et al.*, 1980). In the case of the globins, the segmentation of the molecule in a functional

TABLE II. Nucleotide Sequences of 5′ Flanking Regions of Several Eucaryotic Genes[a]

Eucaryotic gene	Nucleotide sequence	Reference
Human α (1)	GCGCCCCAAGCATAAACCCTGGCGCGCTCGCGGCGCCGGCAC	Michelson and Orkin (1980)
Mouse α	TGGAGGGCATATAAAGTGCTACTTGCTGCAGGTCCAAGACAC	Nishioka and Leder (1979)
Mouse β$_{major}$	GGCAGAGCATATAAAGGTGAGGTAGGATCAGTTGCTCCTCAC	Konkel et al. (1978)
Mouse β$_{minor}$	GCGTATATAAAGCCTGAGCAGGGTCAGTTGCTTCTTAC	Konkel et al. (1979
Human β	GGCTGGGCATAAAAGTCAGGGCAGAGCCATCTATTGCTTAC	Maniatis (personal communication)
Rabbit β	AGGCTTGGGCATAAAAGGCAGAGCAGGCAGCTGCTGCTTAC	Hardison et al. (1979)
Chicken ovalbumin	GCTATATATTCCCCAGGGCTCAGCCAGTGTCTGTAC	Benoist et al. (1980)
Chicken ovomucoid	TTTGTATATATTTGCAGGCAGGCTCGGGGGGACCAT	Lai et al. (1979)

[a] The dotted vertical line indicates the junction of the 5′ flanking regions and the first nucleotide of the mRNA. The underlined sequences roughly 30 nucleotides before this position represent the AT rich, putative promoter regions.

sense is less clear, but the central coding blocks of the α- and β-like genes (from amino acids 31–99 and 30–104, respectively) encode the portion of globin responsible for heme binding (Eaton, 1980; Craik *et al.*, 1980). The specificity of heme binding to the isolated protein "cores" of globin chains has recently been demonstrated, lending some support for this view.

In accord with the model of intervening sequences dividing genes into functional domains, Gilbert (1978) has proposed that intervening sequences may serve to speed up evolution by assembly of novel genes (and, consequently, their protein products) from preexisting ones. If intervening sequences were required to "bring" genes together during evolution, we might later expect the loss of intervening sequences from "mature" genes. The apparent loss of one intervening sequence from the duplicated rat proinsulin genes may represent this phenomenon (Lomedico *et al.*, 1979). Another concept of the "evolutionary" impact of intervening sequences has been proposed by Leder and co-workers and suggests that intervening sequences operate to modulate the target size for genetic recombination among closely related genes (Tiemeier *et al.*, 1978). Their initial ideas stemmed from comparison of mouse immunoglobulin and globin gene systems (Seidman *et al.*, 1978). In the former, the necessity to generate antibody diversity might benefit from recombination events prompted by mechanisms directed toward increasing homology of related gene sequences. It was observed that homology between related immunoglobulin genes extended well beyond the coding sequences into flanking areas. In the case of globin genes, where recombination events would be detrimental, divergence of intervening sequences among closely related, linked genes would serve to reduce recombination. The insertion of intervening sequences, as well as the presence of nonhomologous flanking areas, would reduce the target size for recombination. How general these notions are for other gene systems is uncertain at this time. Comparison of homologous, yet divergent, genes demonstrates that coding regions diverge largely by single nucleotide substitutions whereas intervening sequences evolve via insertions, deletions, and (possibly) crossing-over (Wahli *et al.*, 1980; Van Ooyen *et al.*, 1979; Konkel *et al.*, 1979).

An additional facet of gene organization of potential significance is that of DNA modification (methylation) throughout globin gene clusters. Such methylation of DNA appears to occur differentially among tissues (Ginder *et al.*, 1979; Van der Ploeg *et al.*, 1980*a*). In addition, methylation of specific regions may be correlated with globin gene inactivity (Ginder *et al.*, 1979). It is apparent that methylation *per se* is not a sufficient correlate of gene function but it may be a necessary concomitant (Van der Ploeg *et al.*, 1980*a*). Further work must be done in numerous systems before the significance of methylation is understood.

The Organization and Structure of Human Globin Genes

With the above principles of gene organization and structure as a background we shall now consider in more detail the results of studies on globin gene structure in man. To approach this discussion it is necessary to review the hemoglobin types found in human blood samples at various periods of life and the globin chain composition of these hemoglobins. The hemoglobin types are summarized in Table III. Each is a tetramer of two α-like and two β-like chains. Hemoglobin A, the predominant hemoglobin of normal adults, is a tetramer of α- and β-globin chains. A minor hemoglobin, A_2, present in small amounts in adult blood samples, has δ-, rather than β-chains in the tetramer. The δ-globin chain is quite homologous to the β-chain, indicating that they are the products of gene duplication. The δ- and β-globins differ at eight amino acids out of 146 total. Nucleic acid hybridization has indicated that the 3' end of the δ-globin gene, more specifically the untranslated region, is quite non-homologous with β. Fetal hemoglobin has γ-globin replacing β in the tetramer. γ-Globin, another member of the β-like cluster (see below), is related to δ- and β-globins on the basis of protein sequence but yet more distant (39 amino acids changed from β and 42 from δ). The temporal sequence of globin gene expression is shown in Fig. 2.

TABLE III. Human Hemoglobins

Type of hemoglobin	Chain composition	When present
Hb A	$\alpha_2\beta_2$	Adult life (~95%); small amount during fetal life
Hb F	$\alpha_2\gamma_2$	Fetal life (predominant); adult life (~1–2%)
Hb A_2	$\alpha_2\delta_2$	Adult life (~3%)
Gower-1	$\zeta_2\epsilon_2$	<12 weeks gestation
Gower-2	$\alpha_2\epsilon_2$	<12 weeks gestation
Portland	$\zeta_2\gamma_2$	<12 weeks gestation

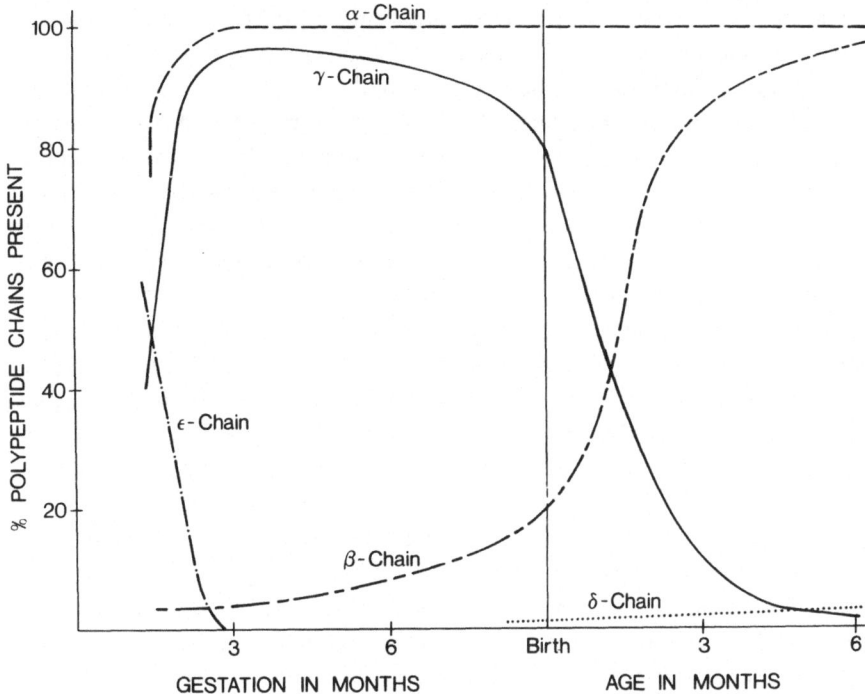

Fig. 2. Temporal pattern of globin chain synthesis during human ontogeny. Adapted from Nathan *et al.* (1979).

Less is known regarding the embryonic human hemoglobins, products found only during early fetal life (before 12 weeks, generally). The hemoglobin compositions are shown in Table III. Molecular cloning studies promise to provide further insights into the precise structure of the embryonic globin chains and their relationships to the adult globin chain members. The availability of a cultured leukemic cell line (k562) that produces embryonic and fetal globins before and after hemin treatment (Rutherford *et al.*, 1979) may help develop these studies.

The nucleotide sequences of the human α, β, and γ mRNAs have been determined in their entirety by analyses of mRNAs and cloned cDNAs (Baralle, 1977; Forget *et al.*, 1980; Kan *et al.*, 1980*b*; Marotta *et al.*, 1977; Proudfoot *et al.*, 1977; Wilson *et al.*, 1977, 1980). In addition, the availability of genomic clones of human globin species has now permitted determination of the complete DNA sequences of human globin species. These sequences reveal common features of globin gene organization cited above and common apparent "signals" for promoter and splicing functions previously described for mouse and rabbit genes.

Study of families with hemoglobinopathies initially revealed that the α- and β-like globin genes were unlinked in the human genome (Weatherall and Clegg, 1979). Furthermore, pedigree studies of families with specific α-globin chain single amino acid mutations indicated that the α structural genes must be duplicated in normal humans (Hollan et al., 1972; Meloni et al., 1980). Although there has been some controversy over the generality of this concept, it is now apparent that the presence of duplicated α-globin genes is, in fact, representative of normal humans and that individuals with only one α gene per haploid genome are thalassemic (see below) (Old et al., 1978a). In addition, hemoglobins that arose by unequal crossing-over between related globin genes (Hb Lepore and Hb Kenya) demonstrated that the γ-, δ-, and β-globin genes resided on a single chromosome and were ordered from N–C terminal (or 5′–3′ in DNA) as $^G\gamma$-$^A\gamma$-δ-β, where $^G\gamma$ and $^A\gamma$ refer to nonallelic γ-globin genes (Bernards et al., 1979; Fritsch et al., 1979).

The chromosomal representation of the human globin genes was a matter of some controversy until recently. Initial attempts to assign those chromosomes carrying the globin genes relied on use of chromosome deletion syndromes and (rather imprecise) molecular hybridization of globin cDNA probes to chromosome spreads. Somatic cell hybridization experiments of Deisseroth et al. (1977, 1978) in which mouse–human hybrid cells segregating human chromosomes were used for analysis, provided convincing evidence for the localization of the α genes on chromosome 16 (Deisseroth et al., 1977) and the β-like genes on chromosome 11 (Deisseroth et al., 1978). More recently, with the advent of gene mapping the sensitivity of these analyses have been developed further. Use of somatic hybrids carrying either deleted human chromosomes or translocations and sorting of specific human chromosomes has confirmed the assignments of Deisseroth and demonstrated that the β-like cluster is present on the short arm of chromosome 11 (Lebo et al., 1979; Gusella et al., 1979; Scott et al., 1979; Sanders-Haigh et al., 1980).

Gene mapping using restriction endonucleases first provided evidence for the tight physical linkage of related globin genes in the DNA. Linkage of β- and δ-globin genes was first demonstrated by Flavell et al. (1978), the linkage of γ genes (Little et al., 1979) with δ and β by Fritsch et al. (1979), and of the duplicated α-globin genes by Orkin (1978). The physical maps and linkage arrangements of globin genes identified by gene mapping have been conclusively established by gene cloning experiments.

The appearance of the α-like cluster of human DNA is depicted in the physical map in Fig. 3 based on the restriction mapping of Orkin (1978), Embury *et al.* (1979), and the cloning experiments of Lauer *et al.* (1980). The duplicated normal α-globin structural genes are located approximately 3.7 kb apart in the DNA (center-to-center). They encode identical polypeptide chains. Upstream from the $5'$-α-globin gene is a nonfunctional α-like gene, the $\psi\alpha1$ gene, a remnant of gene duplication events (Lauer *et al.*, 1980). Further upstream are the duplicated ζ genes, coding for the embryonic α-like chains. Within the region of the duplicated α-globin structural genes there is a broad region of DNA sequence homology (Lauer *et al.*, 1980). First, the coding segments of the α-globin genes are very nearly identical. Direct sequencing of cellular genes, however, suggests that there are substantial differences in DNA sequence between the $\alpha(2)$ and $\alpha(1)$ globin genes in the $3'$-untranslated region (Michelson and Orkin, 1980). The intervening sequences of the two α genes appear similar. The intervening sequences of the $5'$- and $3'$-α-globin genes appear highly homologous, the major difference being the insertion of about nine nucleotides in the large intervening sequence of the $3'$ gene relative to the $5'$ gene of the cluster (Lauer *et al.*, 1980). This suggests that the intervening sequences of the α genes may be evolving by deletion/insertion events, as suggested for the origin of more extreme divergence of the large intervening sequences of the rabbit and mouse β-globin genes (Konkel *et al.*, 1979; Van Ooyen *et al.*, 1979). Second, in addition to extensive homology between the duplicated genes in their coding and intervening sequences, there appears to be extensive homology in the upstream flanking regions of the two structural genes, as determined by

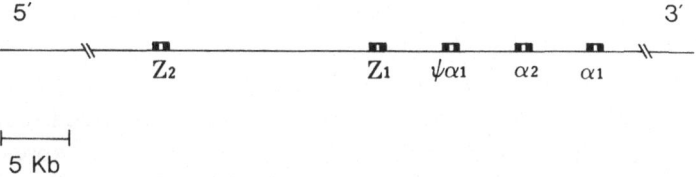

Fig. 3. Arrangement of α-like globin genes on chromosome 16. $\alpha(1)$ and $\alpha(2)$ are the adult α-globin genes. $\psi\alpha1$ is the pseudo-α-gene; $\zeta1$ and $\zeta2$ are the embryonic α chains. Adapted from Lauer *et al.* (1980).

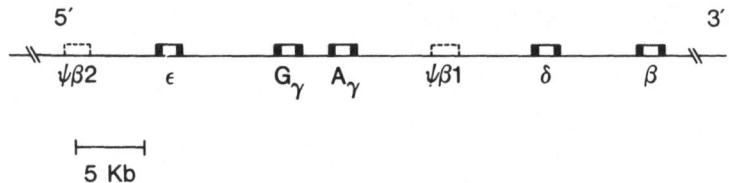

Fig. 4. Arrangement of β-like globin genes on chromosome 11. ψβ1 and ψβ2 are nonfunctional β-like sequences within the cluster. Adapted from the studies of Fritsch *et al.* (1980).

restriction mapping and electron microscopy by Lauer *et al.* (1980). This wide stretch of homology promotes unequal crossing-over events and instability within the α-globin region of normal human DNA, and contrasts with the stability of the β-like globin gene region maintained, perhaps in part, by divergence of nucleotide sequences within intervening sequences and in flanking regions among the γ-, δ-, and β-globin genes (see below).

The β-like globin genes in human DNA are arranged as shown in Fig. 4, spanning a region of more than 30 kb (Fritsch *et al.*, 1980). The single β- and δ-globin genes form the 3′ end of the cluster. The nonallelic γ genes are located about 14 kb 5′ to δ and the ε gene about 13 kb 5′ to γ. The ordering of these genes, embryonic-fetal-adult, suggests possible functional significance to this arrangement. Each globin gene of this cluster is considerably larger than the genes in the α cluster, primarily due to a much greater length of the large intervening sequence (located between codons 104 and 105 of the sequence). The intervening sequences of the β- and δ-globin genes are divergent (Lawn *et al.*, 1978), whereas the γ genes have nearly identical intervening sequences, formally analogous to the relationship between the α genes themselves. Sequences of DNA repeated elsewhere in the human genome are distributed throughout the β-like cluster (Duncan *et al.*, 1979; Fritsch *et al.*, 1980). Evidence for the presence of more than one type of repeated sequence in this region of the DNA has been obtained. One type of repeated sequence, a member of the so-called Alu family of human repeats (Jelinek *et al.*, 1980), is transcribed *in vitro* by RNA polymerase III into discrete transcripts (Duncan *et al.*, 1979). Whether this is a property of repeated sequences of this kind or in any way related to selective globin gene expression in this region is completely unknown at present. But due to the prevalence of repeated sequences throughout the human genome, it would be unwise at this time to ascribe specific functions to these sequences identified in this cluster of globin genes.

Globin Gene Expression and mRNA Biogenesis

Globin genes appear to be transcriptionally active in erythroid precursor cells and dormant in non-hemoglobin-producing tissues. The selective expression of individual members of globin gene families is unexplained, and remains a formidable and important problem in the entire area of hemoglobin molecular genetics. Chromatin digestion experiments have indicated that active and inactive genes appear to have different configurations (that is, susceptibilities to digestion) (Weintraub and Groudine, 1976). However, the configuration of fetal globin genes in chromatin appears not to change in the switch from fetal to adult globin gene expression (Young et al., 1980). These findings imply that more than one feature of chromatin structure is required for maintenance of the "active state."

Study of the synthesis and processing of transcripts from individual globin genes has provided valuable information regarding the biogenesis of mature globin messenger RNAs. Prior to the availability of pure probes for globin sequences, several investigators had suggested that globin mRNAs were synthesized initially as transcripts larger than mature (cytoplasmic) messenger RNA and that processing of these precursors within the nucleus occurred. Bastos and Aviv (1977) and Ross (1976) provided convincing evidence for globin RNA precursors in the mouse system. Curtis et al. (1977) and Ross and Knecht (1978) first demonstrated size differences between α- and β-globin RNA precursors. Soon after the discovery of intervening DNA sequences within globin genes it became apparent that these regions are transcribed into the initial globin RNA products and subsequently excised in splicing maneuvers. Most evidence favors the concept that a globin "transcriptional unit" extends from the 5'-capped nucleotide of globin mRNAs to the 3'-poly(A) addition site. In the case of β-globin genes the greater size of the large intervening sequence compared to α accounts for the greater size of β-RNA precursors. That the splicing reactions involved in processing these globin RNA precursors may be more involved than originally anticipated has been shown by the elegant experiments of Ross and his co-workers in which multiple step processing of mouse globin precursors has been demonstrated (Kinniburgh et al., 1979). The exact DNA sequence signals involved in these complex reactions are unknown. Recently, it has been proposed that small nuclear RNAs may function to align appropriate regions for RNA "spligases" (Lerner et al., 1980; Rogers and Wall, 1980). As originally proposed by Leder (1978), the complexity of the processing steps involved in the biogenesis of globin RNAs suggests the possibility that specific mutations in the DNA may affect adversely any step in the pathway and lead to defective hemoglobin synthesis, the hallmark of thalassemia.

THE NATURE OF THALASSEMIAS

Thalassemias are hypochromic, microcytic anemias that are characterized by defective globin chain synthesis (Weatherall and Clegg, 1972). It is important to stress the heterogeneity of these disorders (Orkin and Nathan, 1976). Specific forms of thalassemia involve the impaired synthesis of specific globin chains, and are so named. Thus, α-thalassemias refer to impaired production of α-globin chains; β-thalassemias to impaired β chain production; $\delta\beta$-thalassemia to decreased β and δ production. Furthermore, the β-thalassemias are formally divided into β^0 and β^+ forms, depending on whether β production is absent or merely reduced from an affected locus.

Study of families affected with thalassemias and variant hemoglobins has indicated that thalassemia mutations are inherited as alleles of the globin genes. Therefore, mutations that lead to thalassemias either affect the globin genes themselves or very closely associated (linked) DNA regions that are of functional importance. In addition, thalassemia mutations act only *in cis,* meaning they affect only the chromosome on which they reside. These findings account for the recessive inheritance of major thalassemia syndromes, in that two "doses" of thalassemia mutations are required to impair significantly and clinically the production of a specific globin chain.

Impaired synthesis of globin chains at the level of protein may be the end result of mutation at different levels. For example, mutations might grossly affect the structure of a globin gene, the transcription of a globin gene, the processing of its precursor RNAs, or the translatability (biological activity) of the mature mRNA. In fact, mutations at almost every level have now been defined. Thus, the thalassemia syndromes are heterogeneous at the molecular level, a finding that is not so surprising in view of the complexity of globin gene structure and function. In the next sections we will specifically detail the molecular basis of various thalassemia syndromes and expand these conclusions.

α-*Thalassemias*

α-Thalassemias are frequent conditions, distributed worldwide. Clinically they can be classified on the basis of the degree of α-globin synthesis impairment (Kan *et al.*, 1968). Available evidence suggests that all four α-globin structural genes are normally active in developing erythroid cells. Dysfunction of one, two, three, or four α genes is associated with clinical states referred to as "silent carrier," α-thalassemia trait, hemoglobin H disease, and homozygous α-thalassemia (hydrops fetalis with hemoglobin Bart's), respectively (Lehmann, 1970). The most severe variety, homozygous α-thalassemia, is a lethal disease in which the fetus cannot synthesize any α-globin chains. In view of the profound nature of this defect and the "quantum" nature of α-thalassemias clinically, it was considered likely that deletion of α structural genes was the genetic mechanism responsible for α-thalassemias (Lehmann, 1970). Studies of hybridization kinetic analysis of globin cDNA with sheared cellular DNAs initially indicated that deletion of most (if not all) α-globin gene sequences occurred in homozygous α-thalassemias (Taylor *et al.*, 1974; Ottolenghi *et al.*, 1974). According to this model and similar studies in other clinical forms of α-thalassemias, it became apparent that the silent carrier state was associated with deletion of one of the four normal α genes; α-thalassemia trait with deletion of two of the four normal genes per cell; and Hb H disease with the deletion of three of the four normal genes (Kan *et al.*, 1975*b*). Although most patients with α-thalassemia fit this scheme, several additional aspects became apparent. First, Kan *et al.* (1977) identified a Chinese family in which Hb H disease was associated with two rather than one structural gene per cell. This implied that one of the two genes detected by hybridization kinetic analysis was in fact "thalassemic" (or carried a "nondeletion" thalassemia defect). Second, clinical differences among α-thalassemias seen in various racial groups were difficult to explain by the deletion model alone. For example, homozygous α-thalassemia is almost exclusively seen among Asian populations, whereas other α-thalassemias are distributed among many different groups.

Application of restriction enzyme mapping procedures to the study of α-thalassemias has provided insights into many areas of both normal and abnormal gene organization, and the molecular basis of these disorders. Gene mapping permits convenient direct quantitation of the number of α-globin genes per cell, especially when used in conjunction with hybridization kinetic analysis (gene copy number assessment) (Orkin *et al.*, 1979*a*). In homozygous α-thalassemia α-globin structural gene sequences appear to be entirely deleted in the DNA when examined by gene mapping (Orkin, 1978; Orkin *et al.*, 1979*a*; Embury *et al.*, 1979; Surrey *et al.*, 1978). These results confirm and extend the previous finding of gene deletion based on less precise techniques. After digestion with the restriction enzyme Eco RI the duplicated α-globin structural genes are contained in a single DNA fragment, 22.5 kb in length (Orkin, 1978); this fragment is missing in homozygous α-thalassemia. Most individuals with Hb H disease have a single α structural gene per cell (Orkin *et al.*, 1979*a*; Embury *et al.*, 1979; Phillips *et al.*, 1980*b*; Kan *et al.*, 1979). This structural gene is generally contained in an 18- to 20-kb Eco RI DNA fragment (Orkin *et al.*, 1979*a*; Embury *et al.*, 1979). The silent carrier state of α-thalassemia is usually associated with the presence of one chromosome containing the 22.5-kb α fragment and the other with the 18- to 20-kb fragment. α-Thalassemia trait (α-thal-1), usually associated with two structural genes per cell, can be characterized by structural genes either in *cis* or *trans* positions, that is, either both on the same chromosome or one each on the chromosome 16 homologs. Kan and his co-workers showed that the former situation is typical of Asians and the latter typical of Blacks (Dozy *et al.*, 1979) and, perhaps, Mediterraneans (Kan *et al.*, 1979). The low frequency at which the chromosome 16 in which both α genes are deleted is found among Blacks (Dozy *et al.*, 1979; Phillips *et al.*, 1979) and, possibly Mediterraneans (Orkin *et al.*, 1979*a*) with α-thalassemias, provides at least one convenient explanation for the near absence of homozygous α-thalassemia among these groups, merely on statistical grounds.

Gene mapping provides a sensitive means for detecting α-globin gene sequences in the DNA that may not contribute to α-globin production (Orkin *et al.*, 1979*a*). These are generally referred to as defective or nondeletion α genes. In our studies of Mediterraneans with Hb H disease we detected putative defective α-globin genes in two different restriction fragments, 22.5 kb and 2.6 kb in size (Orkin *et al.*, 1979*a*). The 22.5-kb fragment has the same physical structure as the normal α-globin gene region. The 2.6-kb fragment appeared to contain no more than a single α-globin gene sequence (Orkin *et al.*, 1979a; Kan *et al.*, 1979). These restriction fragments bearing defective α-globin genes appear to be more common in non-Asians than Asians and may contribute to the absence of homozygous α-thalassemia in the former populations. The nature of the mutation(s) leading to gene dysfunction in the 22.5-kb restriction fragment is unknown. It is possible that the defect involves both structural genes (or their expression) or merely one of the genes in the cluster. Gene cloning, structural analysis, and correlation with gene function will be required to sort out these alternatives.

The defect in the 2.6-kb fragment has been determined in our laboratory by molecular cloning (Orkin and Michelson, 1980). The α-globin gene contained in this fragment is deleted at its 5' end to the position of the sequence corresponding to amino acid 57. The portion of the α gene that is present in this DNA fragment is identical in DNA sequence to that of the more 3' of the normal α-globin genes (Michelson and Orkin, 1980). Therefore, the 2.6-kb α-specific DNA fragment arose by a deletion event originating 5' to the normal duplicated α genes and terminated within the 3' gene in the structural sequence.

New molecular techniques have shed additional insights into the molecular genetics of α-thalassemias. First, comparison of the restriction endonuclease maps of the normal 22.5-kb α-DNA fragment and the 18- to 20-kb α-thalassemia DNA fragment revealed striking homology that could best be explained by an unequal crossing-over event in which two α structural genes are effectively reduced to one in the process (Orkin et al., 1979a). Further work has indicated that an unequal crossing-over event within the region of the duplicated α-globin genes is the usual mechanism for the deletion of one structural gene, characteristic of the α-thalassemia-2 chromosome (Phillips et al., 1980b; Embury et al., 1980). However, in some instances the deletion appears to have removed the 5' α-globin structural gene (Embury et al., 1979, 1980). Therefore, two distinguishable deletions appear to result in the formation of an α-thalassemia-2 (single α gene) chromosome (Embury et al., 1980). Bacteriophage containing the cloned normal duplicated α-globin gene region are very unstable and give rise on propagation to at least two different phage populations in which an α sequence has been deleted (Lauer et al., 1980). It appears that extensive regions of homology in flanking as well as coding and intervening sequences promote these recombination events. The in vitro results (that is, in phage) can mimic in vivo events in this regard (Lauer et al., 1980; Arnheim and Kuehn, 1979).

If unequal crossing-over between two normal α-globin gene regions occurs, we would predict the generation of a triple α gene chromosome in addition to the deleted chromosome. Two different laboratories have recently identified individuals with three α genes in tandem on one chromosome (Goossens *et al.*, 1980; Higgs *et al.*, 1980). Therefore, the cross-ing-over model for α-thalassemia deletions appears to be well founded at this time. The incidence of the triplicated α gene chromosome may differ among racial groups and approach approximately 1 per 200. Although the published evidence is scanty at this time, it seems that all three α loci may be active in these unusual chromosomes.

Whether unequal crossing-over events are responsible for the dele-tion of both α-globin genes on one chromosome (the α-thalassemia-1 defect) must also be considered now in view of recent DNA mapping by Pressley *et al.* (1980). The nature of the DNA deletion in two different fetuses with hydrops fetalis (homozygous α-thalassemia) was examined using a probe for ζ-gene sequences. In a typical hydrops fetalis specimen of Asian origin (from Thailand), in which both α-globin genes on each chromosome 16 were deleted, the ζ-globin region was unaffected by the deletion. However, in a rare hydrops specimen of Mediterranean ances-try, the ζ(1) gene was included in the deletion. These data are depicted in Fig. 5 in addition to the typical crossover leading to the single gene α-thal-2 chromosome and the deletion proposed for the 2.6-kb Eco RI α-thalassemia DNA fragment. Pressley *et al.* (1980) suggested that the hy-drops fetalis deletions that they examined could have been produced by unequal crossing-over mediated by alignment of the ζ(2) and ψ-α-1 genes and actual crossovers downstream from the α(1) locus. Whether this

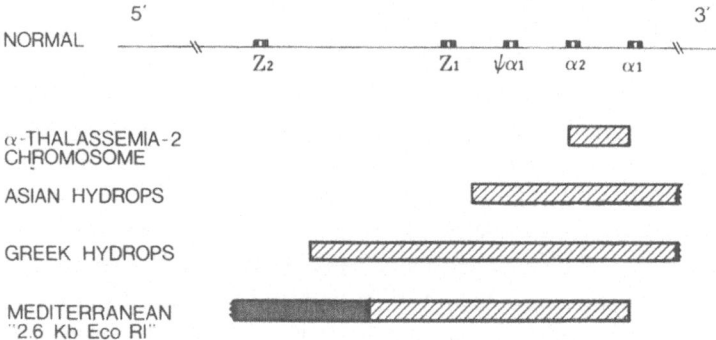

Fig. 5. Deletions within the α-like globin gene cluster in α-thalassemias. Summarized from the studies of Orkin *et al.* (1979a), Embury *et al.*, (1980), Orkin and Michelson (1980), and Pressley *et al.*, (1980). Whether the α-thal-1 chromosome has an α-gene sequence that is, in fact, a "fusion" of the α2 and α1 loci is as yet uncertain, as the crossover might occur in several positions throughout the flanking regions, particularly 5' to the genes (Lauer *et al.*, 1980). Striped area indicates regions that are deleted; stippled area indicates regions that are probably deleted.

model is correct is yet uncertain. Further restriction mapping of these cases and additional hydrops fetalis DNA specimens may help distinguish among the possibilities. Pressley *et al.* (1980) also draw an interesting analogy between deletions in the α-complex and those occurring in the β-globin complex. Comparison of the deletions summarized in Figs. 5 and 6 and in the text makes this point.

The frequency with which α-thalassemias exist indicates that recombination within this segment of the human genome is exceedingly high. In fact, frequent crossing-over tends to "homogenize" genes and has led to the concept of coincident (or concerted) evolution of such members of a gene family (Lauer *et al.*, 1980; Zimmer *et al.*, 1980). In effect, these processes may have served to make the duplicated α and γ genes more similar to each other than the δ- and β-globin loci.

Another important mechanism for α-thalassemia is the presence of abnormal, elongated α-globin chains. The best described example of this class of defects is Hb Constant Spring (Clegg *et al.*, 1971). In this hemoglobin, found in a minority of patients with α-thalassemia, an abnormal α-globin chain in which an additional 31 residues are present on the C-terminal is found. This abnormal chain results from mutation in the terminator for protein synthesis and allows the 3'-untranslated region of the mRNA to be translated into protein. The published mRNA sequences for the 3'-untranslated region match the predicted sequence based on the amino acid sequence of the Constant Spring chain (Wilson *et al.*, 1977, 1980; Proudfoot *et al.*, 1977). Additional elongated α variants, such as Hb Icaria, have been described (Clegg and Tsevrenis, 1974). It is not well understood why the Hb Constant Spring chain is synthesized at a very low rate in individuals possessing this gene. In effect, the presence of a Constant Spring gene mimics the presence of an α-thalassemia-2 defect. DNA sequencing of the thalassemic gene contained in the 2.6-kb fragment described above and the corresponding normal α gene [α(1)] has revealed a 3'-untranslated segment that does not match previously published mRNA sequences or the nucleotide sequence predicted by the Constant Spring chain (Michelson and Orkin, 1980). These findings indicate that the duplicated α-globin genes differ in their DNA sequences in the 3'-untranslated region and that the elongated α-globin chain mutations occurred in the more 5' of the duplicated α-globin genes.

One additional form of α-thalassemia deserves mention, as it is the only acquired form of thalassemia and is as yet poorly understood. In some individuals with myeloproliferative disorders, Hb H disease may appear. In the few cases studied molecular analysis has revealed a profound impairment of α-globin synthesis, approaching that seen in hydrops fetalis states. Molecular hybridization has demonstrated α-mRNA deficiency and apparently no rearrangements in the α-globin gene region (that is, no deletion) (Old *et al.*, 1977). What is particularly curious about this entity is the apparent *trans* effect as both chromosomes are inactivated with respect to α-globin expression. The nature of the alterations responsible for this unusual condition is not understood. Inroads into this problem may provide clues to the understanding of abnormal gene expression in various malignancies.

β-*Thalassemias*

β-Thalassemias refer to those conditions in which β chain synthesis is impaired. We will deal specifically here initially only with those disorders in which β chain synthesis alone is impaired. In a subsequent section we will deal with those conditions associated with defective synthesis of both β and other β-like globin chains.

The typical β-thalassemias may be subdivided simply on the basis of whether any β chains are produced from an affected locus. If no β chains are synthesized, the lesion is of the β^0 type. If some β-globin (albeit a reduced amount) is synthesized, the defect is called β^+. Prior to the use of restriction enzyme mapping and gene cloning, it was apparent that β-thalassemias were generally associated with a quantitative deficiency of β-globin mRNA (Kacian *et al.*, 1973; Housman *et al.*, 1973) and that β-globin gene sequences were largely present in the cellular DNA (Tolstoshev *et al.*, 1976). It was apparent as well that some individuals with β^0-thalassemias had nonfunctional β-mRNA sequences in their reticulocytes and bone marrow (Kan *et al.*, 1975a; Temple *et al.*, 1977; Benz *et al.*, 1978; Old *et al.*, 1978b).

In general, patients with either β^0- or β^+-thalassemias have normal appearing β-globin gene regions in the DNA by restriction mapping experiments (Orkin *et al.*, 1979b; Flavell *et al.*, 1979; Senno *et al.*, 1979). However, one type of rare exception to this statement will be described below. The "normalcy" of the β-globin gene region in β-thalassemias indicates that mutations responsible for β-thalassemias must be limited in size (if in fact, they are very close to or within the β genes themselves). We might readily envision mutations that might affect the transcription of the β-globin gene, the processing or stability of its mRNA transcripts, or the biological activity of the mRNA produced. We fully expect that well-documented examples of each class will soon be recognized at the current pace of work in this area.

β⁰-Thalassemias

The molecular defects in two varieties of β⁰-thalassemia have been well documented at this time. Chang and Kan (1979) characterized the β⁰-thalassemia of a Chinese individual in which biologically inactive β-mRNA was identified. By DNA sequence analysis of transcripts from his β⁰-mRNA, they demonstrated that a single nucleotide change in the codon for amino acid number 17 of the β chain created a terminator for translation (a nonsense mutation). In this instance, normal β-globin chains cannot be generated from this mRNA. *In vitro* suppression of this terminator was demonstrated using a yeast tRNA suppressor (Chang *et al.*, 1979). How common this form of β⁰-thalassemia is uncertain, but several positions within the β-globin sequence could be mutated to a chain terminator in theory. Since approximately 50% of β⁰-thalassemias may have some detectable, biologically inactive mRNA, many examples of this class of defect may be observed in further work.

A second, defined abnormality associated with β⁰-thalassemia is a partial deletion of the β-globin structural gene, so far characterized in only three Asian Indians. Among surveys of β⁰-thalassemia DNA samples using restriction enzyme mapping, Orkin *et al.* (1979b) and Flavell *et al.* (1979) observed these rare examples by noticing a shortened β-specific DNA fragment (Fig. 6). Restriction mapping suggested that a 0.6-kb deletion near the end of the β-globin gene was present in such patients. Interestingly, all patients studied to date are heterozygous for this defect; that is, the defect is seen on only one chromosome 11. The other chromosome in each instance has another β⁰-thalassemia gene that gives a normal restriction pattern. These studies provided direct evidence for heterogeneity of β⁰-thalassemia at the molecular level within a single individual. Further characterization of this molecular defect was achieved by gene cloning experiments. The structurally abnormal β-globin gene of one patient was cloned in bacteriophage λ and then examined directly by electron microscopy after hybridization to either β-globin mRNA or to restriction fragments of a normal β-globin gene (Orkin *et al.*, 1980a). In

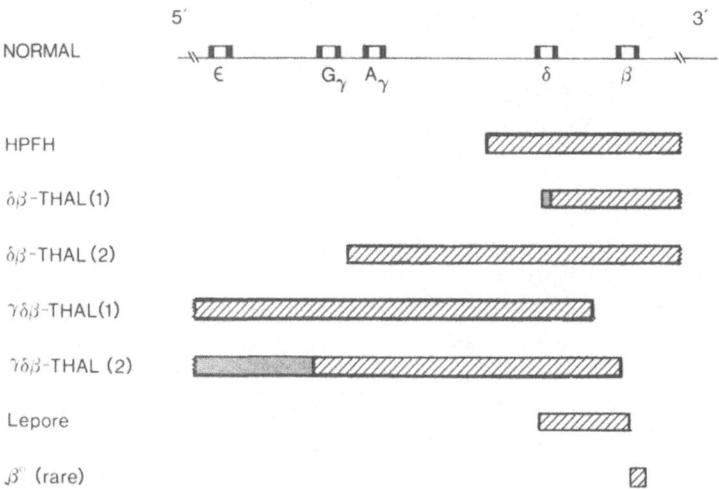

Fig. 6. Deletions within the β-like globin gene cluster in various thalassemia syndromes. Striped area indicates regions that are deleted; stipped area indicates regions that are probably deleted but for which precise data are lacking. Summarized from the studies of Fritsch *et al.* (1979); Orkin *et al.* (1979b,c, 1980a,b); Van der Ploeg *et al.* (1980b); and Flavell *et al.* (1978, 1979).

this manner, it was demonstrated that the last coding block (beyond amino acid 105) is entirely missing in this abnormal β-globin gene and that the deletion is entirely internal within the β-globin region. The deletion removes the terminal third of the last intervening sequence, the entire last coding segment, and (presumably) signals for termination of transcription and polyadenylation of mRNA transcripts located at the end of the β gene. Two patients with this form of β^0-thalassemia have no detectable β-globin mRNA in their reticulocytes by molecular hybridization indicating that neither the structurally normal nor the abnormal β gene is capable of producing mRNA. In view of other experiments in several systems that suggest that intervening sequences must be appropriately processed in order for RNA transcripts to exit from the nucleus and remain stable (Hamer and Leder, 1979c), it appears that deletion of intervening sequence splice junctions in this instance would prevent the formation of processed β-RNA transcripts in developing erythroid cells. It has not been ruled out that this deletion obliterates gene transcription itself by some unknown mechanism.

It is extremely likely that other defects will be discovered in β^0-thalassemias. We expect that these may include defects in gene tran-

scription, severe abnormalities of RNA processing probably due to mutations within intervening sequences, and mutations in DNA signals normally used for transcription termination or polyadenylation of RNA
transcripts. It is also conceivable that some forms of β^0-thalassemia result
from distant mutations, perhaps upstream from the β-globin gene, that
might modulate gene expression in, as yet, unexplained ways.

β^+-Thalassemias

In β^+-thalassemias the deficiency of β-globin production is not absolute. The β-globin chains that are synthesized appear normal. In addition, the impairment of β-globin chain production is correlated closely
with the relative deficiency of β-globin RNA measured in erythroid cells
(Benz et al., 1978). These findings suggest that the underlying defect in
β^+-thalassemia relates either to impaired transcription of the β gene or
to impaired processing of its transcripts. By measurement of steady-state
nuclear and cytoplasmic globin RNA levels in erythroid precursors, Nienhuis et al. (1977) provided some support for the notion that either impaired
processing or impaired transcript stability was the principal problem in
β^+-thalassemia.

Recent technical advances have provided the means for a fresh attack
on this problem. Both gene mapping and gene cloning experiments so far
have indicated that the structure of the β-globin gene is largely normal
in β^+-thalassemia. It is generally believed that specific nucleotide changes
within or around the β-globin gene will be found in association with β^+-
thalassemia. The formal proof of this contention is a formidable problem
in view of the size of the β-globin gene (approximately 1600 base pairs)
and the probable nucleotide sequence polymorphisms that exist among
individuals normally (about $\frac{1}{100}$) (Jeffreys, 1979). The finding by Ross and
colleagues that processing of globin RNA precursor molecules may occur
in multiple steps (Kinniburgh and Ross, 1979) suggests that all nucleotides
within the intervening sequences of thalassemic genes must be scrutinized
carefully and not just nucleotides near the intervening sequence/coding
region junctions.

Recently, several laboratories have obtained evidence to indicate
that the processing of β-globin RNA precursors is abnormal in at least
some patients with β^+-thalassemias. In the studies of Ross and his colleagues (Maquat et al., 1980), it appears that processing of the intranu-

clear β-globin RNA precursors is slow and inefficient (relative to α-globin RNA processing) in β⁺-thalassemia bone marrow cells. Similar data have been obtained by Kantor *et al.* (1980), and Benz *et al.* (1980). In some instances abnormal RNA-processing intermediates may be detectable by gel electrophoresis of erythroid precursor cell RNA samples. Furthermore, these studies have suggested that the δ-globin gene is transcribed at a lower rate than the β-globin gene, accounting largely for the differences in abundance of them in RNA products.

These recent studies have provided rather direct evidence that β⁺-thalassemia is most likely a "processing" disease (Leder, 1978), and suggest (but do not prove as yet) that the responsible mutations lie within intervening sequences. We believe that combined structure–function analyses of cloned β⁺-thalassemia genes, perhaps utilizing systems for the reintroduction of cloned genes into cells, will be required to identify the specific mutations that are the cause of these conditions. Again, in view of the clinical heterogeneity of thalassemias, we can expect that several mutations in various locations in or around the β-globin gene will be discovered in association with β⁺-thalassemia. In addition, although many patients with this entity may have a "processing" disease, it is also likely that some patients may have "transcription" diseases, perhaps related to mutations nearer promoter regions or perhaps upstream from them.

β-Thalassemias Associated with Deficiencies of Other β-like Chains

Less common forms of β-thalassemias are those in which the synthesis of more than one globin chain is affected (Weatherall and Clegg, 1972, 1979). These can be grouped into hereditary persistence of fetal hemoglobin syndromes, δβ-thalassemias and γδβ-thalassemias. These rare lesions are associated with deletions of DNA within the β-like globin gene cluster. It is worthwhile to consider these entities in detail, as further study of individuals with variations on the theme may permit location of intergenic regulators of globin gene expression or fetal switching, if they exist.

δβ-Thalassemia refers to absence of β- and δ-globin chain production from an affected locus. This defect is found distributed among all racial groups. Homozygotes for δβ-thalassemia are rare. Liquid hybridization of globin cDNA and sheared cellular DNA previously demonstrated a deficiency of β and β-like sequences in several samples (Ottolenghi *et al.*, 1976; Ramirez *et al.*, 1976). Prior to gene mapping methods it was not possible to determine the full extent of the deletion and, particularly, whether any residual β or δ sequences were present and whether all cases were similar. Since homozygotes for this condition do not synthesize β or δ chains, only fetal hemoglobin is present in their blood. This fetal hemoglobin could be composed of either or both γ chains ($^G\gamma/^A\gamma$). Most homozygotes for δβ-thalassemia are of the $^G\gamma/^A\gamma$ type (Amin *et al.*, 1979); that is, both γ-globin chains are present in the Hb F. Several homozygotes have been identified of the $^G\gamma$ type (i.e., no $^A\gamma$ production) (Amin *et al.*, 1979; Orkin *et al.*, 1979c).

A similar, and perhaps overlapping entity, is hereditary persistence of fetal hemoglobin syndrome (HPFH). In its best characterized variety, no β or δ chains are produced from the affected chromosome. Homozygotes for HPFH have only Hb F in their blood, of the $^G\gamma/^A\gamma$ type. Homozygotes have all been of Black ancestry. Like δβ-thalassemia liquid hybridization studies have revealed deficient β sequence hybridization (Forget *et al.*, 1976; Kan *et al.*, 1975c). The phenotypes of δβ-thalassemia and HPFH are distinguishable very easily. In HPFH individuals there is no clinical disease. The affected homozygotes produce only Hb F, but are polycythemic, rather than anemic. Globin chain synthetic studies, however, have revealed mild globin chain imbalance (γ/α synthesis equal to 0.65) and the erythrocytes are variably microcytic. In δβ-thalassemia, however, usually the affected homozygotes have mild to moderately severe disease, sometimes referred to as thalassemia intermedia. These patients are anemic, microcytic, and may intermittently require blood transfusions. Globin chain synthetic studies generally reveal greater globin chain imbalance. Therefore, in δβ-thalassemia γ-globin synthesis, although markedly increased over that present in most individuals with more typical β-thalassemia lesions, is not sufficiently augmented to compensate fully for absent β-globin production. In HPFH the compensation by γ-globin gene expression is more nearly complete. Among δβ-thalassemias there is considerable clinical variation, such that in some instances the distinction between HPFH and δβ-thalassemia may be quite difficult (Wood *et al.*, 1979; Clegg *et al.*, 1979).

The differences in compensatory increases in γ-globin synthesis in HPFH and δβ-thalassemia syndromes have suggested that differences in DNA deletions might be responsible if regulatory elements in intergenic regions are important in controlling relative globin gene expression. Huisman *et al.* (1974) have advanced the hypothesis that a region upstream from the δ locus might be required for turning off γ-globin expression postnatally. Deletion of this region in HPFH, but not δβ-thalassemia, might explain the clinical differences. The available data on gene deletions within the β-globin cluster do not permit definitive conclusions with respect to this model, as is discussed below.

An even rarer condition, γδβ-thalassemia, has been described in two families (Kan *et al.*, 1972; Van der Ploeg *et al.*, 1980*b*). In this syndrome, the affected chromosome produces no γ-, δ-, or β-globin chains. In the homozygous state, this condition would be lethal *in utero*. In the heterozygotes identified there is a hemolytic anemia in the neonatal period that evolves into β-thalassemia trait after the first few months of life. Globin chain synthetic studies reveal the expected globin chain imbalance. Originally it was hypothesized that this condition resulted from inactivation of the whole γδβ-globin complex, most likely due to a deletion of the DNA (Kan *et al.*, 1972).

Restriction endonuclease mapping of the DNAs of individuals with these complex forms of β-thalassemia has permitted some distinctions at the molecular level and delineation of the extents of DNA deletions in specific situations. In HPFH, the β- and δ-globin structural genes are deleted in their entirety (Orkin *et al.*, 1978; Mears *et al.*, 1978*a*; Fritsch *et al.*, 1979; Bernards and Flavell, 1980; Tuan *et al.*, 1980; Ottolenghi *et al.*, 1979). The 5′ or upstream end of the deletion is approximately 2–3 kb before the δ-globin gene position in two patients (Fritsch *et al.*, 1979) and 7–9 kb upstream in another (Bernards and Flavell, 1980; Tuan *et al.*, 1980). The end of deletion past the β-globin locus is unknown. The γ-globin gene region is normal in HPFH individuals. δβ-Thalassemia homozygotes so far fall into two general classes. In δβ-thalassemia of the $^{G}\gamma$/$^{A}\gamma$ type (δβ-1), a deletion originating 3′ to the β-globin gene terminates within the large intervening sequence of the δ-globin gene (Mears *et al.*, 1978*a*; Fritsch *et al.*, 1979; Ottolenghi *et al.*, 1979). Again, the γ-globin region is normal. In δβ-thalassemia of the $^{G}\gamma$ type (δβ-2) (in one patient studied to date) a deletion removes the β-, δ-, and $^{A}\gamma$-globin genes in their entirety (Fritsch *et al.*, 1979; Orkin *et al.*, 1979*c*). The deletion terminates near to the $^{G}\gamma$ gene. In a Dutch family with γδβ-thalassemia gene mapping demonstrated a deletion that involved the γ-globin genes and the δ-globin gene, but terminated upstream (about 2 kb) from the β-globin locus (Van der Ploeg *et al.*, 1980*b*). This was particularly curious, as the β-globin gene region was grossly normal yet gave rise to no β-globin chain production. Either this was due to the effects of such an upstream deletion on globin gene expression or due to coincident presence of a β-thalassemia defect on the same chromosome. As yet, these possibilities have not been distinguished. Finally, in the original family reported with γδβ-thalassemia, gene mapping revealed deletion of γ- and some δ-globin sequences, but could not unequivocably identify the structure of the β-globin gene region (Orkin *et al.*, 1980*b*). To ascertain the limits of the deletion near the β-globin gene, the β-globin gene region of this family was cloned and studied directly. This approach demonstrated that the deletion in this family with γδβ-thalassemia terminated with the β-globin coding sequence itself at the position for amino acid 65 of the β sequence (Orkin *et al.*, 1980*b*). These deletions, as well as one crossover form of β-thalassemia (see below) are reviewed in Fig. 6.

Less dramatic forms of the HPFH syndrome are more common. One type, commonly referred to as Swiss HPFH, probably results from a genetic alteration that increases the progenitors programmed to production of Hb F (F-cells) (Wood *et al.*, 1979). Presumably, no deletion in the γδβ-complex is present in the DNA of individuals with this entity.

An additional form of HPFH, called Greek HPFH, found only in the heterozygous state to date, has no detectable deletion in the γδβ region by gene mapping (Bernards and Flavell, 1980; Tuan *et al.*, 1980). The molecular defect probably involves a very limited alteration in an important controlling or modulating state.

Several points emerge from these studies of deletions within the β-globin gene region. First and foremost, there is remarkable genetic heterogeneity in these conditions at the molecular level. Similar phenotypes may have quite different DNA lesions (note γδβ-thalassemias). Second, it is not immediately evident how and if the current picture of these deletions explains the differences in gene expression compensation. For example, if a regulator responsible for shutting off γ gene expression were located upstream from the δ locus (Huisman *et al.*, 1974), we would anticipate that ^{G}γδβ-thalassemia (δβ-thal-2) should resemble HPFH. In fact, the clinical features of δβ-thalassemia of the ^{G}γ and ^{G}γ/^{A}γ types are very similar (Amin *et al.*, 1979). One can argue that deletion of the ^{A}γ genes in the ^{G}γδβ-thalassemia reduces γ gene compensation for β deficiency. This may be the case. Nevertheless, it is not possible to use the current mapping results to support very strongly these regulator hypotheses. It is important to mention that the 3' extents of the HPFH and δβ-thalassemia deletions are largely uncharted yet. It is conceivable that regulators might be located 3' to the β-globin gene, if we are willing to accept downstream regulators of γ gene expression. It appears that further study of additional individuals with these rare conditions will be required to help sort out the various possibilities. An alternate, interesting model proposed by Bernards and Flavell (1980) invokes the notion of chromosomal domains and suggests that deletions in the γδβ region may alter transcription by changing the conformation of chromatin quite widely. We expect that *in vitro* systems for the study of globin gene switching using large genomic segments of the β complex may be required if selected hypotheses are to be subjected to formal testing.

Thalassemias Associated with Structurally Abnormal Globins

Specific structurally abnormal globin chains are associated with thalassemia. The Lepore chain is an abnormal β-like globin chain that results from unequal crossing-over within the δ- and β-globin genes (Weatherall and Clegg, 1979) (see Fig. 6). This fusion globin chain is produced at a reduced rate and represents a β-thalassemia defect, as normal β-globin is not produced from the affected chromosome. In combination with a β-thalassemia defect on the other chromosome a severe variety of β-thalassemia is generally seen.

A different type of abnormal globin chain is found frequently among individuals with α-thalassemia. These α-globin chains are elongated products that result from mutation of a normal terminator codon of the α sequence. During translation the 3'-untranslated region is translated into an additional peptide segment at the C-terminal end of the α chain. Hb Constant Spring and Icaria are of this type (Clegg *et al.*, 1971; Clegg and Tsevrenis, 1974). Each mimics an α-thalassemia lesion in that these abnormal α chains are synthesized at very low rates (for reasons that are not entirely clear). In association with α-thalassemia trait on the other chromosome (α-thal-1), Hb H disease occurs.

Finally, as exemplified by Hb Indianapolis, a single amino acid change in the β chain may rarely lead to a thalassemia phenotype due to unusual instability of the affected product (Adams *et al.*, 1979). Although this unstable hemoglobin is unique in this regard, it further underscores the heterogeneity at the molecular level for thalassemia defects. Individuals heterozygous for Hb Indianapolis have the phenotype of severe β-thalassemia trait. In fact, other unstable β-chain variants may behave in a similar fashion, although not to the same degree.

DNA POLYMORPHISMS IN THE β-LIKE GLOBIN GENE CLUSTER

Although the DNA of each individual is different from that of another, most DNA sequences appear to be conserved. This is reflected in consistent restriction enzyme patterns for most gene sequences among unrelated individuals. However, we might expect that wide surveys of human populations would reveal sequence differences concentrated in specific groups, and perhaps randomly distributed as well. Such is indeed the case. In their initial cloning studies of the β-globin complex, Maniatis and co-workers noted a difference between two cloned δ-globin regions with respect to digestion with the enzyme Pst I (Lawn *et al.*, 1978). In one instance, the normal δ-specific fragment was found. In the other clone, Pst cleaved within the large intervening sequence of the δ-globin gene. These findings implied that the two clones represented different chromosomes and that a polymorphism for Pst cleavage might be found among humans. Subsequent to these findings, Jeffreys (1979) identified this polymorphism for Pst cleavage among normal human DNA samples. Polymorphisms for Hind III cleavage within the γ-globin genes and for Kpn I neighboring the β-globin gene have been described. The Hind III polymorphism appears particularly common among unrelated individuals and may be potentially useful in prenatal diagnosis of thalassemias and sickle cell anemia (see below).

Of particular interest are polymorphisms that might be linked to particular normal or abnormal genes. Kan and Dozy (1978a) provided the pivotal discovery in this area when they identified a restriction site polymorphism for the enzyme Hpa I 3′ to the β-globin gene. In their initial survey the presence of a 13-kb Hpa I β fragment was very often found in association with the sickle β gene, whereas normal β-globin genes were located in a 7.6-kb fragment. This linkage disequilibrium could be used in prenatal diagnosis of sickle cell anemia by fetal DNA study (see below) in families in which this polymorphism existed. Further studies have shown that the 13-kb fragment is only associated with the sickle β gene in about 50–60% east coast American blacks, as opposed to about 75–80% west coast blacks (Feldenzer *et al.*, 1979). Furthermore, the 13-kb fragment may rarely contain a normal β-globin gene. The presence of this polymorphism appears to be useful in defining the origins of the sickle globin gene in that selected populations in which the sickle gene is found vary with respect to the frequency of the Hpa I polymorphism. For example, Asian Indians with the sickle gene do not have the polymorphism, whereas West Africans do (Kan and Dozy, 1979; Kan and Dozy, 1980). The Hpa I polymorphisms also appear to be associated with the Hb C gene in black individuals (Kan and Dozy, 1979; Feldenzer *et al.*, 1979). These fascinating differences now permit examination of evolution of human populations by DNA analysis.

Kan *et al.* (1980*a*) also identified a polymorphism for a Bam HI site 3′ to the β-globin gene that can be used to exclude β-thalassemia in a minority of Sardinian individuals. Although one-third of the normal Sardinian population has a 22.0-kb β-specific Bam fragment, all patients with β⁰-thalassemia have a 9.3-kb fragment.

PRENATAL DIAGNOSIS OF HEMOGLOBIN DISORDERS BY DNA ANALYSIS

In view of the severity of some thalassemias and hemoglobinopathies there has been considerable interest and effort directed toward providing and improving the prenatal diagnosis of hemoglobin disorders. Hemoglobinopathies (sickle cell anemia and thalassemias) can be detected prenatally by globin chain synthesis studies of fetal blood samples obtained by fetoscopy or placental aspiration. As these procedures and the results of these approaches have been extensively reviewed (Nathan *et al.*, 1979), we will not deal with them further except to say that prenatal diagnosis by these means is now well established and in widespread use. To date over 1000 cases have been analyzed during pregnancies at risk for hemoglobinopathies (Alter, personal communication, 1980).

Analysis of DNA obtained from amniotic fluid cells (amniotic fibroblasts) offers great promise for the prenatal detection of hemoglobin disorders (Orkin *et al.*, 1978). Acquisition of these fetal fibroblasts is safe and avoids the necessity to employ specially trained obstetricians to obtain fetal blood samples. Kan and co-workers first utilized fetal DNA samples to detect α-thalassemias by DNA–DNA solution hybridization.

They demonstrated that deletions associated with hydrops fetalis and Hb H disease states could be distinguished by that method. A disadvantage of these approaches, however, was the need for rather large quantities of DNA (about 100–200 μg) in order to obtain accurate hybridization data. The advent of restriction endonuclease gene mapping methods increased the sensitivity of prenatal diagnosis of α-thalassemia and extended its usefulness to other hemoglobin disorders. In the case of gene mapping studies only about 10–20 μg DNA at a minimum is required. In fact, it is often possible to centrifuge the fetal cells present in amniotic fluid samples directly to recover sufficient material for one or two assays. In other situations, it is advisable to culture the fibroblast cells for three weeks or so to obtain greater amounts of DNA for repeated analyses. By restriction endonuclease analysis it is possible to detect α-thalassemias and δβ-thalassemias prenatally, as well as Lepore defects and rare β^0-thalassemias associated with partial gene deletion. Furthermore, the single nucleotide change in the codon for amino acid 121 of the β chain that results in $Hb^{0\text{-Arab}}$ can be detected by the absence of an intragenic Eco RI cleavage site (Flavell et al., 1978; Little et al., 1980b). Except for α-thalassemias among Asians deletion forms of thalassemia are quite rare and prenatal diagnosis by DNA analysis would have only limited application. However, the use of DNA polymorphisms (discussed above) now permit much wider use of these new technologies. In the case of sickle cell anemia use of the Hpa I polymorphism alone should be useful about 25–50% of the time in providing definitive prenatal diagnoses (Kan and Dozy, 1978b). In theory application of additional polymorphisms would be expected to increase the success rate in this approach. In fact, Phillips et al. (1980a) have recently shown that use of the Hind III polymorphism in the γ-globin genes in conjunction with the Hpa I polymorphism allows accurate diagnosis in a greater percentage of families at risk for sickle cell anemia.

As mentioned above, a polymorphism for Bam cleavage 3′ to the β-globin gene can sometimes be used to ascertain the presence or absence of β-thalassemia (Kan et al., 1980a). We should expect that application of this marker, perhaps in conjunction with the frequent Hind III polymorphism, will allow prenatal diagnosis of β-thalassemias by DNA analysis. Use of the Hind III polymorphism for prenatal diagnosis of β-thalassemia has already been reported by Little et al. (1980a).

It is apparent that identification of other polymorphisms in the β-like globin cluster will extend further the application of restriction mapping in prenatal diagnosis. In view of the numerous possibilities in each clinical situation it will be especially important to study families at risk for sickle cell anemia and thalassemia syndromes as early as possible, most favorably prior to pregnancy. In this manner, appropriate marker linkage can be established. This process will generally necessitate study of other family members and offspring in order to be certain of the restriction sites on each chromosome. Once these data are acquired, the actual prenatal diagnosis can be accomplished with greater confidence. In many instances DNA analysis may not be able to provide definitive genotype assignments, but may allow exclusion of homozygosity and, therefore, obviate the need for fetal blood samplings.

ADDITIONAL APPROACHES TO THE CORRELATION OF GENE STRUCTURE AND FUNCTION

New molecular and genetic procedures may be expected to yield important information regarding normal and abnormal gene expression in the very near future. These include various methods for the reintroduction of cloned genes into living cells and assays for transcription of cloned genes *in vitro*. Introduction of cloned genes into cells via SV40 viral DNA vectors should permit assessment of essential features of gene expression and RNA processing (Hamer and Leder, 1979*a,b*). When normal mouse globin chromosomal genes are introduced into monkey kidney cells in culture in this manner, normal mouse globin chains are produced. Gene transcription and RNA biogenesis occur normally. We expect that this system will be particularly useful in the analysis of mutant globin genes, and may permit controlled study of transcriptional and processing mutations. Apparent integration of foreign genes in fibroblast DNA can be achieved by DNA transformation (Wigler *et al.*, 1979). The introduction of rabbit and human β-globin genes into fibroblasts can be accomplished. It appears, however, that expression of the transformed β genes may not be faithful in these cells as transcription appears not to start at the appropriate site in the globin gene (Wald *et al.*, 1979). Although this is somewhat disappointing, it is conceivable that further developments in this area may permit more directed introduction of genes stably into cells. Further speculation regarding genetic engineering as a cure for hemoglobin disorders is premature until the directed insertion of genes into proper environments in the DNA is regularly possible.

Finally, another important new development sure to have impact on analysis of hemoglobin disorders is establishment of *in vitro* assays of gene transcription. Previous attempts to transcribe chromatin *in vitro* using exogenous RNA polymerases met with numerous artifacts and pitfalls. Recently, two laboratories have independently designed *in vitro* systems capable of accurate initiation of transcription of cloned genes (Weil *et al.*, 1979; Manley *et al.*, 1980). In the case of mouse and human globin genes these systems synthesize RNA transcripts that are initiated normally. Termination of transcription and RNA processing do not appear to occur in these systems. It is anticipated that use of these assays will be useful, perhaps in combination with chemical mutagenesis *in vitro*, in defining important nucleotides in putative promoters and neighboring controlling regions and in identifying thalassemias resulting from mutations affecting gene transcription.

SUMMARY AND CONCLUSIONS

Over the past two or three years extensive progress has been made in determining the structure of human globin genes and their organization in the cellular DNA. Coincident with these developments has been rapid application of this new knowledge to study of hemoglobin disorders, primarily the thalassemia syndromes. At present, many different genetic lesions associated with thalassemias have been identified (see Table IV) and, certainly, more are to be documented in the near future. Without a doubt the approaches already employed will serve as models for the study of other human diseases. Fortunately, the basic findings in the molecular genetics of human hemoglobins have had rapid impact in the prenatal diagnosis of both thalassemias and sickle cell anemia. Again, at the current rate of progress we should expect that DNA analysis may supplant fetal blood sampling in the not too distant future as the standard method for prenatal diagnosis of these conditions.

When new insights into old problems are gained unexpectedly they are especially appreciated. This has been the case in the study of hemoglobin molecular genetics as well. For example, consideration of the composition of the α gene region of human DNA and the defects in α-thalassemias has provided insights into the instability of particular regions of the human genome. It is likely that such situations are not so rare and that many parts of the human genome are constantly evolving at a high

TABLE IV. Examples of Genetic Lesions Established in
Thalassemias

1. Deletion of entire globin structural genes
   α genes—α-thalassemias
   δ and β genes—HPFH syndrome and $\delta\beta$-thalassemia-1
   $^A\gamma$, δ, β genes—$^G\gamma$-$\delta\beta$-thalassemia
   γ and δ genes—$\gamma\delta\beta$-thalassemias
2. Partial deletion of globin structural genes
   α gene—α-thalassemia gene in "2.6-kb Eco RI" fragment
   β gene—β^0-thalassemia in Asian Indians
3. Biologically inactive globin mRNA
   β-mRNA—nonsense mutation in β^0-thalassemia in a Chinese
   patient
4. Inefficient globin mRNA precursor processing
   β-mRNA precursor—β^+-thalassemias (? intervening sequence
   mutations)
5. Thalassemias associated with abnormal globins
   Lepore chain—$\delta\beta$ crossover chains
   Elongated α chains—Hb Constant Spring
   Unstable β chain—Hb Indianapolis

rate due to unequal crossing-over mechanisms. In addition, the recognition of restriction endonuclease cleavage sites as markers of chromosome regions and interesting genes now allows reconsideration of population evolution and dynamics. We expect that further application of the new genetic techniques to the study of human globin genes will provide additional, unsuspected insights and stimulate analysis of other human loci.

ACKNOWLEDGMENTS. We thank Ms. Catherine Lewis for the preparation of this manuscript. Studies in our laboratory have been supported by grants from the National Institutes of Health and the March of Dimes-National Foundation. Dr. Orkin is the recipient of a Research Career Development Award of the NIH.

REFERENCES

Adams, J. G., III, Boxer, L. A., Baehner, R. L., Forget, B. G., Tsistrakis, G. A., and Steinberg, M H., 1979, Hemoglobin Indianapolis (β112[G14]arginine), *J. Clin Invest.* **63**:931–938.

Alwine, J. C., Kemp, D. J., and Stark, G. R., 1977, Method for detection of specific RNAs in agarose gels by transfer to diazobenzyloxymethyl-paper and hybridization with DNA probes, *Proc. Natl. Acad. Sci. USA* **74**:5350–5354.

Amin, A. B., Pandya, N. L., Diwen, P. P., Darbre, P. D., Kattamis, C., Metaxaton-Mavromati, A., White, J. M., Wood, W. G., Clegg, J. B., and Weatherall, D. J., 1979, A comparison of the homozygous states for $^G\gamma$ and $^G\gamma^A\gamma\delta\beta$-thalassaemia, *Br. J. Haematol.* **43**:537–548.

Arnheim, N., and Kuehn, M., 1979, The genetic behavior of a cloned mouse ribosomal DNA segment mimics mouse ribosomal gene evolution, *J. Mol. Biol.* **134**:743–765.

Baralle, F. E., 1977, Complete nucleotide sequence of the 5' noncoding region of human α- and β-globin mRNA, *Cell* **12**:1085–1095.

Bastos, R. N., and Aviv, H., 1977, Globin mRNA precursor molecules: Biosynthesis and processing in erythroid cells. *Cell* **11**:641–650.

Benoist, C., O'Hare, K., Breathnach, R., and Chambon, P., 1980, The ovalbumin gene-sequence of putative control regions, *Nucleic Acids Res.* **8**:127–142.

Benton, W. D., and Davis, R. W., 1977, Screening λgt recombinant clones by hybridization to single plaques *in situ, Science* **196**:180–182.

Benz, E. J., Jr., Forget, B. G., Hillman, D. G., Cohen-Solal, M., Pritchard, J., Cavallesco, C., Prensky, W., and Housman, D., 1978, Variability in the amount of β-globin mRNA in β^0-thalassemia, *Cell* **14**:299–312.

Bernards, R., and Flavell, R. A., 1980, Physical mapping of the globin gene deletion in hereditary persistence of foetal haemoglobin (HPFH), *Nucleic Acids Res.* **8**:1521–1534.

Bernards, R., Little, P. F. R., Annison, G., Williamson, R., and Flavell, R. A., 1979, Structure of the human $^G\gamma$-$^A\gamma$-δ-β-globin gene locus, *Proc. Natl. Acad. Sci. USA* **76**:4827–4831.

Blattner, F. R., Williams, B. G., Blechl, A. E., Thompson, K. D., Faber, H. E., Furlong, L. A., Grunwald, D. J., Kiefer, D. O., Moore, D. D., Schumm, J. W., Shelton, E. L., and Smithies, O., 1977, Charon phages: Safer derivatives of bacteriophage lambda for DNA cloning, *Science* **196**:161–169.

Blattner, F. R., Blechl, A. E., Denniston-Thompson, K., Faber, H. E., Richards, J. E., Slighton, J. L., Tucker, P. W., and Smithies, O., 1978, Cloning human fetal γ globin and mouse α-type globin DNA: Preparation and screening of shotgun collections, *Science* **202**:1279–1284.

Breathnach, R., Benoist, C., O'Hare, K., Gannon, F., and Chambon, P., 1978, Ovalbumin gene: Evidence for a leader sequence in mRNA and DNA sequences at the exon-intron boundaries, *Proc. Natl. Acad. Sci. USA* **75**:4853–4857.

Chang, J. C., and Kan, Y. W., 1979, β^0-thalassemia: A nonsense mutation in man. *Proc. Natl. Acad. Sci. USA* **76**:2886–2889.

Chang, J. C., Temple, G. F., Trecartin, R. F., and Kan, Y. W., 1979, Suppression of the nonsense mutation in homozygous β^0-thalassaemia, *Nature (London)* **281**:602–603.

Clegg, J. B., and Tsevrenis, H., 1974, Haemoglobin Icaria, a new chain termination mutant which causes α-thalassaemia, *Nature (London)* **251**:245–247.

Clegg, J. B., Weatherall, D. J., and Milner, P. F., 1971, Haemoglobin Constant Spring—a chain termination mutant?, *Nature (London)* **234**:337–340.

Clegg, J. B., Metaxaton-Mavromati, A., Kattamis, C., Sofroniadou, K., Wood, W. G., and Weatherall, D. J., 1979, Occurrence of $^G\gamma$ HbF in Greek HPFH: Analysis of heterozygotes and compound heterozygotes with β-thalassaemia, *Br. J. Haematol.* **43**:521–536.

Craik, C. S., Buchman, S. R., and Beychok, S., 1980, Characterization of globin domains: Heme binding to the central exon product, *Proc. Natl. Acad. Sci. USA* **77**:1384–1388.

Curtis, P. J., Mantei, N., van den Berg, J., and Weissmann, C., 1977, Presence of a putative

15S precursor to β-globin mRNA but not to α-globin mRNA in Friend cells, *Proc. Natl. Acad. Sci. USA* **74**:3184–3188.

Deisseroth, A., Nienhuis, A., Turner, P., Velez, R., Anderson, W. F., Ruddle, F., Laurence, J., Creagen, R., and Kucherlapati, R., 1977, Localization of the human α-globin structural gene to chromosome 16 in somatic cell hybrids by molecular hybridization, *Cell* **12**:205–218.

Deisseroth, A., Nienhuis, A., Laurence, J., Giles, R., Turner, P., and Ruddle, F., 1978, Chromosomal location of human β-globin gene on human chromosome 11 in somatic cell hybrids, *Proc. Natl. Acad. Sci. USA* **75**:1456–1460.

Denhardt, D. T., 1966, A membrane-filter technique for the detection of complementary DNA, *Biochem. Biophys. Res. Commun.* **23**:641–646.

Dodgson, J. B., Strommer, J., and Engel, J. D., 1979, Isolation of the chicken β-globin gene and a linked embryonic β-like globin gene from a chicken DNA recombinant library, *Cell* **17**:879–887.

Doolittle, W. F., 1978, Genes in pieces, were they ever together?, *Nature* **272**:581–582.

Dozy, A. M., Kan, Y. W., Embury, S. H., Mentzer, W. C., Wang, W. C., Lubin, B., Davis, J. R., Jr., and Koenig, H. M., 1979, α-Globin gene organisation in blacks precludes the severe form of α-thalassaemia, *Nature (London)* **280**:605–607.

Duncan, C., Biro, P. A., Choudary, P. V., Elder, J. T., Wang, P. R. C., Forget, B. G., deRiel, J. K., and Weissman, S. M., 1979, RNA polymerase III transcriptional units are interspersed among human non-α globin genes, *Proc. Natl. Acad. Sci. USA* **76**:5095–5099.

Eaton, W. A., 1980, The relationship between coding sequences and function in haemoglobin, *Nature (London)* **284**:183–185.

Efstratiadis, A., Kafatos, F. C., and Maniatis, T., 1977, The primary sequence of rabbit β-globin mRNA as determined from cloned DNA, *Cell* **10**:571–585.

Embury, S. H., Lebo, R. V., Dozy, A. M., and Kan, Y. W., 1979, Organization of the α-globin genes in the Chinese α-thalassemia syndromes, *J. Clin. Invest.* **63**:1307–1310.

Embury, S. H., Miller, J. A., Dozy, A. M., Kan, Y. W., Chan, V., and Todd, D., 1980, Two different molecular organizations account for the single α-globin gene of the α-thalassemia-2-genotype, *J. Clin. Invest.* **66**:1319–1325.

Engel, J. D., and Dodgson, J. B., 1980, Analysis of the closely linked adult chicken α-globin genes in recombinant DNAs, *Proc. Natl. Acad. Sci. USA* **77**:2596–2600.

Feldenzer, J., Mears, G., Burns, A. L., Natta, C., and Bank, C., 1979, Heterogeneity of DNA fragments associated with the sickle-globin gene, *J. Clin. Invest.* **64**:751–755.

Flavell, R. A., Kooter, J. M., DeBoer, E., Little, P. F. R., and Williamson, R., 1978, Analysis of the human β-δ-globin gene loci in normal and Hb Lepore DNA: Direct determination of gene linkage and intergene distance, *Cell* **15**:25–41.

Flavell, R. A., Bernards, R., Kooter, J. M., DeBoer, E., Little, P. F. R., Annison, G., and Williamson, R., 1979, The structure of the human β-globin gene in β-thalassemia, *Nucleic Acids Res.* **6**:2749–2760.

Forget, B. G., Hillman, D. G., Lazarus, H., Borell, E. F., Benz, E. J., Jr., Caskey, C. T., Huisman, T. H. J., Schroeder, W. A., and Housman, D., 1976, Absence of messenger RNA and gene DNA for beta globin chains in hereditary persistence of fetal hemoglobin, *Cell* **7**:323–329.

Forget, B. G., Wilson, J. T., Wilson, L. B., Cavallesco, C., Reddy, V. B., deRiel, J. K., Biro, P. A., Ghosh, P. K., and Weissman, S. M., 1980, Globin mRNA structure: General features and sequence homology, in: *Cellular and Molecular Regulation of Hemoglobin Switching* (G. Stamatoyannopoulos and A. W. Nienhuis, eds), pp. 569–593, Grune and Stratton, New York.

Fritsch, E. F., Lawn, R. M., and Maniatis, T., 1979, Characterisation of deletions which affect the expression of fetal globin genes in man, *Nature* (*London*) **279:**598–603.

Fritsch, E. F., Lawn, R. M., and Maniatis, T., 1980, Molecular cloning and characterization of the human β-like globin gene cluster, *Cell* **19:**959–972.

Gilbert, W., 1978, Why genes in pieces?. *Nature* (*London*) **271:**501–502.

Ginder, G. D., Wood, W. I., and Felsenfeld, G., 1979, Isolation and characterization of recombinant clones containing the chicken adult β-globin gene, *J. Biol. Chem.* **254:**8099–8102.

Goossens, M., Dozy, A. M., Embury, S. H., Zachariades, Z., Hadjiminas, M. G., Stamatoyannopoulos, G., and Kan, Y. W., 1980, Triplicated α-globin loci in humans, *Proc. Natl. Acad. Sci. USA* **77:**518–521.

Gough, N. M., Kemp, D. J., Tyler, B. M., Adams, J. M., and Cory, S., 1980, Intervening sequences divide the gene for the constant region of mouse immunoglobulin μ chains into segments, each encoding a domain, *Proc. Natl. Acad. Sci. USA* **77:**554–558.

Grosschedl, R., and Birnstiel, M. L., 1980, Identification of regulatory sequences in the prelude sequences of an H2A histone gene by the study of specific deletion mutants *in vivo, Proc. Natl. Acad. Sci. USA* **77:**1432–1436.

Gusella, J., Varsanyi-Breiner, A., Kao, F.-T., Jones, C., Puck, T. T., Keys, C., Orkin, S., and Housman, D., 1979, Precise localization of human β-globin gene complex on chromosome 11, *Proc. Natl. Acad. Sci. USA* **76:**5239–5243.

Hamer, D., and Leder, P., 1979*a*, Expression of the chromosomal mouse β^{maj}-globin gene cloned in SV40, *Nature* (*London*) **281:**35–40.

Hamer, D., and Leder, P., 1979*b*, SV40 recombinants carrying a functional RNA splice junction and polyadenylation site from the chromosomal mouse β^{maj}-globin gene, *Cell* **17:**737–747.

Hamer, D., and Leder, P., 1979*c*, Splicing and the formation of stable RNA, *Cell* **18:**1299–1302.

Hamer, D. H., Smith, K. D., Boyer, S. H., and Leder, P., 1979, SV40 recombinants carrying rabbit β-globin gene coding sequences, *Cell* **17:**725–735.

Hardison, R. C., Butler, E. T., III, Lacy, E., Maniatis, T., Rosenthal, N., and Efstratiadis, A., 1979, The structure and transcription of four linked rabbit β-like globin genes, *Cell* **18:**1285–1297.

Higgs, D. R., Old, J. M., Pressley, L., Clegg, J. B., and Weatherall, D. J., 1980, A novel α-globin gene arrangement in man, *Nature* (*London*) **284:**632–635.

Hollan, S. R., Szelenyi, J. G., Brimhall, B., Duerst, M., Jones, R. T., Koler, R. D., and Stocklenz, A., 1972, Multiple alpha chain loci for human haemoglobins: Hb J-Buda and Hb G-Pest, *Nature* (*London*) **235:**47–50.

Housman, D., Forget, B. G., Skoultchi, A., and Benz, E. J., Jr., 1973, Quantitative deficiency of chain specific globin messenger ribonucleic acids in the thalassemia syndromes, *Proc. Natl. Acad. Sci. USA* **70:**1809–1813.

Huisman, T. H. J., Schroeder, W. A., Etremer, G. D., Dume, H., Miller, A., Brodie, A., Shelton, J. R., Shelton, J. B., and Apell, G., 1974, The present status of the heterogeneity of fetal hemoglobin in β-thalassemia: An attempt to unify some observations in thalassemia and related disorders, *Ann. N.Y. Acad. Sci.* **232:**107–122.

Hutchison, C. A., III, Phillips, S., Edgell, M. H., Gillam, S., Jahnke, P., and Smith, M., 1978, Mutagenesis at a specific position in a DNA sequence, *J. Biol. Chem.* **253:**6551–6560.

Jeffreys, A. J., 1979, DNA sequence variants in the ^{G}γ-, ^{A}γ-, δ-, and β-globin genes in man, *Cell* **18:**1–10.

Jeffreys, A. J., and Flavell, R. A., 1977a, A physical map of the DNA regions flanking the rabbit β-globin gene, *Cell* **12**:429–439.

Jeffreys, A. J., and Flavell, R. A., 1977b, The rabbit β-globin gene contains a large insert in the coding sequence, *Cell* **12**:1097–1108.

Jelinek, W. R., Toomey, T. P., Leinwand, L., Duncan, C. H., Biro, P. A., Chondary, P. V., Weissman, S. M., Rubin, C. M., Houck, C. M., Deininger, P. L., and Schmid, C. W., 1980, Ubiquitous, interspersed repeated sequences in mammalian genomes, *Proc. Natl. Acad. Sci. USA* **77**:1398–1402.

Kacian, D. L., Gambino, R., Dow, L. W., Grossbard, E., Natta, C., Ramirez, F., Spiegelman, S., Marks, P. A., and Bank, A., 1973, Decreased globin messenger RNA in thalassemia detected by molecular hybridization, *Proc. Natl. Acad. Sci. USA* **70**:1886–1890.

Kan, Y. W., and Dozy, A. M., 1978a, Polymorphism of DNA sequence adjacent to the human β-globin structural gene: Relationship to sickle mutation, *Proc. Natl. Acad. Sci. USA* **75**:5631–5635.

Kan, Y. W., and Dozy, A. M., 1978b, Antenatal diagnosis of sickle cell anemia by DNA analysis of amniotic fluid cells, *Lancet* **ii**:91–93.

Kan, Y. W., and Dozy, A. M., 1979, The evolution of the S and C genes in the world populations (Abstract), *Clin. Res.* **27**:274A.

Kan, Y. W., and Dozy, A. M., 1980, Evolution of the hemoglobin S and C genes in world populations, *Science* **209**:388–390.

Kan, Y. W., Schwartz, E., and Nathan, D. G., 1968, Globin chain synthesis in the alpha thalassemia syndrome, *J. Clin. Invest.* **47**:2515–2522.

Kan, Y. W., Forget, B. G., and Nathan, D. G., 1972, γβ-Thalassemia: A cause of hemolytic disease of newborns, *N. Engl. J. Med.* **286**:129–134.

Kan, Y. W., Holland, J. P., Dozy, A. M., and Varmus, H. E., 1975a, Demonstration of nonfunctional β-globin mRNA in homozygous β⁰-thalassemia, *Proc. Natl. Acad. Sci. USA* **72**:5140–5144.

Kan, Y. W., Dozy, A. M., Varmus, H. E., Taylor, J. M., Holland, J. P., Lie-Injo, L. E., Ganesan, J., and Todd, D., 1975b, Deletion of α-globin genes in haemoglobin-H disease demonstrates multiple α-globin structural loci, *Nature (London)* **255**:255–256.

Kan, Y. W., Holland, J. P., Dozy, A. M., Charache, S., and Kazazian, H. H., Jr., 1975c, Deletion of the β-globin structural gene in hereditary persistence of foetal haemoglobin, *Nature (London)* **258**:162–163.

Kan, Y. W., Dozy, A. M., Trecartin, R., and Todd, D., 1977, Identification of a nondeletion defect in α-thalassemia, *N. Engl. J. Med.* **297**:1081–1084.

Kan, Y. W., Dozy, A. M., Stamatoyannopoulos, G., Hadjiminas, M. G., Zacharaides, Z., Furbetta, M., and Cao, A., 1979, Molecular basis of hemoglobin-H disease in the Mediterranean population, *Blood* **54**:1434–1438.

Kan, Y. W., Lee, K. Y., Furbetta, M., Angius, A., and Cao, A., 1980a, Polymorphism of DNA sequence in the β-globin gene region: Application to prenatal diagnosis of β⁰-thalassemia in Sardinia, *N. Engl. J. Med.* **302**:185–188.

Kan, Y. W., Chang, J. C., and Poon, R., 1980b, Nucleotide sequences of the untranslated 5' and 3' regions of human α, β, and γ-globin mRNAs, in: *Cellular and Molecular Regulation of Hemoglobin Switching* (G. Stamatoyannopoulos, and A. W. Nienhuis, eds.), pp. 595–606, Grune and Stratton, New York.

Kantor, J. A., Turner, P. H., and Nienhuis, A. W., 1980, Beta thalassemia: Mutations which affect processing of the β-globin in RNA precursor, *Cell* **21**:149–157.

Kinniburgh, A. J., and Ross, J., 1979, Processing of the mouse β-globulin mRNA precursor: At least two cleavage-ligation reactions are necessary to excise the larger intervening sequence, *Cell* **17**:915–921.

Kinniburgh, A. J., Mertz, J. E., and Ross, J., 1978, The precursor of mouse β-globin messenger RNA containing two intervening RNA sequences, *Cell* 14:681–693.

Konkel, D. A., Tilghman, S. M., and Leder, P., 1978, The sequence of the chromosomal mouse β-globin major gene: Homologies in capping, splicing, and poly (A) sites, *Cell* 15:1125–1132.

Konkel, D. A., Maizel, J. V., Jr., and Leder, P., 1979, The evolution and sequence comparison of two recently diverged mouse chromosomal β globin genes, *Cell* 18:865–873.

Lacy, E., Hardison, R. C., Quon, D., and Maniatis, T., 1979, The linkage arrangement of four rabbit β-like globin genes, *Cell* 18:1273–1283.

Lai, E. C., Stein, J. P., Catterall, J. F., Woo, S. L., Mace, M. L., Means, A. R., and O'Malley, B. W., 1979, Molecular structure and flanking nucleotide sequences of the natural chicken ovomucoid gene, *Cell* 18:629–642.

Lauer, J., Shen, C.-K. J., and Maniatis, T., 1980, The chromosomal arrangement of human α-like globin genes: Sequence homology and α-globin gene deletions, *Cell* 20:119–130.

Lawn, R. M., Fritsch, E. F., Parker, R. C., Blake, G., and Maniatis, T., 1978, The isolation and characterization of linked δ- and β-globin genes from a cloned library of human DNA, *Cell* 15:1157–1174.

Lebo, R. V., Carrans, A. V., Burkhart-Schultz, K., Dozy, A. M., Yu, L.-C., and Kan, Y. W., 1979, Assignment of human β-, γ-, and δ-globin genes to the short arm of chromosome 11 by chromosome sorting and DNA restriction enzyme analysis, *Proc. Natl. Acad. Sci. USA* 76:5804–5808.

Leder, A., Miller, H. I., Hamer, D. H., Seidman, J. G., Norman, B., Sullivan, M., and Leder, P., 1978, Comparison of cloned mouse α- and β-globin genes: Conservation of intervening sequence locations and extragenic homology, *Proc. Natl. Acad. Sci. USA* 75:6187–6191.

Leder, P., 1978, Discontinuous genes, *N. Engl. J. Med.* 298:1079–1081.

Leder, P., Tiemeier, D., and Enquist, L., 1977, EK2 derivatives of bacteriophage lambda useful in the cloning of DNA from higher organisms: The λgtWES system, *Science* 196:175–177.

Lehmann, H., 1970, Different types of alpha thalassaemia and significance of haemoglobin Bart's in neonates, *Lancet* ii:78–80.

Lerner, M. R., Boyle, J. A., Mount, S. M., Wolin, S. L., and Steitz, J. A., 1980. Are snRNPs involved in splicing?, *Nature (London)* 283:220–224.

Little, P. F. R., Curtis, P., Contelle, C., von den Berg, J., Dalgleish, R., Malcolm, S., Courtney, M., Westaway, D., and Williamson, R., 1978, Isolation and partial sequence of recombinant plasmids containing human α-, β-, and γ-globin cDNA proponents, *Nature (London)* 273:640–643.

Little, P. F. R., Flavell, R. A., Kooter, J. M., Annison, G., and Williamson, R., 1979, Structure of the human fetal globin gene locus, *Nature (London)* 278:227–231.

Little, P. F. R., Annison, G., Darling, S., Cleamson, R. W., Camba, L., and Modell, B., 1980a, Model for antenatal diagnosis of β-thalassaemia and other monogenic disorders by molecular analysis of linked DNA polymorphisms, *Nature (London)* 285:144–147.

Little, P. F. R., Whitelaw, E., Annison, G., Williamson, R., Kooter, J. M., Flavell, R. A., Goossens, M., Sergeant, G. R., and Montgomery, D., 1980b, The detection and use of hemoglobin mutants in the direct analysis of human globin genes, *Blood* 55:1060–1062.

Lomedico, P., Rosenthal, N., Efstratiadis, A., Gilbert, W., Kolodner, R., and Tizard, R., 1979, The structure and evolution of two nonallelic rat preproinsulin genes, *Cell* 18:545–558.

Maniatis, T., Kee, S. K., Efstratiadis, A., and Kafatos, F. C., 1976, Amplification and characterization of a β-globin gene synthesized *in vitro*, *Cell* 8:163–182.

Maniatis, T., Hardison, R. C., Lacy, E., Lauer, J., O'Connell, C., Quon, D., Sim, G. K., and Efstratiadis, A., 1978, The isolation of structural genes from libraries of eucaryotic DNA, *Cell* **15**:687–701.

Manley, J. L., Five, A., Cano, A., Sharp, P. A., and Gefter, M. L., 1980, DNA-dependent transcription of adenovirus genes in a soluble whole-cell extract, *Proc. Natl. Acad. Sci. USA* **77**:3855–3859.

Mantei, N., Boll, W., and Weissmann, C., 1979, Rabbit β-globin mRNA production in mouse L cells transformed with cloned rabbit β-globin chromosomal DNA, *Nature (London)* **281**:40–46.

Maquat, L. E., Kinniburgh, A. J., Beach, L. R., Honig, G. R., Lazerson, J., Ershler, W. B., and Roos, J., 1980, Processing of the human β-globin mRNA precursor to mRNA is defective in three β⁺-thalassemias, *Proc. Natl. Acad. Sci. USA* **77**:4287–4291.

Marotta, C. A., Wilson, J. T., Forget, B. G., and Weissman, S. M., 1977, Human β-globin messenger RNA. III. Nucleotide sequences derived from complementary DNA, *J. Biol. Chem.* **252**:5040–5053.

Maxam, A. M., and Gilbert, W., 1977, A new method for sequencing DNA, *Proc. Natl. Acad. Sci. USA* **74**:560–564.

Mears, J. G., Ramirez, F., Leibowitz, D., Nakamura, F., Bloom, A., Konotey-Abulu, F., and Bank, A., 1978a, Changes in restricted human cellular DNA fragments containing globin gene sequences in thalassemias and related disorders, *Proc. Natl. Acad. Sci. USA* **75**:1222–1226.

Mears, J. G., Ramirez, F., Leibowitz, D., and Bank, A., 1978b, Organization of human δ- and β-globin genes in cellular DNA and the presence of intragenic inserts, *Cell* **15**:15–23.

Meloni, T., Pilo, G., Camardella, L., Cancedda, F., Lania, A., Pepe, G., and Luzzatto, L., 1980, Coexistence of three hemoglobins with different α-chains in two unrelated children (with family studies indicating polymorphism in the number of α-globin genes in the Sardinian population), *Blood* **55**:1025–1032.

Michelson, A., and Orkin, S. H., 1980, The 3'-untranslated regions of the duplicated human α-globin genes are unexpectedly divergent, *Cell* **22**:371–377.

Mulligan, R. C., Howard, B. H., and Berg, P., 1979, Synthesis of rabbit β-globin in cultured monkey kidney cells following infection with a SV40 β-globin recombinant genome, *Nature (London)* **277**:108–114.

Nathan, D. G., Alter, B. P., and Orkin, S. H., 1979, Prenatal diagnosis of hemoglobino-pathies, *Clin. Perinatol.* **6**:275–291.

Nathans, D., 1979, Restriction endonucleases, simian virus 40, and the new genetics, *Science* **206**:903–909.

Nienhuis, A. W., Turner, P., and Benz, E. J., Jr., 1977, Relative stability of α- and β-globin messenger RNAs in homozygous β-thalassemia, *Proc. Natl. Acad. Sci. USA* **74**:3960–3964.

Nishioka, Y., and Leder, P., 1979, The complete sequence of a chromosomal mouse α-globin gene reveals elements conserved throughout vertebrate evolution, *Cell* **18**:875–882.

Nishioka, Y., Leder, A., and Leder, P., 1980, Unusual α-globin-like gene that has clearly lost both globin intervening sequences, *Proc. Natl. Acad. Sci. USA* **77**:2806–2809.

Old, J., Longley, J., Wood, W. G., Clegg, J. B., and Weatherall, D. J., 1977, Molecular basis for acquired haemoglobin H disease, *Nature (London)* **269**:524–525.

Old, J. M., Clegg, J. B., Weatherall, D. J., and Booth, P. B., 1978a, Haemoglobin J. Tongariki is associated with α-thalassaemia, *Nature (London)* **273**:319–320.

Old, J. M., Proudfoot, N. J., Wood, W. G., Longley, J. I., Clegg, J. B., and Weatherall, D. J., 1978b, Characterization of β-globin mRNA in the β⁰-thalassemias, *Cell* **14**:289–298.

Orkin, S. H., 1978, The duplicated human α-globin genes lie close together in cellular DNA, *Proc. Natl. Acad. Sci. USA* **75**:5950–5954.

Orkin, S. H., and Michelson, A., 1980, Partial deletion of the α-globin structural gene in human α-thalassaemia, *Nature (London)* **286**:538–540.

Orkin, S. H., and Nathan, D. G., 1976, The thalassemias, *N. Engl. J. Med.* **295:**710–714.

Orkin, S. H., Alter, B. P., Altay, C., Mahoney, M. J., Lazarus, H., Hobbins, J. C., and Nathan, D. G., 1978, Application of restriction endonuclease mapping to the analysis and prenatal diagnosis of thalassemias caused by globin gene deletion, *N. Engl. J. Med.* **289:**166–172.

Orkin, S. H., Old, J., Lazarus, H., Altay, C., Gurgey, A., Weatherall, D. J., and Nathan, D. G., 1979a, The molecular basis of α-thalassemias: Frequent occurrence of dysfunctional α loci among non-Asians with Hb H disease, *Cell* **17:**33–42.

Orkin, S. H., Old, J. M., Weatherall, D. J., and Nathan, D. G., 1979b, Partial deletion of β-globin gene DNA in certain patients with β⁰-thalassemia, *Proc. Natl. Acad. Sci. USA* **76:**2400–2404.

Orkin, S. H., Alter, B. P., and Altay. C., 1979c, Deletion of the ^Aγ globin gene in ^Cγ-δβ-thalassemia, *J. Clin. Invest.* **64:**866–869.

Orkin, S. H., Kolodner, R., Michelson, A., and Husson, R., 1980a, Cloning and direct examination of a structurally abnormal human β⁰-thalassemia globin gene, *Proc. Natl. Acad. Sci. USA* **77:**3558–3562.

Orkin, S. H., Goff, S., and Nathan, D. G., 1980b, Molecular heterogeneity in the deletion in γδβ-thalassemia, *J. Clin. Invest.* (in press).

Ottolenghi, S., Lanyon, W. G., Paul, J., Williamson, R., Weatherall, D. J., Clegg, J. B., Pritchard, J., Pootrakul, S., and Boon, W. H., 1974, The severe form of α-thalassaemia is caused by a haemoglobin gene deletion, *Nature (London)* **251:**389–392.

Ottolenghi, S., Comi, P., Giglioni, B., Tolstoshev, P., Lanyon, W. G., Mitchell, G. J., Williamson, R., Russo, G., Musumeci, S., Schiliro, G., Tsistrakis, G. A., Charache, S., Wood, W. G., Clegg, J. B., and Weatherall, D. J., 1976, δβ-Thalassemia is due to a gene deletion, *Cell* **9:**71–80.

Ottolenghi, S., Giglioini, B., Comi, P., Gianni, A. M., Polli, E., Acquaye, C. T. A., Oldham, J. H., and Masera, G., 1979, Globin gene deletion in HPFH, δ⁰β⁰-thalassemia and Hb Lepore disease, *Nature (London)* **278:**654–657.

Phillips, J. A., III, Scott, A. F., Smith, K. D., Young. K. E., Lightbody, K. L., Jiji, R. M., and Kazazian, H. H., Jr., 1979, A molecular basis for hemoglobin H disease in American Blacks, *Blood* **54:**1439–1445.

Phillips, J. A., III, Panny, S. R., Kazazian, H. H., Jr., Boehm, C., Scott, A. F., And Smith, D. D., 1980a, Prenatal diagnosis of sickle cell anemia by restriction endonuclease analysis: Hind III polymorphism in γ-globin genes extend test applicability, *Proc. Natl. Acad. Sci. USA* **77:**2853–2856.

Phillips, J. A., III, Vik, T. A., Scott, A. F., Young, K. E., Kazazian, H. H., Jr., Smith, K. D., Fairbanks, V. F., and Koenig, H. M., 1980b, Unequal crossing-over: A common basis of single α-globin genes in Asians and American Blacks with hemoglobin H disease, *Blood* **55:**1066–1069.

Pressley, L., Higgs, D. R., Clegg, J. B., and Weatherall, D. J., 1980, Gene deletions in α-thalassemia prove that 5′ ζ locus is functional, *Proc. Natl. Acad. Sci. USA* **77:**3586–3589.

Proudfoot, N. J., Gillam, S., Smith, M., and Longley, J. I., 1977, Nucleotide sequence of the 3′-terminal third of rabbit α-globin messenger RNA: Comparison with human α-globin messenger RNA, *Cell* **11:**807–818.

Ramirez, F., O'Donnell, J. V., Marks, P. A., Bank, A., Musumeci, S., Schiliro, G., Pizzarelli, G., Russo, G., Luppis, R., and Gambino, R., 1976, Abnormal or absent β-mRNA in β⁰-Ferrara and gene deletion in δβ-thalassemia, *Nature (London)* **263:**471–475.

Roberts, R., 1980, Directory of restriction endonucleases, in: *Methods in Enzymology*, Vol. 65 (L. Grossman and K. Moldane, eds.), pp. 1–15, Academic Press, New York.

Rogers, J., and Wall, R., 1980, A mechanism for RNA splicing, *Proc. Natl. Acad. Sci. USA* **77:**1877–1879.

Ross, J., 1976, A precursor of globin messenger RNA, *J. Mol. Biol.* **106**:402–420.

Ross, J., and Knecht, D. A., 1978, Precursors of alpha and beta globin messenger RNAs, *J. Mol. Biol.* **119**:1–20.

Rutherford, T. R., Clegg, J. B., and Weatherall, D. J., 1979, K562 human leukemic cells synthesise embryonic haemoglobin in response to haemin, *Nature (London)* **280**:164–165.

Sakano, H., Rogers, J. H., Huppi, K., Brack, C., Traunecher, A., Maki, R., Wall, R., and Tonegawa, S., 1979, Domains and the hinge region of an immunoglobulin heavy chain are encoded in separate DNA segments, *Nature (London)* **277**:627–633.

Sanders-Haigh, L., Anderson, W. F., and Franche, U., 1980, The β-globin gene is on the short arm of human chromosome 11, *Nature (London)* **283**:683–686.

Sanger, F., and Coulson, A. R., 1975, A rapid method for determining sequences in DNA by primer synthesis with DNA polymerase, *J. Mol. Biol.* **94**:441–448.

Scott, A. F., Phillips, J. A., III, and Migeon, B. R., 1979, DNA restriction endonuclease analysis for localization of human β- and δ-globin genes on chromosome 11, *Proc. Natl. Acad. Sci. USA* **76**:4563–4565.

Seidman, J. G., Leder, A., Nau, M., Norman, B., and Leder, P., 1978, Antibody diversity, *Science* **202**:11–17.

Senno, L. D., Conconi, F., Little, P. F. R., and Williamson, R., 1979, Restriction enzyme analysis of the β-globin gene in DNA from β⁰-thalassemic subjects from Ferrara, *Biophys. Biochem. Res. Commun.* **91**:548–553.

Shen, C.-K. J., and Maniatis, T., 1980, The organization of repetitive sequences in a cluster of rabbit β-like globin genes, *Cell* **19**:379–391.

Shortie, D., and Nathans, D., 1978, Local mutagenesis: A method for generating viral mutants with base substitutions in preselected regions of the viral genome, *Proc. Natl. Acad. Sci. USA* **75**:2170–2174.

Smithies, O., Blechl, A. E., Denniston-Thompson, K., Newell, N., Richards, J. E., Slighton, J. L., Tucker, P. W., and Blattner, F. R., 1978, Cloning human foetal γ-globin and mouse α-type globin DNA: Characterization and partial sequencing, *Science* **202**:1284–1289.

Southern, E. M., 1975, Detection of specific sequences among DNA fragments by gel electrophoresis, *J. Mol. Biol.* **98**:503–577.

Sternberg, N., Tiemeier, D., and Enquist, L., 1977, *In vitro* packaging of a λ-Dam vector containing Eco RI DNA fragments of *E. coli* and phage P1, *Gene* **1**:255–280.

Surrey, S., Chambers, J. S., Muni, D., and Schwartz, E., 1978, Restriction endonuclease analysis of human globin genes in cellular DNA, *Biochem. Biophys. Res. Commun.* **83**:1125–1131.

Swanstrom, R., and Shank, P. R., 1978, X-ray intensifying screens greatly enhance the detection by autoradiography of the radioactive isotopes ^{32}P and ^{125}I, *Anal. Biochem.* **86**:184–192.

Taylor, J. M., Dozy, A., Kan, Y. W., Varmus, H. G., Lie-Injo, L. E., Ganesan, J., and Todd, D., 1974, Genetic lesion in homozygous α-thalassaemia (hydrops fetalis), *Nature (London)* **251**:392–393.

Temple, G. F., Chang, J. C., and Kan, Y. W., 1977, Authentic β-globin mRNA sequences in homozygous β⁰-thalassemia, *Proc. Natl. Acad. Sci. USA* **74**:3047–3051.

Tiemeier, D. C., Tilghman, S. M., Polsky, F. I., Seidman, J. G., Leder, A., Edgell, M. H., and Leder, P., 1978, A comparison of two cloned mouse β-globin genes and their surrounding and intervening sequences, *Cell* **14**:237–245.

Tilghman, S. M., Tiemeier, D. C., Polsky, F., Edgell, M. H., Seiiman, J. G., Leder, A., Enquist, L. W., Norman, B., and Leder, P., 1977, Cloning specific segments of the mammalian genome: Bacteriophage λ containing mouse globin and surrounding gene sequences, *Proc. Natl. Acad. Sci. USA* **74**:4406–4410.

Tilghman, S. M., Tiemeier, D. C., Seidman, J. G., Peterlin, B. M., Sullivan, M., Maizel, J. V., and Leder, P., 1978a, Intervening sequences of DNA identified in the structural portion of a mouse β-globin gene, *Proc. Natl. Acad. Sci. USA* **75**:725–729.

Tilghman, S. M., Curtis, P. J., Tiemeier, D. C., Leder, P., and Weissmann, C., 1978b, The intervening sequence of a mouse β-globin gene is transcribed within the 15S β-globin mRNA precursor, *Proc. Natl. Acad. Sci. USA* **75**:1309–1313.

Tolstoshev, P., Mitchell, J., Lanyon, G., Williamson, R., Ottolenghi, S., Comi, S., Giglioni, B., Masera, G., Modell, B., Weatherall, D., and Clegg, J. B., 1976, Presence of gene for β-globin in homozygous β⁰-thalassaemia, *Nature (London)* **259**:95–98.

Tuan, D., Biro, P. A., deRiel, J. K., Lazarus, H., and Forget, B. G., 1979, Restriction endonuclease mapping of the human γ-globin gene loci, *Nucleic Acids Res.* **6**:2519–2544.

Tuan, D., Murnane, M. J., deRiel, J. K., and Forget, B. G., 1980, Heterogeneity in the molecular bases of hereditary persistence of fetal haemoglobin, *Nature (London)* **285**:335–337.

Van den Berg, J., van Ooyen, J., Mantei, A., Schambock, N., Grosveld, A., Flavell, R. A., and Weissmann, C., 1978, Comparison of cloned rabbit and mouse β-globin genes showing strong evolutionary divergence of two homologous pairs of introns, *Nature (London)* **276**:37–44.

Van der Ploeg, L. H. T., and Flavell, R. A., 1980a, DNA methylation in the human γδβ-globin locus in erythroid and nonerythroid tissues, *Cell* **19**:947–958.

Van der Ploeg, L. H. T., Konings, A., Oort, M., Roos, D., Bernini, L., and Flavell, R. A., 1980b, γ-β-Thalassaemia studies showing that deletion of the γ- and δ-genes influences β-globin gene expression in man, *Nature (London)* **283**:637–642.

Van Ooyen, A., van den Berg, J., Mantei, N., and Weissmann, C., 1979, Comparison of total sequence of a cloned rabbit β-globin gene and its flanking regions with a homologous mouse sequence, *Science* **206**:337–344.

Wahli, W., David, I. B., Wyler, T., Weber, R., and Ryffel, G. U., 1980, Comparative analysis of the structural organization of two closely related vitellogenin genes in *X. laevis, Cell* **20**:107–117.

Wald, B., Wigen, M., Lacy, E., Maniatis, T., Silverstein, S., and Arnel, R., 1979, Introduction and expression of a rabbit β-globin gene in mouse fibroblasts, *Proc. Natl. Acad. Sci. USA* **76**:5634–5688.

Wasylyk, B., Kedinger, C., Corden, J., Brison, O., and Chambon, P., 1980, Specific *in vitro* initiation of transcription on conalbumin and ovalbumin genes and comparison with adenovirus-2 early and late genes, *Nature (London)* **285**:367–373.

Weatherall, D. J., and Clegg, J. B., 1972, *The Thalassaemia Syndromes*, 2nd edn, Blackwell, Oxford.

Weatherall, D. J., and Clegg, J. B., 1979, Recent developments in the molecular genetics of human haemoglobin, *Cell* **16**:467–479.

Weil, P. A., Luse, D. S., Segall, J., and Roeder, R. G., 1979, Selective and accurate initiation of transcription at the Ad2 major late promoter in a soluble system dependent on purified RNA polymerase II and DNA, *Cell* **18**:469–484.

Weintraub, H., and Groudine, M., 1976, Chromosomal subunits in active genes have an altered conformation, *Science* **93**:848–853.

Wigler, M., Sweet, R., Sim, G. K., Wold, B., Pellicer, A., Lacy, E., Maniatis, T., Silverstein, S., and Axel, R., 1979, Transformation of mammalian cells with genes from procaryotes and eucaryotes, *Cell* **16**:777–785.

Wilson, J. T., deRiel, J. K., Forget, B. G., Marotta, C. A., and Weissman, S. M., 1977, Nucleotide sequence of the 3'-untranslated portion of human alpha globin mRNA, *Nucleic Acids Res.* **4**:2353–2368.

Wilson, J. T., Wilson, L. B., deRiel, J. K., Villa-Komaroff, L., Efstratiadis, A., Forget,

B. G., and Weissman, S. M., 1978, Insertion of synthetic copies of human globin genes into bacterial plasmids, *Nucleic Acids Res.* **5**:563–581.

Wilson, J. T., Wilson, L. B., Reddy, V. B., Cavallesco, C., Ghosh, P. K., deRiel, J. K., Forget, B. G., and Weissman, S. M., 1980, Nucleotide sequence of the coding portion of human α-globin messenger RNA, *J. Biol. Chem.* **255**:2807–2815.

Wood, W. G., Clegg, J. B., and Weatherall, D. J., 1979, Hereditary persistence of fetal haemoglobin (HPFH) and δβ-thalassaemia, *Br. J. Haematol.* **43**:509–520.

Young, N. S., Benz, E. J., Jr., and Nienhuis, A. W., 1980, Structure of the individual globin genes in chromatin, in: *Cellular and Molecular Regulation of Hemoglobin Switching* (G. Stamatoyannopoulos, and A. W. Nienhuis, eds.), pp. 749–760, Grune and Stratton, New York.

Zimmer, E. A., Martin, S. L., Beverley, S. M., Kan, Y. W., and Wilson, A. C., 1980, Rapid duplication and loss of genes coding for the alpha chains of hemoglobin, *Proc. Natl. Acad. Sci. USA* **77**:2158–2162.

Chapter 5

Advances in the Treatment of Inherited Metabolic Diseases

Robert J. Desnick and Gregory A. Grabowski

Division of Medical Genetics
Mount Sinai School of Medicine
New York, New York 10029

INTRODUCTION

Major advances have been made in the elucidation of the molecular pathologies of inherited metabolic diseases during the past two decades. The clinical and pathophysiologic manifestations have been delineated and the metabolic derangements have been characterized in an ever-increasing number of these myriad disorders.[1-2] Sophisticated chemical and enzymatic techniques as well as *in vitro* tissue culture systems have been developed to identify the specific enzymatic defects in more than 150 of the over 400 catalogued, recessively inherited, inborn errors of metabolism.[2] Implementation of these techniques in major centers has made the diagnosis of these disorders a reality. Indeed, the demonstration of the specific enzymatic deficiency has provided for the accurate diagnosis of affected homozygotes or hemizygotes, detection of heterozygous carriers, and the capability to prenatally diagnose and prevent the birth of affected fetuses. However, in spite of these major diagnostic achievements, patients and their families have become increasingly disappointed by the absence of specific therapies for most of these debilitating disorders.

During the past decade, considerable attention has been focused on the development of strategies to treat patients with inherited metabolic diseases (Table I). Theoretically the ideal cure for these inherited dis-

281

TABLE I. Approaches to the Treatment of
Inherited Metabolic Diseases

Metabolic Manipulation
  Dietary restriction
  Substrate depletion techniques
    Chelation enhanced excretion
    Plasmapheresis/affinity binding
    Surgical bypass procedures
  Metabolic inhibition
  Product replacement
Gene Product Therapy
  Cofactor supplementation
  Enzyme induction
  Allotransplantation
  Enzyme replacement therapy
Gene Therapy
  Production of human gene products
  Gene replacement
Preventive Therapy
  Heterozygote screening
  Genetic counseling
  Prenatal diagnosis

orders would be the insertion of the normal segment of DNA coding for the synthesis of the normal gene product. Therapeutic intervention at the level of the primary genetic defect, or gene therapy, is presently precluded by our limited biochemical and cellular technology to insert a gene which will be under proper genetic regulation for normal expression. However, the recent developments in recombinant DNA biochemistry and their rapid application to *in vitro* gene-product production have signaled the future prospects for "gene transfer" as a means to replace defective human genes.

Early therapeutic endeavors primarily involved attempts to alter the disease course by manipulations at the level of the metabolic or biochemical defects (Table I). In selected diseases, investigators have attempted to reduce the levels of the accumulated substrate (or precursors proximal to the metabolic block) by dietary restriction, chelation, or administration of appropriate metabolic inhibitors. Alternatively, the deficient metabolic product has been supplied with documented examples of chemical and clinical success. Therapeutic trials at the level of the biochemical defect have involved direct administration of the appropriate gene product, the specific active enzyme or deficient cofactor, or by the transplantation of allografts capable of producing the normal gene product. The limitations, as well as the encouraging experiences, of various strategies for the treatment of genetic diseases have been the subject of recent symposia and reviews.[3–7]

A major thrust of current research is directed at the aminoacidopathies, organic acidopathies, and the lysosomal storage diseases; these inherited enzymatic deficiencies have become a focus of exploratory therapeutic endeavors. Perhaps the experience and information gained from the study of these disorders, as prototypes, may provide the basis for future therapeutic endeavors specifically modified for other inherited metabolic diseases. The development of these strategies has been and will continue to be dependent on the further elucidation and understanding of the basic molecular pathology of specific inherited enzymatic deficiencies. Thus, it is the purpose of this discussion to provide an overview of the various strategies for the treatment of genetic diseases, examples of new or novel approaches, and a more detailed account of the current status of enzyme replacement for the treatment of selected lysosomal storage diseases.

METABOLIC MANIPULATION

As illustrated in Fig. 1, therapeutic manipulation of the metabolic alterations resulting from an enzymatic defect has been designed either to (1) limit the intake or deplete the accumulation of the toxic substrate and/or its precursors or (2) supply adequate concentrations of crucial metabolic products. In each amenable disorder, the metabolic manipulation is based on an understanding of the specific pathogenic compound, the accumulated substrate, or deficient product. Table II lists the genetic diseases that have been treated by various metabolic manipulations.

Dietary Restriction

Dietary restriction was the first therapeutically successful manipulation for an inborn error of metabolism. In 1953, Bickel et al.[8] demonstrated the value of a low phenylalanine diet to limit the accumulation of the toxic substrate in patients with phenylketonuria (phenylalanine hydroxylase deficiency). The recent identification of phenylketonuria associated with a deficiency of tetrahydrobiopterin has led to the discovery

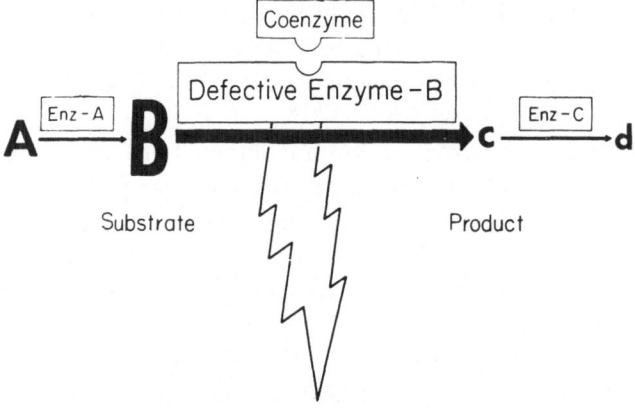

Therapeutic Strategies

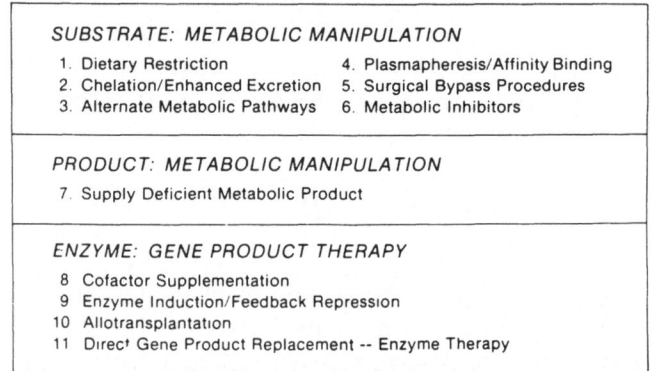

Fig. 1. Treatment of inherited metabolic diseases: therapeutic strategies.

of two new genetic forms, one due to a dihydropteridine reductase deficiency[9] and a second due to a more proximal defect in tetrahydrobiopterin synthesis.[10] Since tetrahydrobiopterin deficiency also results in reduced activities of tyrosine and tryptophan hydroxylases, the amounts of dopa and 5-hydroxytryptophan (precursors of the neurotransmitters dopamine, norepinephrine, epinephrine, and serotonin) are reduced. Attempts to replace these neurotransmitter precursors (carbidopa, 5-hydroxytryptophan, and DOPA) did not prove clinically effective in patients who already had neurologic damage,[9] but appears to be effective if initiated early.[11] However, it has been suggested that the direct administration of tetrahydrobiopterin may be therapeutic,[12] particularly if this compound can cross the blood–brain barrier.[9]

A novel method for dietary restriction in phenylketonuria is the oral administration of gelatin capsules containing the plant enzyme, phenyl-alanine ammonia lyase.[13] The enzyme converts phenylalanine in the gut to *trans*-cinnamic acid and thereby selectively depletes the phenylalanine derived from food protein before absorption. A pilot trial of oral enzyme ingestion in a previously untreated patient with phenylalanine hydrox-ylase deficiency reduced the level of blood phenylalanine about 25%. Further experience with this enzyme will determine its usefulness in re-ducing the blood phenylalanine levels as an adjunct to dietary therapy. In addition, this method may provide a tolerable form of substrate de-pletion for the management of older patients who are not on a low-phen-ylalanine diet, particularly previously treated females who become pregnant.

Subsequent experience with dietary restriction has proven clinically effective not only for patients with phenylketonuria, but also for other inborn errors whose pathology results from toxic substrate accumulation. These include galactosemia,[14,15] hereditary fructose intolerance,[16] lactose intolerance,[17] B_6-unresponsive homocystinuria,[18,19] branched-chain ke-

TABLE II. Genetic Diseases Treated by Metabolic Manipulation

Dietary restriction	Primary gout
Argininemia	β-Thalassemia
Argininosuccinic aciduria	Wilson disease
Branched-chain ketoacidosis	Plasmapheresis/affinity binding
Citrullinemia	Refsum disease
Carbamyl phosphate synthetase	Gaucher disease
deficiency	Fabry disease
Cystinuria	Hypercholesterolemia type II
Galactosemia	Surgical bypass procedures
Gyrate atrophy	Glycogenosis type Ia
Hereditary fructose intolerance	Glycogenosis type III
Histidinemia	Hypercholesterolemia type II
Isovaleric acidemia	Metabolic inhibitors
Lactose intolerance	Hyperlipoproteinemia type III
Methylmalonic acidemia	Lesch–Nyhan syndrome
Ornithine transcarbamylase deficiency	Primary gout
Phenylketonuria	Product replacement
Propionic acidemia	Adrenogenital syndromes
Tyrosinemia	Congenital hypothyroidism
Substrate depletion techniques	Orotic aciduria
Chelation/enhanced excretion	Pituitary dwarfism
Argininosuccinic aciduria	Hemophilias
Cystinuria	Hypo- and agammaglobulinemias
Familial hyperlipoproteinemia	Diabetes mellitus
Nonketotic hyperglycinemia	

toaciduria,[20,21] isovaleric acidemia,[22] and tyrosinemia.[23] However, experimental attempts to treat cystinuria,[24,25] cystinosis,[26] histidinemia,[27,28] and other aminoacidurias by appropriate dietary restriction have met with limited biochemical or clinical success. It should be noted that generalized protein restriction has proven biochemically and clinically effective in urea cycle defects, especially those with partial enzyme defects,[29] and in certain organic acidopathies.[30] Efforts to decrease the hyperammonemia in these disorders (due to deficiency of carbamyl-phosphate synthetase, ornithine transcarbamylase, or argininosuccinic acid synthetase) by the use of low-protein diets supplemented with nitrogen-free α-keto-acid analogs (keto-valine, keto-leucine, keto-isoleucine, phenyllactate, and hydroxy-methionine)[29,31,32] require further evaluation to determine their efficacy.

Gyrate atrophy is a recent addition to the list of disorders in which dietary restriction has proven therapeutic. An arginine-deficient diet corrected the hyperornithinemia, ornithinuria, and lysinuria in patients with B_6-nonresponsive gyrate atrophy of the choroid and retina which results from the deficient activity of ornithine-δ-aminotransferase, a pyridoxine-dependent enzyme.[33] In addition, the excretion of ornithine was enhanced by oral administration of α-aminoisobutyric acid, which increases the renal clearance of dibasic amino acids. Patients receiving this diet for more than 2 years did not experience continued chorioretinal degeneration. Furthermore, improvement was noted in vision, including dark adaptation, visual fields, and electroretinographic studies.[33,34]

Substrate Depletion Techniques

Chelation

Another approach to decrease the concentrations of the noxious substrate and/or precursors and metabolic derivatives is the administration of chelators or other drugs which enhance the excretion of these compounds. In Wilson's disease,[35,36] the accumulated copper may be depleted by the administration of D-penicillamine. This chelating agent binds to, mobilizes, and promotes the excretion of the intracellularly accumulated copper ions. Triethylene tetramine dihydrochloride has proven an effective copper chelator for patients with Wilson's disease who developed late drug tolerance or early penicillamine sensitivity.[37] Penicillamine also has been used for the treatment of cystinuria[38]; the drug participates in a mixed disulfide reaction to solubilize cystine calculi and promote the urinary excretion of the more soluble cysteine–penicillamine mixed disulfide.

In β-thalassemia, the necessity for frequent therapeutic blood transfusions leads to iron overload and eventually hemosiderosis. Desferrioxamine B has been used effectively to chelate the accumulated ferritin and prevent iron loading in transfusion-dependent thalassemia.[39]

Enhanced Excretion

Cholestyramine, a nonabsorbable anion-exchange resin that binds bile acids in the intestinal lumen and thus prevents their absorption, has been administered orally to reduce cholesterol concentrations in familial hypercholesterolemia. The increased fecal excretion of bile salts leads to an increased conversion of cholesterol to bile acids, and results in decreased plasma cholesterol levels.[40] Therapy in primary gout has utilized various uricosuric drugs (e.g., probenecid, sulfinpyrazone) to decrease the systemic uric acid accumulation by increasing its renal excretion and mobilizing the intracellular deposits of uric acid salts.[41,42]

Alternative Metabolic Pathways for Excretion

Recent efforts to treat the urea cycle defects have focused on the use of alternative metabolic pathways to enhance excretion of waste nitrogen in the form of metabolites other than urea, namely, urea cycle intermediates and amino acid acylation products.[43] This maneuver exploits the fact that certain nonurea nitrogen-containing metabolites have high excretion rates; by promoting the synthesis of these urea cycle intermediates, waste nitrogen excretion can be enhanced. For example, argininosuccinic acid, a urea cycle intermediate which contains both nitrogen atoms for urea synthesis, is rapidly excreted and is relatively nontoxic. Dietary supplementation of pharmacologic amounts of arginine (and/or ornithine) in patients with argininosuccinase deficiency increased nitrogen waste excretion (as argininosuccinic acid), decreased plasma levels of ammonium, and maintained other urea precursors in their respective normal ranges.[44] A similar approach, promoting orotic acid excretion, has been suggested for ornithinine transcarbamylase deficiency.[45]

Amino acid acylation provides a second alternative pathway which may increase waste nitrogen excretion.[43] Oral administration of sodium benzoate results in the enhanced urinary excretion of hippuric acid (the glycine conjugate of benzoic acid). Similarly, administration of phenylacetic acid results in the acylation of glutamine to form phenylacetylglutamine, a rapidly excreted nitrogen-containing compound. Preliminary trials of each of these compounds have been carried out in a patient with partial carbamyl phosphate synthetase deficiency with encouraging results.[46] It has been suggested that combined use of both compounds might provide an effective means to promote waste nitrogen excretion in the inborn errors of ureagenesis as well as in other nitrogen accumulation disorders.[43]

Another approach for substrate depletion by an alternative pathway has been used for the treatment of isovaleric acidemia.[47–50] Based on the studies of bovine glycine N-acylase,[50] Tanaka[47] reasoned that the administration of supplementary glycine to patients with isovaleric acidemia with high hepatic isovaleryl-CoA concentrations would promote the production of isovalerylglycine, enhance its excretion and alleviate their ketoacidosis. Indeed, prompt biochemical resolution of acidotic crises and increased isovalerylglycine excretion was observed in two patients when glycine was administered orally or rectally.[47,49] Although Tanaka[47] observed improved growth in one patient during a period of chronic glycine therapy, the use of glycine to prevent acidotic episodes and improve development in these patients will require a controlled, collaborative study.

Plasmapheresis/Affinity Binding

Two mechanical approaches to deplete an accumulated circulating substrate include plasmapheresis and affinity binding. Plasmapheresis simply removes a large volume of plasma containing the toxic compound, whereas affinity binding selectively removes a compound or class of compounds from the circulation which specifically bind to the affinity ligand. Theoretically, the affinity binding method is superior since plasmapheresis removes large amounts of plasma and nonselectively removes other molecules (e.g., trace metals, cofactors, and other essential compounds) which may be crucial for certain metabolic processes. In addition, chronic plasmapheresis requires volume replacement with commercial plasma products or fresh plasma, which both involve the risk of hepatitis.

Plasmapheresis has been attempted in several disorders characterized by high circulating levels of specific lipids. In Refsum disease, chronic plasmapheresis has therapeutic value. This disorder is characterized by high circulating levels of the fatty acid, phytanic acid, due to defective phytanic acid oxidation. Previously, rigorous dietary restriction of phytanic acid and phytol has proven beneficial.[51] More recently, chronic plasmapheresis has been used effectively to markedly reduce the plasma levels of phytanic acid; concomitantly, improvement in the patients' neurologic status, including peripheral nerve function, muscle strength, and stabilized vision, has been noted.[51-53] It is likely that combined dietary restriction and chronic plasmapheresis will enhance therapy in this disease since phytanic acid is entirely of dietary origin and plasmapheresis can be used to further minimize phytanic acid levels and perhaps deplete substrate that is mobilized from tissues.

Another disease in which plasmapheresis has been used effectively for substrate depletion (as an adjunct to dietary restriction) is homozygous familial hypercholesterolemia. In two LDL-receptor-negative patients, decreased plasma cholesterol and regression of xanthomas were observed during chronic plasmapheresis over a 2-year period.[54] These studies and [14C]cholesterol tracer experiments performed in hypercholesterolemic adults undergoing plasmapheresis[55] indicated that accumulated cytosolic lipid substrates can be depleted from tissues by this technique.

Plasmapheresis has been attempted in two lysosomal storage diseases, Fabry and Gaucher diseases,[53,56,57] in which the glycosphingolipids trihexosyl ceramide and glucosyl ceramide accumulate, respectively. The rationale for this approach was based on the fact that these substrates accumulate in the plasma where they are primarily associated with the low- and high-density lipoproteins. Pilot trials in patients with Fabry disease,[53,56] have suggested that this modality may deplete adequate amounts of circulating substrates to warrant further evaluation, particularly since the manifestations of this disease result from lipid deposition in vascular endothelium.[57] The major question to be resolved is whether intervention by chronic plasmapheresis will deplete enough substrate, compared to that newly synthesized, so that the net result is decreased substrate deposition in the target sites of pathology, the vascular endothelium. Thus, further long-term evaluation is required to determine the efficacy of this strategy in Fabry disease.

Pilot trials in Gaucher type 1 disease, however, have indicated that chronic plasmapheresis does not deplete enough substrate to alter the course of the disease.[56] We carried out a study of serial biweekly plasmaphereses over a 4-week period in a severely affected 21-year-old homozygote with Gaucher type 1 disease. During each plasmapheresis approximately 1000 ml of plasma containing about 16 μmoles of glucosyl ceramide was removed. Typically, the plasma levels of glucosyl ceramide 1 hr after plasmapheresis were essentially the same as the pretreatment values, indicating a rapid exchange between the plasma and other pools. The amount of substrate removed with each plasmapheresis was about 3.2% of the estimated daily normal turnover of this substrate (about 400 mg/day) due to the senescence of leukocyte and erythrocyte membranes which are the major sources of the glucosyl ceramide precursors.[58] Thus, the removal of only a small percentage of the daily substrate turnover precludes the feasibility of this approach in Gaucher disease.

Affinity binding has been used to deplete circulating cholesterol in homozygous familial hypercholesterolemia.[59,60] The affinity ligand, heparin-agarose, was used for the extracorporeal binding of low density lipoprotein from the patients' blood. This ligand binds the lipoprotein by the formation of precipitating heparin–lipoprotein complexes. Blood drawn into a transfer bag containing heparin, heparin-agarose beads, and $CaCl_2$ was mixed, and then transfused through a filter to remove the heparin–lipoprotein complexes. Repetitive use of this technique, reduced the plasma cholesterol about 50%. The cholesterol level returned to initial levels in 3–7 days after treatment. Recently, an improved system, using a heparin-agarose column coupled to a blood cell processor, has been developed[61] which may increase the efficiency and effectiveness of this technique; however, this system has not been used in human trials. If this plasma perfusion system proves practical, then use of various affinity ligands may provide an attractive technique for the efficient depletion of toxic, accumulated substrates from the circulation, particularly in selected amino acid, organic acid, or other disorders in which alternative approaches have not proven effective.

Surgical Bypass Procedures

Intriguing surgical strategies have been used for the treatment of various metabolic diseases.[62] In glycogenoses types I[63] and III,[64] the progressive incorporation of absorbed glucose into hepatic glycogen was partially reversed by a surgical anastamosis between the portal vein and inferior vena cava. This portacaval shunt permitted some of the glucose, absorbed by the intestine, to bypass the hepatocyte where glucose is pathologically and irreversibly deposited as glycogen; the glucose-rich blood was shunted systemically to nourish the tissues. In addition to reestablishing normoglycemia, these patients have had documented clinical improvement. Similarly, ileal–jejunal bypass procedures have been successful in reducing the hypercholesterolemia in patients with hyperlipoproteinemia type IIa, by decreasing the absorption of cholesterol from the gut.[65]

Metabolic Inhibition

Metabolic inhibitors have been used to reduce the synthesis of accumulated substrates or precursors. In patients with Lesch–Nyhan syndrome (hypoxanthine-guanine phosphoribosyl transferase deficiency) and primary gout, allopurinol has been used therapeutically to inhibit the enzyme, xanthine oxidase, in order to reduce the uric acid concentrations.[41,66] Clofibrate, which inhibits the synthesis or release of glyceride from the liver, has been found effective in reducing blood lipids to normal levels in patients with hyperlipoproteinemia type III.[67] Although not a specific inhibitor of glycine biosynthesis, strychnine has been used successfully to antagonize the binding of glycine to receptors in the central nervous system and improve the severe depression of respiratory and motor function resulting from the high cerebrospinal fluid levels of glycine[68] in patients with severe infantile glycine encephalopathy (nonketotic hyperglycinemia).

Bartter syndrome, or hyperplasia of the juxtaglomerular apparatus, is characterized by hypokalemic alkalosis, hyperreninemia, aldosteronism, high urinary prostaglandin (PGE_2) excretion, and normal blood pressure.[69,70] Although the basic defect in this disorder is unknown, the recent finding of increased urinary excretion of 6-keto-$PGF_{1\alpha}$ suggested that the overproduction of prostacyclin (PGI_2) mediates both the hyperreninemia and the hyporesponsiveness of blood pressure to pressor agents.[71] Treatment with indomethacin, a prostaglandin endoperoxide synthetase inhibitor,[72] has been reported to correct the hyperreninemia, aldosteronism, and the high PGE_2 excretion,[70,71,73] although the inhibitor did not affect the defect in distal fractional chloride reabsorption.[74] However, after prolonged therapy, it has been reported that the level of plasma renin activity was no longer supressed.[75] Further understanding of the nature of the primary defect should result in the development of specific therapy.

Product Replacement

The clinically most effective metabolic manipulations involve direct product replacement in disorders whose pathogenesis results from the defective enzyme's failure to produce a crucial metabolic product. The administration of appropriate steroids to patients with the congenital adrenal hyperplasia syndromes,[76] thyroxine for hypothyroidism,[77] growth hormone for pituitary dwarfism,[78] and uridine for orotic aciduria[79] have provided therapeutically effective approaches to override the inherited metabolic block.

Past efforts to treat Menkes disease by copper replacement are instructive and underscore the necessity for a fundamental understanding of the primary defect prior to the development and initiation of therapeutic endeavors. These trials were based on the early evidence that the primary defect was due to defective copper absorption.[80] Thus, several investigators administered copper sulfate by oral,[81] subcutaneous,[82] or intravenous[80,81,83,84] routes in an attempt to treat the apparent copper deficiency. Although the serum concentrations of copper and ceruloplasmin were increased to normal levels,[80,83,85] clinical improvement was not observed. More recent studies[86–89] have indicated that the defect in Menkes disease results in elevated levels of intracellular copper secondary to an abnormality in the regulation of metallothionein, a copper binding protein. Thus, future attempts to treat this disease will require elucidation of the molecular defect in metallothionein regulation and the development of strategies to control the levels of metallothionein and intracellular copper in affected males with this X-linked disease.

In contrast to Menkes disease, trace metal replacement has proven clinically effective in acrodermatitis enteropathica. In this disorder, inherited as an autosomal recessive trait, the systemic zinc deficiency has been shown to result from a defective zinc-binding factor in the intestine.[90] The inability to absorb and/or transport zinc leads to bullous skin lesions, alopecia, and deficient activities of several zinc-dependent enzymes.[91–93] Administration of human breast milk, which contains the zinc-binding factor, has been shown to ameliorate the clinical manifestations.[90,94,95] In addition, oral zinc supplementation has proven therapeutic in this disease; when serum zinc levels reach normal range, clinical improvement ensues.[94,95]

Another disorder in which a binding or transport protein may be defective is X-linked hypophosphatemia. In this disease, a primary renal defect in phosphate resorption leads to poor bone mineralization (rickets) and hypocalcemia. Recently, efforts directed at symptomatic improvement have involved oral administration of phosphate and 1,25-dihydroxycholecalciferol.[96] Although bone mineralization and hypocalcemia improved, the primary defect in renal wastage of phosphate did not change. In addition, this experimental regimen was associated with an increased risk for hypercalcemia and, thus, requires careful monitoring of serum calcium levels.

GENE PRODUCT THERAPY

Therapeutic endeavors at the level of the biochemical defect, a functionally defective enzyme or gene product, have become the focus of recent efforts to treat inherited metabolic diseases. Characterization of the molecular nature of a specific enzymatic defect provides the biochemical rationale for the development of effective therapeutic strategies.

Cofactor Supplementation

Many enzymatic reactions require specific cofactors, often a vitamin or its derivative, for normal catalytic activity. In certain inborn errors, the enzymatic defect may involve (1) the binding site for a specific cofactor or vitamin (resulting in an altered affinity for apoenzyme–coenzyme binding) or (2) the abnormal transport or biosynthesis of the active form of the cofactor. The coenzyme may normally form part of the active site of the holoenzyme and thus participate directly in catalysis. The apoenzyme–coenzyme interaction may decrease the K_m of the reaction by altering the conformation of the enzyme. Alternatively, the coenzyme may render the holoenzyme less susceptible to intracellular degradation and thus enhance the number of catalytically active molecules. It is notable that even a small increase in enzymatic activity, resulting from supplementation with the proper cofactor, may have a significant effect on the metabolic defect. Indeed, experience with the vitamin-dependent enzymatic deficiency diseases has indicated that for certain mutations, cofactor supplementation may increase the residual activity of certain mutant holoenzyme complexes and provide biochemical and clinical im-

TABLE III. Gene Product Therapy: Cofactor Supplementation[a]

Cofactor	Metabolic disorder	Biochemical defect
Cobalamin	Anemia	Deficiency of intrinsic factor
	Anemia	Inactive intrinsic factor
	Anemia	Ileal transport
	Methylmalonic aciduria	L-Methylmalonyl-CoA mutase
	Methylmalonic aciduria[b]	Adenosylcobalamin synthesis
	Methylmalonic aciduria with homocystinuria[b]	Adenosylcobalamin and methylcobalamin synthesis
	Transcobalamin II deficiency	Transport into cells
Folate	Congenital malabsorption of folate	Intestinal absorption
	Dihydrofolate reductase deficiency	Dihydrofolate reductase
	Formiminoglutamic aciduria	Formiminotransferase (and cyclodeaminase)
	Homocystinuria	Methylenetetrahydrofolate reductase
Biotin	β-Methylcrotonylglycinuria	β-Methylcrotonyl-CoA carboxylase
	Mixed carboxylase deficiency[b]	Holocarboxylase synthetase
	Propionic acidemia	Propionyl-CoA carboxylase
Thiamine	Branched-chain ketoaciduria	Branched-chain ketoacid decarboxylase
	Pyruvic acidemia	Pyruvate decarboxylase
	Subacute necrotizing encephalomyelopathy (Leigh)	Pyruvate carboxylase (?)
Ascorbate	Ehlers–Danlos (VI) syndrome	Collagen lysyl hydroxylase
Pyridoxine	Cystathioninuria	γ-Cystathionase
	Gyrate atrophy	Ornithine ketoacid aminotransferase
	Homocystinuria	Cystathionine β-synthase
	Hyperoxaluria I	Glyoxylate: α-ketoglutarate carboligase
	Hyperoxaluria II	D-Glyceric dehydrogenase
	Infantile convulsions	Glutamate decarboxylase
	Pyridoxine-responsive anemia	Unknown
	Xanthurenic aciduria	Kynureninase
Vitamin D	Familial hypophosphatemia	Defective phosphate transport/reabsorption?

[a] Modified from Fleisher and Gaull. [97]
[b] Prenatal cofactor supplementation accomplished or possible.

provement. This strategy for the "biochemical manipulation" of the abnormal, residual enzymatic activity has been applied to the treatment of over 25 cofactor or vitamin-responsive disorders (Table III).[97–99]

In several of these disorders, vitamin-responsive and nonresponsive subtypes involving the same deficient enzymatic activity have been identified (e.g., methylmalonic aciduria,[100] cystathioninuria[101]). Based on *in vivo* and *in vitro* studies, it has been hypothesized that responsive and nonresponsive forms of a particular disease represent different mutations of the apoenzyme or a mutation in a gene responsible for the synthesis of the cofactor.

In the responsive mutation, the apoenzyme's cofactor binding site may be altered to a form that is still capable of binding when large concentrations of the cofactor are provided; the nonresponsive mutation may alter irreversibly the enzyme's conformation so that the residual enzymatic activity is not increased with the administration of large doses of the appropriate cofactor, or may result in an unstable protein or no enzyme synthesis. Alternatively, if the absorption, transport, or synthesis of the required cofactor is impaired, supplementation with the proper cofactor derivative may partially reconstitute enzymatic activity.

Overall, the clinical effectiveness of cofactor supplementation therapy, often accompanied by other dietary manipulations, has been variable. It is anticipated, however, that even slight enhancement of enzyme activity may prove beneficial, and follow-up studies are still required in these disorders. For example, cofactor supplementation has been attempted recently in both bovine and human mannosidosis.[102,103] Trials of oral zinc sulfate supplementation were undertaken based on the findings that zinc cations stimulated normal plant, mammalian, and human acid α-mannosidase activity[104-106] as well as the residual acid α-mannosidase in tissues and fluids from bovine and human mannosidosis.[102,103] Following oral zinc supplementation, a modest increase in the activity of the residual acid α-mannosidase was observed in bovine liver, kidney and pancreas (organs in which zinc accumulates). A concomitant decrease in the levels of mannosyl-oligosaccharides also was observed in these tissues. However, in the brain of the treated calf, the residual enzymatic activity and oligosaccharide content were not changed.[102] A controlled human trial was undertaken in four sibs with mannosidosis whose residual acidic activity also was stimulated by zinc sulfate *in vitro*.[103] However, oral zinc supplementation did not result in detectable increases in the residual acid α-mannosidase activities in plasma, leukocytes, or tears. A significant decrease in the excretion of urinary mannosyl-rich oligosaccharides was observed during the treatment periods, suggesting an effect in renal tissue. At least two possible mechanisms may account for these results: (1) the low concentrations of zinc attained in the blood and/or (2) the preferential tissue deposition of zinc.[107] *In vitro*, a zinc concentration of 0.4 mM was required to stimulate the residual acidic α-mannosidase activity to 140% of initial activity. *In vivo*, maximal zinc levels in serum were between 0.05 and 0.1 mM. These levels may have been insufficient for detectable activation of the residual activity. Alternatively, the serum zinc concentration may not have reflected the levels of zinc in tissues. As a consequence, the activation of residual acidic activity may be detectable only in those tissues with adequate intracellular zinc concentration. Thus, an increased kidney content of zinc may have preferentially lowered the urinary content of oligosaccharides. These findings correspond to those in the bovine animal model and indicate that the effect of zinc supplementation may be confined to tissues which accumulate zinc and that zinc uptake by tissues of the nervous system may be inadequate. Therefore, it is unlikely that long-term supplementation would be therapeutic.

In contrast, preliminary *in vivo*[108] and *in vitro*[109] studies indicate that ascorbate may prove beneficial to individuals afflicted with Ehlers–Danlos type VI syndrome, hydroxylysine-deficient collagen disease. A significant increase in collagen lysyl hydroxylase activity was observed when fibroblasts from an affected individual were incubated in the presence of ascorbate.[109] In addition, oral ascorbate supplementation altered the urinary hydroxyproline/hydroxylysine ratio,[108] indicating an *in vivo* metabolic response. Further controlled clinical trials in vitamin responsive patients must be designed to evaluate this treatment strategy.

It is notable that cofactor supplementation has been successfully used for the *in utero* treatment of a fetus prenatally diagnosed as having B_{12}-responsive methylmalonic acidemia.[110] This child, now six years old, has maintained normal psychomotor development with daily oral cofactor supplementation (M. J. Mahoney, personal communication). The therapeutic value of cofactor supplementation as well as a theoretical discussion of the nature of the molecular pathologies in this group of diseases have been recently reviewed.[3,97,99]

Enzyme Induction/Feedback Repression

Another approach at the level of the enzymatic defect involves the use of drugs which are capable of increasing a residual enzymatic activity. Phenobarbital and related drugs apparently stimulate the production of smooth endoplasmic reticulum as well as the synthesis of specific enzymes of the endoplasmic reticulum, including hepatic UDP-glucuronyl transferase. These findings have provided the basis for administering phenobarbital to patients with Gilbert syndrome and Crigler–Najjar syndrome.[111] Although enzyme modification or stabilization has not been ruled out, the drug presumably induces smooth endoplasmic reticulum synthesis, and concomitantly, increases the amount of UDP-glucuronyl transferase, resulting in increased conjugation of unconjugated bilirubin and a decrease in the plasma bilirubin levels. This approach may be of value in selected enzymatic defects resulting from a decreased synthetic rate of a specific enzyme located on the smooth endoplasmic reticulum.

Enzyme induction has been used as a therapeutic modality in hereditary angioneurotic edema[112] and α_1-antitrypsin deficiency.[113] Both disorders result from the defective synthesis of glycoproteins produced and secreted by the liver. Angioneurotic edema is an autosomal dominant disorder characterized by half-normal levels of functionally active serum C1 esterase inhibitor. In this disorder, the administration of danazol, an androgen-related compound (a derivative of ethynyltestosterone) resulted in three- to fivefold-increased serum levels of C1 inhibitor (as well as increased C4) in most patients. Although the mechanism of induction has not been characterized, it has been suggested that the androgen may induce synthesis of C1 inhibitor mRNA.[112] Prophylactic administration of oral danazol has been proven effective in decreasing or preventing the acute attacks of angioneurotic edema with minimal virilization and hepatotoxicity. Recently, intravenous administration of a plasma concentrate containing C1 inhibitor has been used effectively during acute abdominal or laryngeal attacks to abate the symptoms and increase serum C4 activity.[114]

Danazol also has been used to increase the levels of the serum antiprotease, α_1-antitrypsin, in individuals with this recessively inherited deficiency.[113] After thirty days of androgen therapy, the levels of functional α_1-antitrypsin in PiZZ, PiM$_{Duarte}$Z, and PiSZ individuals increased 37%, 85%, and 87%, respectively. Analysis of the increased glycoprotein revealed the same electrophoretic patterns as those observed pretreatment. Since PiZZ, PiM$_{Duarte}$Z, and PiSZ individuals are at risk for development of severe emphysema due to the deficiency of this antiprotease in their circulation, danazol has been suggested as a preventive approach to improve the protease–antiprotease imbalance and impede the progression of the lung disease.[113]

Enzyme repression has been used as a strategy for the treatment of the acute porphyrias. For example, acute intermittent porphyria (AIP) is a dominantly inherited disorder resulting from the half-normal activity of the enzyme, uroporphyrinogen I synthase. This primary enzymatic deficiency results in the accumulation of its substrate, porphobilinogen (PBG) and its immediate precursor, δ-aminolevulinic acid (ALA). In addition, the defect in heme biosynthesis results in decreased heme production. Heme is known to repress the activity of the first and rate-limiting enzyme in the pathway, δ-aminolevulinate synthetase (ALAS), by an end-product feedback inhibition.[115-119] During acute attacks in AIP, the levels of PBG and ALA are markedly increased. It has been shown that hepatic ALAS is induced by various drugs, dietary factors, or other metabolites as well as by the decreased production of heme secondary to the deficiency of uroporphyrinogen I synthase.[120,121] On the assumption that heme would repress the hepatic ALAS activity and thereby decrease the toxic levels of PBG and ALA, intravenous hematin was infused in patients during acute attacks of AIP,[120-122] porphyria variegata,[121,122] and coproporphyria.[121,122] To date, over 30 acute attacks have been treated with hematin.[120-121] "Prompt and often dramatic recovery" was observed in over 80% of the treated attacks.[121] The high serum concentrations of ALA and PBG were markedly reduced, often to zero, following hematin infusions.[121] Thus, the use of hematin to repress hepatic ALAS activity has proven to be of therapeutic value. Further studies of hepatic ALAS, δ-aminolevulinate dehydrase, and uroporphyrinogen I synthase activities before and after hematin infusions in patients with AIP, are required to document the biochemical mechanisms underlying the hematin effect. Also, it should be noted that infusions of glucose or levulose have been effective in the treatment of acute attacks[123,124] and the use of hematin in conjunction with glucose should be considered for the treatment of severe attacks.

Allotransplantation

An intriguing means for transferring normal genetic information into patients with selected structural and metabolic defects is allotransplantation.[125-127] This approach exploits the grafting of cells, tissues, or organs containing the normal DNA for the production of active enzymes or other gene products in the recipient.[126] For structural gene defects with pathology limited to specific organs or tissues, successful transplantation of the appropriate allograft may provide effective treatment. For inherited metabolic defects, allotransplantation may provide a strategy for the grafting of appropriate tissues for the continuous synthesis of active gene products. Figure 2 illustrates the concept of "enzyme transplantation"

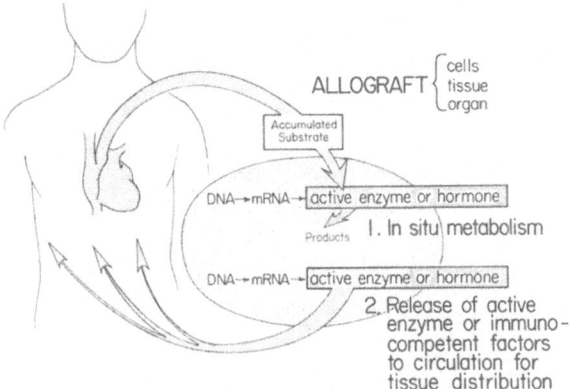

Fig. 2. Mechanism of enzyme transplantation.

and indicates two potential mechanisms by which the allograft might therapeutically metabolize the accumulated substrate in recipients with selected inherited enzymatic deficiencies. In disorders characterized by substrate accumulation in the plasma, the active enzyme in the allograft may metabolize or clear the accumulated substrate which is delivered to the transplanted tissue by the circulation. As the accumulated substrate is cleared from the plasma, a concentration gradient would be established between the plasma and the tissue sites of substrate deposition, allowing for the continuous resaturation of the plasma and continual clearance of the systemic substrate load. *In situ* metabolism would probably be the major mechanism of substrate metabolism by transplanted organs such as liver, spleen, and kidney.

Alternatively, the normal allograft may synthesize active enzyme, an essential cofactor, hormone, or immunocompetent factor, which is either released by the turnover of allograft cells or by direct secretion into the circulation. The active enzyme or gene product is then distributed to the tissues where it may gain access to cells for substrate metabolism. The release and distribution of normal gene products conceivably would be a therapeutic mechanism of transplanted pancreas, bone marrow, thymus, and, to a lesser degree, liver and kidney. The selection of the allograft in a particular inherited disorder must be based on the specific nature of the defective gene product, the pathophysiology of the disease, as well as the probable mechanism by which the allograft might provide the normal gene product.

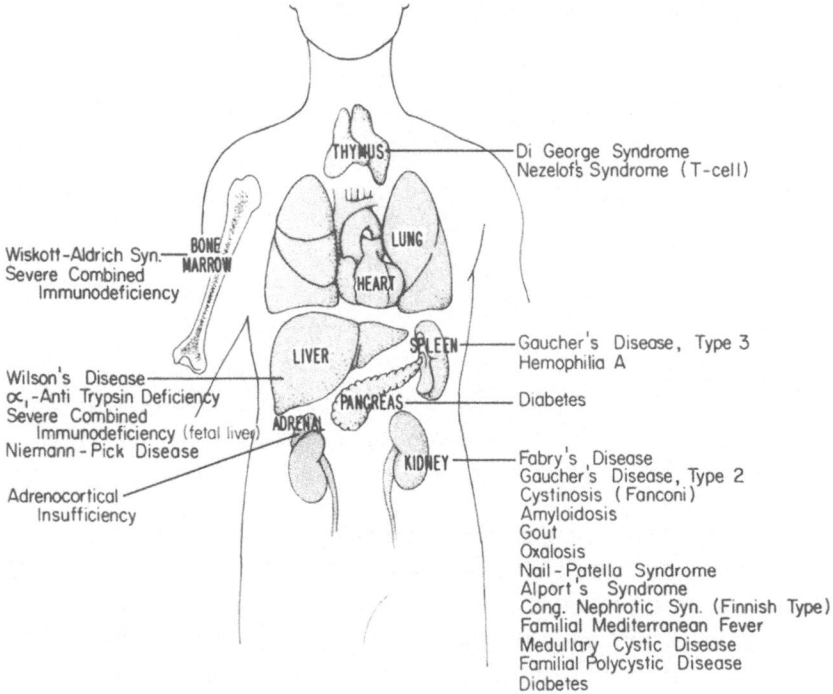

Fig. 3. Allotransplantation in genetic diseases.

Figure 3 summarizes the various genetic diseases in which allotransplantation has been accomplished. Several of these endeavors were specifically designed to be therapeutic—to continuously replace defective enzymes, hormones, or immunologic factors or effectively restore the functional alterations resulting from structural gene defects.

Transplantation of Bone Marrow, Thymus, and Fetal Liver

Bone marrow, thymus, and fetal liver have been successfully transplanted for the cellular, metabolic, and immunologic correction of various congenital and inherited disorders.[128] Bone marrow has been effective in reconstituting immunocompetency in over 20 patients with severe combined immunodeficiency disease[127,129] and in several patients with the Wiskott–Aldrich syndrome.[130–132] Recently, bone marrow transplantation has been used to correct the cellular and metabolic defects in severe infantile osteopetrosis.[133–135] Differentiation of donor marrow stem cells into osteoclasts appears to provide the mechanism for the reversal of this connective tissue disease. It is notable that bone marrow cells are also the source of other differentiated mesenchymal cells (e.g., hepatic Kupffer cells[136]), suggesting that this tissue may be useful to correct other diseases resulting from defective cellular and metabolic functions of mesenchymal cell derivatives. The current status and problems of bone marrow transplantation have been the subject of a recent review.[137]

Fetal liver cells also corrected the immunologic defects in severe combined immunodeficiency disease.[138] Fetal thymus allografts have reconstituted the immunologic and thymic deficiencies in the Di George syndrome (congenital absence of the thymus)[139] and corrected the immunologic defect in the Nezelof syndrome, a rare T-cell immunodeficiency.[140]

Orthotopic Liver Transplantation

Total liver allografts have been attempted in patients with Wilson disease,[141–143] α_1-antitrypsin deficiency,[144–147] and Niemann–Pick disease.[148] Although the specific defect has not been identified in Wilson disease, the levels of serum ceruloplasmin and copper increased from low to normal levels 2 weeks posttransplantation. Urinary copper excretion also increased posttransplant, and homograft biopsies have shown no copper accumulation. In one recipient, the Kaiser–Fleischer rings disappeared.[143] Hepatic transplantation in α_1-antitrypsin deficiency resulted in normal levels of circulating α_1-antitrypsin of the donor phenotype shortly after transplantation; however, most recipients have not survived longer than one year.[146,147]

An intriguing, but preliminary, report describes the successful orthotopic liver graft in a patient with Niemann–Pick type A disease.[148] Posttransplantation, the sphingomyelinase activity in plasma, urine, and most notably, cerebrospinal fluid, appeared to be increased, suggesting that normal hepatic tissue is capable of secreting active enzyme which can be distributed to various tissues of the recipient's body, including cerebrospinal fluid and neural tissue.

Pancreatic Transplantation

Transplantation of the pancreas has been accomplished in more than 20 patients with juvenile diabetes mellitus.[149,150] Almost immediately posttransplantation, the insulin levels became homeostatic and the glucose levels returned to normal. Several patients have had normal glucose and hormone levels for more than one year posttransplantation.

Splenic Transplantation

Splenic allotransplantation has been accomplished in patients with hemophilia A[151] and juvenile or type 3 Gaucher disease.[152,153] The levels of Factor VIII were essentially unchanged; however, the level of plasma glucosyl ceramide, the Gaucher substrate, was decreased posttransplantation. Unfortunately, these grafts were unsuccessful due to graft rejection.

Renal Transplantation

Extensive experience with renal transplantation has demonstrated that the kidney is one of the most successfully transplanted organs. Renal transplantation has been accomplished in over a dozen genetic diseases. The majority of these disorders involve primary renal pathology, and the allograft corrects the abnormal renal function. Renal transplantation in familial polycystic disease, medullary cystic disease, familial Mediterranean fever, congenital nephrotic syndrome, Alport syndrome, nail–patella syndrome, and amyloidosis results in excellent renal function and no apparent recurrence of the renal disease. The protein and lipid levels in patients with the congenital nephrotic syndrome became normal after transplantation.[154] Recipients with primary gout have normal renal function, but the hyperuricemia and gouty symptoms persist.[155] Transplantation in familial lecithin:cholesterol acyltransferase (LCAT) deficiency corrected the renal insufficiency but had no effect on the metabolic disease. The abnormal lipid metabolites slowly reaccummulate, but years of normal allograft function are expected.[156] Kidney transplantation in several patients with primary oxalosis and familial Mediterranean fever with amyloidosis has been unsuccessful due to the rapid reaccumulation of calcium oxalate crystals and amyloid in the respective allografts.[157]

Renal transplantation has been accomplished in more than 20 children with cystinosis.[158] Most recipients have received allografts from their parents who were obligate heterozygotes for the cystinotic gene. The cystine levels in cornea, bone marrow, and peripheral leukocytes have not decreased posttransplantation. Cystine has not reaccumulated in the proximal renal epithelium of the allografts, but reaccumulation has occurred in the renal interstitial cells, presumably due to infiltration by the recipients's macrophages. However, no cystine reaccumulation was demonstrated 2 years after transplantation in a patient who received a cadaver allograft; this finding suggests that the unrelated, normal allograft is more capable of handling cystine than heterozygous kidneys.[158]

Renal transplantation in patients with Fabry disease (defective α-galactosidase A) was undertaken to monitor the biochemical and clinical effectiveness of enzyme transplantation. Since patients with Fabry disease develop renal failure, and since renal tissue contains active α-galactosidase A, it was hypothesized that a renal allograft might provide active enzyme to correct the metabolic defect of Fabry disease. Although biochemical and clinical improvement following successful renal transplantation has been reported in several recipients,[159-162] no biochemical effect could be demonstrated in others.[163-165]

The mechanism by which the renal allograft is responsible for the chemical and clinical observations in recipients with Fabry disease is the focus of current investigations. At present, the data support the mechanism of renal filtration and *in situ* metabolism; the allograft may filter the accumulated lipid substrate from the plasma and catabolize the lipid within the kidney. Then as the accumulated lipid is cleared from the plasma, a concentration gradient would be established between the plasma and tissue sites of lipid deposition, allowing for the continuous resaturation of the plasma and eventual clearance of the systemic lipid accumulation. However, further studies are in progress to determine if active enzyme is released into the circulation by the allograft and taken up by the recipient's tissue.

Fibroblast Transplantation

Recently, the subcutaneous transplantation of cultured skin fibroblasts has been accomplished in patients with mucopolysaccharidoses (MPS) types I-H,[166] II,[167-169] III A,[169] and VI.[170] The rationale for fibroblast transplantation was based on the prior demonstration that cultured normal fibroblasts secreted enzymes which corrected the abnormal mucopolysaccharide metabolism in fibroblasts from patients with these disorders.[171-173] Thus, it was reasoned that the grafted normal fibroblasts would continuously release enzyme for distribution and uptake by the recipient's connective-tissue cells; in addition, the grafted cells would replicate and possibly reach other tissue sites, thus providing an ever-expanding source of the normal enzyme. To date five patients with MPS I-H, three with MPS II, three with MPS IIIA, and one with MPS VI have received these cellular allografts. In most cases, donor fibroblasts obtained from histocompatible siblings with enzymatic activity were transplanted and the recipients received immunosuppressive drugs. However, several recent recipients received no immunosuppression.[166]

In an attempt to assess the biochemical effects of these fibroblast grafts, the urinary mucopolysaccharide excretion patterns have been monitored. Several reports[166–169] have indicated an increase in shorter-chain glycosaminoglycan fragments, consistent with mucopolysaccharide degradation, while another did not.[170] Immunosuppression alone was found to increase the excretion of shorter-chain fragments.[166] Notably, fibroblasts obtained from the sites of administration up to 6 months after transplantation were cultured and found to have low levels of enzymatic activity suggesting that at least some of the transplanted cells had survived.[166] In addition, small increases in the levels of enzymatic activity were found in sera and/or circulating leukocytes of some recipients.[169] However, the biochemical (and clinical) effectiveness of this approach has not been proven and requires further evaluation and long-term follow-up of the fibroblast recipients. Future experimental animal model studies using the recently discovered cats with MPS I-H[174] and MPS VI[175,176] should provide important data concerning the efficacy of fibroblast transplantation prior to further human trials.

Experimental Transplantation in Animal Model Systems

Several recent reports have described the successful transplantation of heterotopic allografts and cell implants for the correction of metabolic deficiencies in animal model systems. Pancreatic islet cell implants have reestablished normal insulin and glucose levels in alloxan-induced diabetic rats.[177] Bone marrow from congenic mice corrected the enzymatic defect in acatalasemic mice.[178]

Heterotopic liver allografts have corrected the UDP-glucuronyl transferase deficiency in Gunn rats[179]; analogously, isolated hepatocytes also corrected the defective bilirubin metabolism in the Gunn rat.[178] Hepatocellular transplantation may provide a unique method to correct metabolic diseases due to defective hepatocyte metabolism or secretion. Partial hepatectomy of the histocompatible donor (i.e., removal of the left lobe) will provide adequate cells for transplantation. The metabolic capacity of the donor's liver eventually will be regained due to the organ's capacity to regenerate. Following injection of the isolated donor hepatocytes into the portal vein of the recipient, the donor cells will lodge in the hepatic sinusoids. Further evaluation of the viability and ability of these hepatocellular grafts to correct metabolic defects in animal models will determine the efficacy of this approach for human trials.

Thus, the grafting of enzyme-producing tissues for the continuous production of the normal gene products may provide another strategy for the treatment of appropriate recessively inherited metabolic diseases. It must be emphasized that this approach is exploratory, and further study of experimental animal transplantation and the long-term results in patients transplanted to date must be evaluated before allotransplantation is undertaken in other inherited metabolic diseases. However, histocompatibility restrictions remain the major current obstacle to transplantation. It is anticipated that further advances in our understanding of histocompatibility barriers[180,181] may provide methods to overcome this limitation and render the transfer of enzyme-producing cells and/or tissues more practical.

Erythrocyte Transfusion Therapy

Periodic erythrocyte transfusions have provided a novel approach to replace the deficient adenosine deaminase (ADA) in patients with severe combined immunodeficiency disease. In this disorder, the enzymatic defect leads to the accumulation of adenosine, deoxyadenosine, and several of their metabolites, dATP, cAMP, and S-adenosyl-homocystine. These compounds accumulate in the plasma, erythrocytes, and lymphocytes of affected patients; several of these metabolites have been shown to inhibit normal immune function as well as interfere with normal lymphocyte differentiation and function in vitro.[182] Matched, normal, irradiated erythrocytes containing normal ADA activity have been infused biweekly by partial exchange transfusions (reaching heterozygous levels of ADA) in over 10 patients with this disease.[127] Although not as metabolically or clinically effective as bone marrow transplantation,[127,129] the erythrocyte transfusions have resulted in decreased plasma deoxyadenosine levels and the decreased urinary excretion of adenosine and deoxyadenosine. The dATP concentrations became near normal in the recipients' erythrocytes and lymphocytes. Concomitantly, humoral immune function improved markedly although cellular immune function remained somewhat impaired in the transfused patients.[127,182,183]

Erythrocyte transfusion therapy also has proven effective for purine nucleoside phosphorylase deficiency, which is associated with severely defective T-cell immunity.[184] Repeated erythrocyte transfusions resulted in decreased urinary levels of inosine, deoxyinosine, guanosine, and deoxyguanosine with concomitantly increased levels of urine and serum uric acid. In addition, erythrocytic 2,3-diphosphoglycerate increased, causing a shift to the right of the oxygen dissociation curve. Importantly, the recipient's immunologic status was partially improved; it was hypothesized that the residual dGTP levels prevented complete immunologic reconstitution.[184]

ENZYME REPLACEMENT THERAPY

The thrust of current research is directed at the treatment of inherited metabolic diseases at the level of the primary defect—the specific enzymatic lesion. Lacking the technology to modify or replace defective genes, investigators have turned to enzyme replacement as a potential means to treat selected inborn errors of metabolism, particularly the lysosomal storage diseases (for reviews see references 3–7). Since these disorders are caused by a critical mutation in a segment of DNA whose resultant gene product is a defective enzyme, investigators hypothesized that the metabolic defect in each disorder could be corrected by replacement of the defective enzymatic activity, analogous to the success of hormone,[185–190] Factor VIII,[191,192] and gamma globulin[193,194] replacement therapy.

Recent developments in enzyme and cellular engineering technologies have provided new strategies which can be exploited for enzyme replacement endeavors. Affinity chromatographic techniques can be utilized to provide sufficient quantities of the appropriate active enzyme for experimental and clinical trials in selected disorders. These methods can be modified for the large-scale production of human enzyme for long-term replacement trials. With adequate amounts of enzyme available, the major focus of current research has been directed at the development and evaluation of methods to deliver the enzyme to the crucial tissue and subcellular sites of metabolic pathology. Therefore, in this section, we review the concept and current status of enzyme replacement in genetic diseases, and in particular, the application of receptor- and carrier-mediated strategies for the delivery of enzymes to target sites of pathology. Although these recent studies have focused on enzyme replacement in selected metabolic diseases, these strategies may have broader relevance for the target delivery of hormonal, antimicrobial, antineoplastic, and other potent therapeutic agents.

Rationale

The rationale for enzyme replacement therapy in selected lysosomal storage diseases evolved from two fundamental observations: (1) the identification of lysosomes as the subcellular site of pathology and (2) the elucidation of the basic role of the lysosome in cellular catabolism. Thus, it was reasoned that after endocytosis and fusion of the various components of the lysosomal apparatus,[195] exogenous enzyme would be brought into contact with the accumulated substrate for hydrolysis.

The working hypothesis that exogenously supplied enzymes can be delivered to lysosomes for effective substrate catabolism has been supported by the *in vitro* "correction" of substrate accumulation in cultured fibroblasts. When the appropriate active enzyme was supplied in the media of cultured fibroblasts obtained from individuals with various lysosomal storage diseases, the exogenous enzyme gained access to the accumulated intracellular substrates and normalized substrate turnover. For example, the defective heparan sulfate degradation in fibroblasts from patients with mucopolysaccharidosis type IIIB was corrected by the addition of partially purified N-acetyl-α-D-glucosaminidase to the culture medium; the uptake of exogenous enzyme to intracellular levels of less than 5% of normal mean activity resulted in the catabolism of 70% of the accumulated mucopolysaccharide.[196] Similarly, human β-glucuronidase taken up by deficient fibroblasts was localized in the lysosomes and 58% of the exogenous activity remained in these cells after 19 days in culture.[197] Similar uptake and/or correction studies have been accomplished in enzyme-deficient fibroblasts from individuals with various glycosphingolipidoses,[198–200] mucopolysaccharidoses,[201–203] glycogenosis,[204] and glycoproteinoses.[205] Significantly, these studies indicated the feasibility of enzyme replacement and, in particular, that low levels of exogenous enzyme could gain access to intracellular lysosomal sites and effect normalization of substrate metabolism. Furthermore, they provided the rationale for clinical trials of enzyme replacement in the lysosomal storage diseases.

Early Trials of Enzyme Replacement

The early clinical trials of enzyme replacement are summarized in Table IV. The earliest trials used commercially available or partially purified fungal[204,206-208] or bovine[209,210] enzymes administered intravenously or intrathecally to severely debilitated patients. Although these studies suggested that exogenously administered enzymes could gain access to visceral tissues, reactions to the heterologous proteins[207,211] clearly indicated the need for nonimmunogenic, sterile enzymes, preferably from human sources.

Since quantities of highly purified, human lysosomal hydrolases were unavailable, investigators turned to the use of readily available human preparations which contained the active enzyme. Fresh normal plasma, plasma concentrates, and leukocytes were administered to patients with a variety of lysosomal disorders.[212-224] Although small amounts of enzymatic activity were administered, some investigators reported evidence for substrate catabolism[204,206,207,212,216,218,220,221,223,224] while others did not.[208,209,213,216,218,219,222] It was concluded that this approach was inadequate and even long-term administration of these preparations was unlikely to have significant therapeutic effect.

Requisites for Enzyme Replacement Therapy

The experiences of the early clinical trails were instructive and identified the obstacles that must be overcome if enzyme replacement is to become an effective therapeutic modality. The difficulties in the pilot human trials included (1) the short circulating and intracellular half-lives of the administered activity, (2) the inability to target enzymes to specific tissue and subcellular sites of pathology, (3) the impracticality and difficulty of serial evaluation of the physiologic and biochemical factors affecting the fate of administered enzymes, and (4) immunologic complications. Thus, the following requisites for effective enzyme replacement

TABLE IV. Early Human Trials with Crude Enzyme Preparations

Disease	Enzyme deficiency	Year	Enzyme source	Route	Organ uptake	Evidence for substrate catabolism	Reference
Glycogenosis type II	Acid α-glucosidase	1964	A. niger	IV[a]	Liver	+ Liver	204, 206
		1967	A. niger	IV	Liver	+ Liver by E. M.	207
		1968	A. niger	IM[a]	Liver/leukocytes	− Liver	208
Metachromatic leukodystrophy	Arylsulfatase A	1967	Human urine	IT[a]			211
		1969	Bovine brain	IV+IT	Liver	− CNS by E. M.	209
Hurler syndrome (MPS I-H)	α-L-Iduronidase	1971	Plasma	IV		+ Urine	212
		1972	Plasma	IV			213
		1973	Plasma	IV			212, 214
		1974	Bovine hyaluronidase	IV	Liver	+/− Urine	210
Hunter syndrome (MPS II)	α-L-Induronate sulfatase	1971	Plasma	IV		+ Urine	212
		1971	Leukocytes	IV		+ Urine	215
		1972	Blood	IV		− Urine	213
		1972	Plasma	IV		− Urine	216
		1974	Bovine hyaluronidase	IV	Liver	+/− Urine	210
Sanfilippo A (MPS IIIA)	Heparan N-sulfate sulfatase	1972	Plasma	IV		+ Urine	217
		1973	Plasma	IV		− Urine	213
		1973	Plasma	IV		− Urine	218
Sanfilippo B (MPS IIIB)	N-Acetyl-α-glucosaminidase	1972	Blood	IV		− Urine	213
MPS VII	β-Glucuronidase	1974	Plasma	IV		− Urine	219
G$_{M2}$ gangliosidosis type 2	β-Hexosaminidase A and B	1972	Plasma	IV		+ Plasma	220
Fabry	α-Galactosidase A	1970	Plasma	IV		+ Plasma	221
		1970	Plasma	IV		− Plasma	222
		1973	Plasma, leukocytes and platelets	IV		+ Plasma	223

[a] IM, Intramuscular injection; IV, intravenous injection; IT, intrathecal injection.

therapy were formulated.[6] Each of these requisites have been discussed in detail elsewhere.[5-7]

1. Availability of large quantities of stable, nonimmunogenic, sterile enzymes with high specific activities.
2. Protection of the administered activity from bioinactivation and immunologic surveillance and delivery to the major tissue and subcellular sites of pathology.
3. Mammalian model systems to evaluate and maximize strategies for enzyme therapy.
4. Appropriately designed serial biochemical and clinical evaluations of human trials.

Human and Animal Trials—Replacement with Purified Enzymes

In the early 1970s, investigators recognized the need to use partially purified enzymes from human sources to avoid or minimize potential immune reactions (requisite 1). At about the same time, other investigators developed animal model systems to experimentally evaluate and maximize enzyme replacement strategies prior to human trials (requisite 3). Our initial efforts were directed to systematically determine the *in vivo* fate of an intravenously administered lysosomal enzyme in a mammalian model system.[225] The results of these human and animal studies are summarized below. In retrospect, it is notable, and instructive, that similar findings were obtained from these two approaches.

Human Trials with Purified Homologous Enzymes

Table V summarizes the results of the clinical trials of direct enzyme replacement in various lysosomal disorders. Pilot intravenous administrations of the appropriate human enzymatic activity were accomplished in patients with G_{M2} gangliosidosis type 2 (Sandhoff disease; urinary β-hexosaminidase A[226]), glycogenosis type II (placental α-glucosidase[227]), Fabry disease (placental α-galactosidase[223]), and most recently, Gaucher disease (placental β-glucosidase[228-232]). In each case, the highly purified human enzyme was shown to hydrolyze its natural substrate *in vitro* prior to *in vivo* trials. The injected enzymes were rapidly cleared from the circulation and exogenous enzymatic activity was recovered in biopsied

TABLE V. Human Trials with Purified Human Enzymes

Disease	Year	Enzyme administered	Source	Route	Organ uptake	Evidence for substrate catabolism (tissue)		Reference
Glycogenosis type II	1973	Acid α-glucosidase	Placenta	IV[b]	Liver	−	Liver, muscle	227
G$_{M2}$ gangliosidosis type 2	1973	β-Hexosaminidase A	Urine	IV	Liver	+	Plasma	226
Fabry	1973	α-Galactosidase A	Placenta	IV	Liver	+	Plasma	223
Gaucher types 1 and 3	1974	β-Glucosidase	Placenta	IV	Liver	+	Liver, RBC, plasma	228
Gaucher type 1	1977	β-Glucosidase	Placenta	IV		+	Lymphocytes, platelets	230
						−	RBC, plasma	
G$_{M2}$ gangliosidosis type 1	1979	β-Hexosaminidase A[a]	Placenta	IT[b] or IV		+	Serum	233

[a] Native and PVP-modified β-hexosaminidase administered.
[b] IV, Intravenous; IT, intrathecal.

liver samples from the recipients. As shown in Table VI, the approximate half-lives of activity in the circulation were between 10 and 20 min. Evidence for concomitant substrate catabolism was demonstrated; decreased concentrations of plasma globoside (G_{M2} gangliosidosis type 2[226]), hepatic glycogen (glycogenosis type II[227]), plasma trihexosyl ceramide (Fabry disease[223]), and plasma, erythrocytic, and hepatic glucosyl ceramide[228,229] or leukocyte and platelet glucosyl ceramide[230] (Gaucher disease) were found after enzyme administration.

These preliminary, but encouraging, results supported the feasibility of enzyme replacement with highly purified human enzymes. However, a most important observation was made relevant to the treatment of disorders with primary neuronal involvement. Following the intravenous administration of enzyme to a patient with G_{M2} gangliosidosis type 2,[226] β-hexosaminidase A activity was rapidly cleared from the circulation, and a significant increase in β-hexosaminidase activity was detected in hepatic tissue biopsied percutaneously 45 min after injection compared to the level in the preinjection biopsy. However, no significant increase of β-hexosaminidase A activity was observed in either lumbar or ventricular cerebrospinal fluid or in biopsied brain tissue, thus demonstrating the inability of an intravenously administered enzyme to gain access to the central nervous system.[226] In an attempt to overcome this obstacle, Von Specht *et al.*[233] intrathecally administered highly purified placental

TABLE VI. Half-Lives for the Plasma Disappearance of Unentrapped, Purified Human Enzyme Intravenously Administered to Patients with Lysosomal Diseases

Disease	Administered activity	Source	Half-life[a] (min)	Reference
G_{M2} gangliosidosis type 2	β-Hexosaminidase A	Urine	10	226
Gaucher type I	β-Glucosidase	Placenta	18	229
	β-Glucosidase	Placenta	19	228
Fabry	α-Galactosidase A	Placenta	10–12	223
		Plasma	70	337
		Spleen	10–12	337
Glycogenosis type II	α-Glucosidase	Placenta	20	227
G_{M2} gangliosidosis type I	β-Hexosaminidase A	Placenta	7[b], 15[c]	233
	PVP-β-hexosaminidase A	Placenta	80	233

[a] Time approximate.
[b] 100% was serum level 2 min after injection.
[c] 100% was CSF level 2 min after intraventricular injection.

β-hexosaminidase A to two patients with G_{M2} gangliosidosis type 1 (Tay–Sachs disease). Case 1 received four intraventricular and 15 lumbar injections of pure enzyme; case 2 received weekly lumbar injections over 10 months. Although they demonstrated decreased levels of G_{M2} ganglioside in plasma after injection, the pathologic substrate accumulation in frontal lobe biopsied pre- and postinfusion appeared unchanged by light and electron microscopy. These studies documented the constraints of the blood–brain barrier as well as the inability of intrathecally administered enzyme to gain access to neural cells. Thus, future replacement endeavors for disorders with severe neurologic involvement, such as Tay–Sachs and Niemann–Pick diseases, will require the development of techniques that permit macromolecular enzymes to gain access to the neuronal sites of substrate deposition.

Animal Model Trials—In Vivo Fate of Purified Enzymes

In order to systematically determine the *in vivo* fate of an intravenously administered glycoprotein lysosomal hydrolase, we developed a murine model system. Partially purified bovine hepatic β-glucuronidase was administered intravenously to C3H/HeJ *Gus^h* β-glucuronidase deficient mice.[225] A selective thermal inactivation assay allowed the sensitive and reliable discrimination of the bovine β-glucuronidase activity from the residual murine tissue activities.[225] The heterologous enzyme was purposely chosen in order to assess potential immune complications.

Figure 4 shows the *in vivo* fate of unentrapped bovine β-glucuronidase administered intravenously to the β-glucuronidase-deficient mice. A rapid clearance from the circulation (half-life $\sim$3 min) and a rapid reciprocal uptake of the injected enzyme in the liver were observed at either dosage administered. Maximal hepatic recovery, 70% of dose, was detected at 30 min, decreasing to nondetectable levels by 24 hr. Subcellular fractionation of hepatic tissue (2 hr after injection) revealed that more than 85% of the recovered activity was in the mitochondrial–lysosomal fraction, indicating the *in vivo* localization of the administered enzyme primarily in hepatic lysosomes. No exogenous bovine activity was detected in any other tissue, including spleen, lung, bone marrow, intestine, muscle, and, especially, brain.

The rapid plasma clearance and preferential hepatic uptake of bovine β-glucuronidase in the murine system were similar to the findings in the

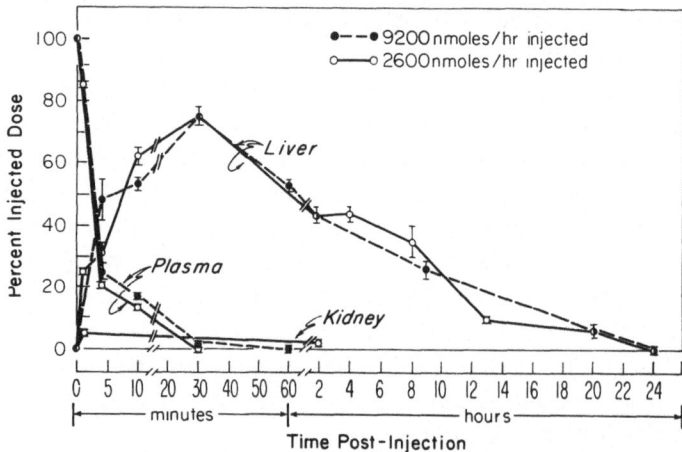

Fig. 4. Fate of unentrapped bovine β-glucuronidase administered intravenously to β-glucuronidase-deficient mice.[225]

human trials using highly purified human enzymes (Tables V and VI).
These observations identified the need to develop strategies designed to
optimize the delivery of enzyme to selected target sites of pathologic
substrate deposition for effective therapy.

Therapeutic Considerations

Target Sites of Lysosomal Pathology

As indicated in Table VII, the primary tissue and cellular sites of
lysosomal substrate accumulation differ among the lysosomal storage
diseases. Therefore, efforts to control or reverse the pathologic substrate
accumulation in each disorder requires enzyme delivery to these target
tissues and/or cells (requisite 2). The results of the pilot human trials and
the animal model studies underscored the limited tissue distribution (primarily hepatic) of these glycoprotein enzymes when injected intravenously. Furthermore, the inability of the enzyme to cross the blood–brain
barrier precludes the treatment of disorders with primary neuronal involvement. Clearly, if an enzyme is to reach sites other than the liver,
delivery strategies must be designed to enhance the uptake of enzyme to
the crucial cell type(s) in each disease. Targeting to specific sites may
require either purification of an isozyme from specific tissue sources and/

or the chemical modification of the enzyme; alternatively, the enzyme preparation may be immobilized in various biodegradable vesicles that can be chemically or physically modified to promote uptake by particular cells or tissues.

Equilibrium-Depletion of Stored Substrate

In certain diseases, the pathophysiologic mechanisms responsible for substrate synthesis, distribution, and accumulation may be therapeutically exploited. For example, the deposition of trihexosyl ceramide in the vascular endothelium, more than in any other cell or tissue, leads to the morbid manifestations of Fabry disease. Although this glycosphingolipid is a plasma-membrane component of most cells, the major sites of synthesis are hepatocytes[234] and erythroid precursors.[235] Presumably, the hepatocyte-synthesized glycosphingolipid is associated and secreted with LDL into the circulation.[236] The LDL-associated substrate is taken up into endothelial lysosomes by the LDL-receptor-mediated process where it is accumulated due to the defective α-galactosidase A activity. This mechanism for the preferential accumulation of substrate in vascular endothelial cells is supported by the following findings: (1) very little, if any, substrate accumulation occurs in hepatocytes of affected individuals,[237] (2) the majority of circulating substrate is associated with low-density lipoproteins,[237–239] (3) the plasma substrate levels are increased 5- to 10-fold in patients with Fabry disease[240] and 3- to 5-fold in patients with familial hypercholesterolemia (due to defective LDL uptake),[241] and, finally, (4) the vascular endothelial lysosomes are a major site of substrate deposition in Fabry disease.[57,237] Thus, efforts to hydrolyze or deplete the accumulated substrate in the circulation may lead to control of the progressive deposition in the vascular endothelium and may provide an alternate strategy to treat patients with Fabry disease. In addition, it is possible that the circulating substrate is in a dynamic equilibrium with that in endothelial as well as other cells. Based on *in vitro* studies,[57] it is possible that depletion of the circulating substrate will result in the resaturation of the plasma by substrate stored in endothelial and other cells. Thus, enzyme replacement endeavors that result in decreased levels of circulating substrate (e.g., by prolonged maintenance in the circulation using modified enzymes[233]) may provide effective therapy in Fabry disease without the necessity to deliver the enzyme to the vascular endothelium.

TABLE VII. Primary Cellular Site of Pathology in the Lysosomal Storage Diseases[a]

Disease	Defective enzyme	Primary site of lysosomal pathology
Sphingolipidoses:		
Fabry disease	α-Galactosidase A	Vascular endothelial cells
Gaucher Disease	β-Glucosidase	
Type 1		RES[a]
Type 2		Neuron
G_{M2} gangliosidosis		
Type 1, 3	β-Hexosaminidase A	Neuron
Type 2	β-Hexosaminidase A and B	Neuron, RES
G_{M1} gangliosidosis	G_{M1}-β-galactosidase	Neuron, RES
Krabbe disease	Galactosylceramide: β-galactosidase	Neuron
Metachromatic leukodystrophy	Cerebroside sulfatase	Schwann and glial cells
Nieman–Pick disease	Sphingomyelinase	
Type A		Neuron
Type B		RES
Farber disease	Ceramidase	Neuron, RES
Mucopolysaccharidoses:		
MPS I	α-L-Iduronidase	
I-H		Neuron, CTC, RES
I-S		CTC, RES
I-H/S		CTC, RES
MPS II	α-L-Iduronidate sulfatase	
Mild		CTC, RES
Severe		Neuron, CTC, RES
		? Neuron
MPS III		
A	Heparan N-sulfatase	Neuron, CTC, RES
B	N-Acetyl-α-D-glucosaminidase	Neuron, CTC, RES
C	Acetyl CoA: α-glucosaminide N-acetyl transferase	Neuron, CTC, RES
D	N-Acetyl glucosamine-6-sulfate sulfatose	Neuron, CTC, RES
MPS IV	Hexosamine 6-sulfatase	Chondrocytes, CTC, RES
MPS VI	N-acetylgalactosamine 4-sulfatase	
Severe		RES, CTC
Mild		RES, CTC
MPS VII	β-Glucuronidase	? Neuron, CTC, RES
Glycoproteinoses:		
Mannosidosis	α-Mannosidase A and B	Neuron
Fucosidosis	α-L-Fucosidase	Neuron
Aspartylglucosaminuria	Aspartylglucosaminidase	? Neuron

(continued)

TABLE VII. (*Continued*)

Disease	Defective enzyme	Primary site of lysosomal pathology
Mucolipidosis II, III	?	Neuron, CTC
Sialidosis	α-Neuraminidase	
With mental retardation		Neuron
Without mental retardation		RES, CTC
Glycogenosis:		
Type II	Acid α-glucosidase	Striated muscle cell

[a] Adapted from Stanbury *et al.*[1]
[b] RES, Reticuloendothelial system; CTC, connective tissue cells.

This "equilibrium-depletion" concept may be extended to other lysosomal storage diseases. In Gaucher type 1 disease, the pathologic substrate glucosyl ceramide, appears to be readily exchanged between the erythrocyte membrane and plasma.[235] If this lipid substrate is in a dynamic equilibrium with other cell types, enzyme replacement which results in substrate catabolism in reticuloendothelial cells, for example, may lead to total body substrate depletion. This mechanism may provide an effective, but not specific, approach for enzyme replacement in disorders which have readily diffusable, lipid-soluble or rapidly equilibrating substrates. The feasibility of this approach should be investigated in appropriate animal models prior to human trials.

Enzyme Delivery Strategies

Two attractive strategies to enhance the target delivery of lysosomal hydrolases (and other glycoconjugates) involve (1) receptor-mediated molecular recognition processes and (2) the use of carrier vesicles including liposomes and autologous erythrocytes. In the former, the binding of the carbohydrate moiety of the administered glycoprotein-hydrolase to a cell-specific membrane-bound receptor signals the internalization and uptake of the molecule by components of the lysosomal apparatus. Thus, the identification of cell-specific receptor systems and the chemical modification and/or selection of appropriate isozymes may be exploited to "address-label" an enzyme to a particular cell type(s). Enzyme entrapment in biodegradable vesicles, such as liposomes or autologous erythrocytes, also provides a strategy to enhance enzyme delivery to critical sites of substrate pathology as well as protect the active enzyme from bioinactivation and immunologic surveillance in the circulation. Furthermore, the surface of these vesicles can be chemically modified or coated with compounds which may facilitate cell-specific uptake.

Receptor-Mediated Uptake Systems

Table VIII lists the known molecular recognition markers involved in glycoprotein uptake by various mammalian cells and tissues. To date, receptor-mediated uptake has been demonstrated for complex circulating glycoproteins,[242–248] lysosomal hydrolases,[247–252] low-density lipoproteins,[253,254] and transcobalamin II.[255] Since most lysosomal hydrolases are glycoproteins, the studies of receptor-mediated glycoprotein clearance and uptake are particularly relevant to the *in vivo* fate of administered hydrolases. Reviews of the current status of these receptor systems are available.[242,248,256,257]

TABLE VIII. Mammalian Recognition Markers for Receptor-Mediated Uptake[a]

Organ	Cell type	Animal	Marker	Reference
Liver	Hepatocyte	Rat, rabbit, mouse	β-Gal	243, 244
	Hepatocyte	Mouse	Fucα(1→3)glcNAc	244, 345
	Kupffer cell	Rat	αMan, GlcNAc	245-249
Lung	Alveolar macrophages	Rat	Manα(1→6) > Manα(1→2) or Manα(1→3)	249
		Calf	β-Gal	267
		Rat	β-Gal	243
Heart		Calf	β-Gal	243
		Rat	β-Gal	243
Kidney		Rat	β-Gal	243
Spleen		Rat	β-Gal	243
Thymus		Rat	β-Gal	243
Brain		Rat	β-Gal	243
	Synaptosomes	Rat	?	336
	Neuroblastoma	Mouse	β-Gal	243
Skin	Fibroblast	Human	Man-6-P	275
		Human	LDL	253
	Artery (smooth muscle)	Human	LDL	254

[a] After Wold.[251]

Recognition Markers for Prolonged Circulation and Hepatocyte Uptake

The discovery by Ashwell and colleagues[250,251] that asialoglycoproteins were selectively cleared from the circulation by hepatocytes was the seminal observation which led to the characterization of carbohydrate-mediated recognition processes. They found that circulating glycoproteins containing nonreducing termini of N-acetylneuraminic acid [(NANA)-β-Gal-X] remained in the circulation when injected into rats; the removal of the terminal sialic acid, exposing the penultimate galactose residue, however, signaled the clearance of the asialoglycoprotein from the circulation and its selective uptake and degradation by hepatocytes. This galactose-specific receptor has been purified from hepatocyte membranes and characterized.[258,259] Recently, another hepatocyte-specific receptor has been identified in which fucose is the recognition marker.[244] These findings provided important information concerning the mechanism(s) by which circulating glycoproteins are maintained in and cleared from the circulation.

The fact that sialylated glycoproteins were retained in the circulation suggests that enzyme replacement for disorders which require the active exogenous enzyme to be retained in the circulation (e.g., α_1-antitrypsin deficiency) should use the most sialylated isozyme or attempt to appropriately conjugate sialic acid residues onto the exogenous enzyme. In addition, disorders in which the hepatocyte is a target cell should exploit the galactose (or fucose) receptor to address administered enzymes or other therapeutic molecules to hepatocytes.

Recognition Markers for Reticuloendothelial Cell Uptake

The *in vivo* plasma clearance and tissue uptake and/or the *in vitro* tissue or cellular uptake of modified glycoproteins and neoglycoproteins has been determined by different laboratories.[245,247,260–267] Using the Ashwell approach, specific carbohydrate residues were sequentially removed from asparagine-linked complex glycoproteins as shown in Fig. 5. Another strategy involved conjugation of a particular carbohydrate residue with a protein such as bovine albumin or orosomucoid.[248] A series of these neoglycoproteins were constructed with different terminal carbohydrate moieties, (e.g., [^{125}I]mannan, [^{125}I]-Glc, [^{125}I]-Gal, etc.). These com-

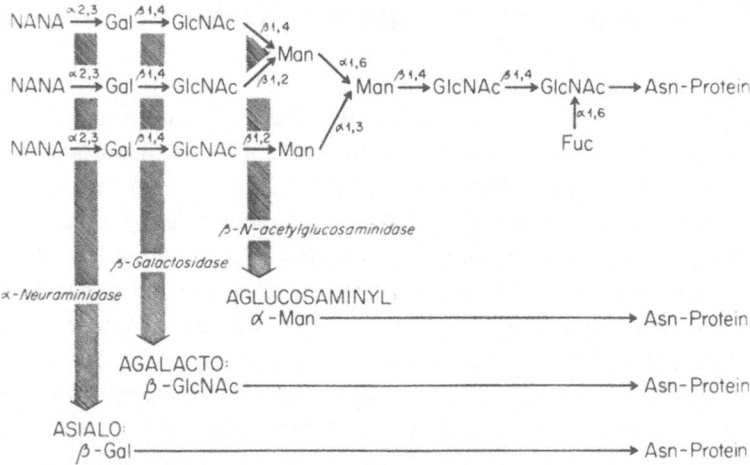

Fig. 5. Modified glycoproteins produced by the sequential enzymatic removal of the terminal carbohydrate residues from asparagine-linked complex glycoproteins. NANA, N-acetyl-neuraminic acid; Gal, galactose; GlcNAc, N-acetyl-glucosamine; Man, mannose; Fuc, L-fucose; Asn, asparagine.

pounds, as well as various purified carbohydrates, were used as receptor probes when injected with another glycoprotein (e.g., a lysosomal hydrolase) to determine which carbohydrate moiety inhibited the receptor-mediated uptake of the administered native or modified glycoprotein. In this way, a series of modified or neoglycoproteins can be used to probe the carbohydrate specificity of the receptor for each of these glycoconjugates.

Using these techniques, the mannosyl/N-acetylglucosamine receptor on reticuloendothelial cells (including Kupffer cells) and alveolar macrophages was discovered. Agalacto-orosomucoid markedly inhibited the plasma clearance of simultaneously injected β-glucuronidase (as well as several other partially purified lysosomal hydrolases) from the circulation of rats.[268] These findings indicated that the terminal GlcNAc residue of agalacto-orosomucoid (Fig. 5) competed with the enzyme for receptor-mediated uptake. Further experiments demonstrated that the clearance of agalacto-orosomucoid was inhibited by mannose, and mannans, as well as N-acetylglucosamine, indicating that mannose was also recognized by the same receptor.[269] The clearance of agalacto-orosomucoid from the circulation was inhibited more by mannans containing α1 → 6 linkages than α1 → 3 or α1 → 2 linkages which further characterized the specificity of the mannose/N-acetylglucosamine receptor.[249] The fact that these compounds did not alter the clearance kinetics of asialo-orosomucoid distinguished this receptor from the galactose-specific receptor of hepatocytes. More recently, the mannose/N-acetylglucosamine receptor has been isolated from rat and rabbit liver and shown to be a mannan-binding protein.[270,271]

The mannose/N-acetylglucosamine glycoprotein receptor may be of particular relevance to enzyme replacement endeavors. As shown in Table VII, there are a number of storage disorders in which the target site of pathology is the reticuloendothelial system. Modification of the appropriate enzyme to expose GlcNAc terminal residues should enhance reticuloendothelial cell uptake.

Recognition Marker for Fibroblast Uptake

The early work of Neufeld and colleagues[272] lead to the characterization of a carbohydrate recognition marker for lysosomal enzyme uptake by cultured skin fibroblasts. They found that hydrolases treated with periodate were not taken up by fibroblasts.[273] Sly and co-workers[203] reported that different tissue sources of human β-glucuronidase were taken up at different rates by β-glucuronidase-deficient fibroblasts. The "high-uptake" forms had a subpopulation of molecules that were more acidic than the less acidic, "low-uptake" forms.[274] Once pinocytosed by the fibroblasts, the "high-uptake" form was converted to the less acidic, "low-uptake" form. Experiments demonstrated that the "high-uptake" property of the enzyme could be abolished by treatment with alkaline phosphatase, implicating phosphate in the recognition process. These observations and the use of sugar and oligosaccharide inhibitors of enzyme uptake, led to the finding that the "high uptake" form had a phosphomannosyl recognition marker, mannose-6-phosphate (Man-6-P), incorporated into its oligosaccharide moiety; this moiety effected the receptor-mediated uptake of the enzyme into fibroblasts. Man-6-P has been found on other lysosomal hydrolases and it has been hypothesized that this moiety is involved in the uptake and trafficking of lysosomal hydrolases in the lysosomal apparatus.[275–277]

It is conceivable that this recognition system may be useful for enzyme replacement in disorders characterized by lysosomal accumulation in fibroblasts and other connective tissue cells (e.g., mucopolysaccharidoses without neural involvement, MPS II mild, IV, and VI). Further characterization of the ubiquity and tissue specificity of this receptor in other differentiated mesenchymal cells (e.g., osteoblasts, osteoclasts, histiocytes, microglia, smooth muscle, vascular endothelium, and macrophages) will determine the potential of the Man-6-P receptor-mediated uptake for targeted enzyme delivery.

LDL Recognition for Uptake by Fibroblasts, Vascular Endothelium, and Smooth Muscle

Goldstein and Brown[235,253,254,278-281] elucidated the mechanisms by which low-density lipoproteins (LDL) regulate the uptake, storage, and synthesis of cholesterol in human cells. The keystone of their work was the finding that normal human fibroblasts possessed a cell-surface receptor which mediated the endocytosis of cholesterol-rich LDL.[282] The receptor-mediated uptake of LDL has been shown for cultured fibroblasts, arterial smooth muscle cells, lymphoid cells, endothelial cells, and glial cells.[283-286] LDL is synthesized in the liver and intestine.[287,288] It is secreted into the circulation and becomes bound to its high-affinity receptor. Following endocytosis and delivery to the lysosomal apparatus, the cholesterol esters, proteins, and other associated lipids are hydrolyzed.[279-281,288]

The LDL receptor appears to be the protein product of a single gene locus.[289] Patients heterozygous for familial hypercholesterolemia have half the normal number of LDL receptors on their fibroblast membranes. Extensive studies of receptor binding to [125I]-LDL have been reported.[279,280,287] The binding is cation dependent and reversed by EDTA. VLDL competes with LDL for receptor binding, however HDL appears to be an extremely poor inhibitor. The receptor is sensitive to pronase, papain, and trypsin, but resistant to neuraminidase and β-galactosidase.[290] Recent studies have indicated that the receptor recognizes lysine and arginine residues on the LDL molecule.[291] The internalization of receptor bound LDL is extremely rapid; half of the [125I]-LDL bound is internalized every 3 min at 37°C.[292,293] The LDL particle is delivered intact to the lysosomal apparatus, where both the protein and lipid components are hydrolyzed. It is notable that the receptor does not appear to be internalized with the LDL. Electron-microscopic studies using ferritin-labeled LDL demonstrated that the LDL receptors were concentrated in fibroblast membrane pits (~0.5 nm) which were ferritin coated on both sides. These coated regions represented only 2% of the total surface of the fibroblast membrane.[293]

The application of the receptor-mediated uptake of LDL for enzyme replacement may be limited because in the steady state (i.e., in the presence of LDL), the number of receptors on normal fibroblasts (and presumably all cells containing the receptor *in vivo*) is low. Only when cultured cells were grown in LDL-depleted media was there a rapid increase in the number of LDL receptors. Thus, efforts to treat patients with various lysosomal disorders by using LDL-conjugated enzymes may require prior depletion of circulating LDL. Although circulating LDL is cleared primarily by nonhepatic cells, it is likely that many cell types have LDL receptors which may dilute efforts to target LDL-conjugated macromolecules to specific cell types.

Application of Receptor-Mediated Uptake for Enzyme Replacement

Receptor-mediated uptake systems offer an attractive and relatively unexploited strategy to target lysosomal enzymes to cells other than the liver. Several aspects of these recognition marker–receptor systems may be used to enhance enzyme delivery to selected tissue or cell types. These include (1) the selective removal of carbohydrate moieties from the enzyme to expose specific recognition markers or the conjugation of specific carbohydrate "address labels" onto hydrolases, (2) the use of nontoxic carbohydrates, oligosaccharides, or glycopeptides to selectively block enzyme binding for uptake by other cell types, (3) the selection and/or modification of specific isozymes with "high-uptake" and stability characteristics, or (4) a combination of the above.

The feasibility of each of these receptor-mediated strategies for enzyme replacement in selected lysosomal storage disease already has been demonstrated *in vitro* and *in vivo*. Brady and colleagues[294] sequentially removed the sialic acid and β-galactosyl residues from purified human placental β-glucosidase. They demonstrated that the modified enzymes were taken up preferentially by isolated hepatocytes (asialoglucosidase) or Kupffer cells (agalactoglucosidase) presumably by the galactose and mannose/N-acetylglucosamine recognition systems, respectively. These observations indicate that specific isozymes may be biologically coded and/or chemically modified to promote their differential distribution and uptake for the controlled and specific delivery of exogenous enzymes.

The coupling of specific carbohydrate residues to glycoprotein enzymes (neoglycoproteins) offers another intriguing possibility for target delivery. Although this approach has been used to investigate carbohydrate receptor specificities, [248,252] neoglycohydrolases have not been synthesized to date. However, this strategy may provide an efficient means of selective cell delivery for enzymes and other therapeutic proteins and is worthy of investigation.

Rattazzi *et al.*[295,296] have achieved some success in minimizing the hepatic uptake of β-hexosaminidase A in cats by injecting mannans prior to enzyme; in this way, enzyme uptake by the mannosyl/*N*-acetylglucosamine (and perhaps Man-6-P) recognition system was partially blocked. Since the mannan infusion did not appear to be toxic, this approach may be useful to enhance the nonhepatic uptake of enzyme in disorders in which hepatocytes and reticuloendothelial cells are not the primary sites of pathology. Further studies are needed to assess the side effects (e.g., hyperosmolality, toxicity) of the blocking compounds. Alternatively, the hyperosmolality may be used to open the blood–brain barrier and gain access to the central nervous system.

The third receptor-mediated strategy involves the selective purification of different molecular forms or tissue isozymes of a specific enzymatic activity for differential uptake and survival.[297,298] For example, rat β-glucuronidase and *N*-acetyl-β-glucosaminidase purified from preputial gland and liver, respectively, were rapidly cleared from the circulation of rats when compared to the slower clearance rates of serum β-glucuronidase and epididymal *N*-acetyl-β-glucosaminidase.[297]. Similarly, a differential intracellular survival of recovered hepatic activity was observed following the intravenous administration of purified bovine splenic, hepatic, and renal β-glucuronidase isozymes into β-glucuronidase-deficient mice.[298] Although all three molecular forms of this enzyme were rapidly cleared from the circulation ($t_{1/2}$ 2–3 min) by the liver, the splenic form remained in liver up to 62 hr whereas the hepatic and renal forms were not detectable by 24 hr. These findings have implications for the target delivery and lysosomal survival of various enzyme forms or isozymes. Specific human isozymes, obtained perhaps from the normal tissues or fluids corresponding to the pathologic target tissue, may further minimize the possibility of these complications (e.g., the specific, normal splenic or hepatic β-glucosidase isozyme for replacement in Gaucher disease type 1 or the specific normal endothelial cell α-galactosidase A isozyme for Fabry disease). In addition, selected isozymes may promote uptake by the pathologic target tissues via recognition of specific tissue and perhaps subcellular receptors and may also provide additional protection from endogenous catheptic destruction.

Carrier Mediated Delivery

The demonstration that enzyme can be entrapped in synthetic lipid spherules, termed liposomes,[299-301] and human erythrocytes[302,316] suggested the intriguing possibility that these biodegradable vesicles may be useful as carriers of exogenous enzyme. Therefore, we explored the feasibility of enzyme entrapment in negatively and positively charged liposomes and in autologous murine erythrocytes and evaluated the effectiveness of these enzyme-containing vesicles as a strategy for efficient enzyme delivery in the murine system.[300,302]

Liposome Entrapment of Enzyme—in Vivo Fate

A procedure for the entrapment of enzyme in negatively- and positively-charged liposomes[300] is illustrated in Fig. 6. Negatively charged liposomes containing bovine β-glucuronidase were prepared with highly purified phosphatidyl choline, cholesterol, and negatively charged phosphatidic acid in a molar ratio of $7:2:1$ by the method of Gregoriadis and Leathwood.[303] To separate unentrapped enzyme, the mixture was centrifuged and the pelleted liposomes were resuspended in buffer and applied to a Sepharose 6B column. The enzyme-loaded liposomes were eluted in the void volume, centrifuged, and resuspended in buffered saline; column fractions containing the unentrapped enzyme were pooled, concentrated, and dialyzed for subsequent liposome entrapment.[300]

Positively charged liposomes were prepared similarly, except that stearylamine, a positively charged lipid, was used instead of phosphatidic acid. Enzyme entrapment was carried out at pH 4.0, below the isoelectric point of 5.1 for bovine β-glucuronidase, to avoid the formation of lipid–enzyme aggregates between the positively charged liposomes and the enzyme which is negatively charged at pH 7.2. The positively charged liposomes were washed extensively by repeated resuspension in buffer and centrifugation; the unentrapped enzyme in the supernatant was recovered for reuse.[300] Both positively and negatively charged liposomes were resuspended in buffered saline prior to intravenous injection into β-glucuronidase-deficient mice. Approximately 5% of the enzyme added

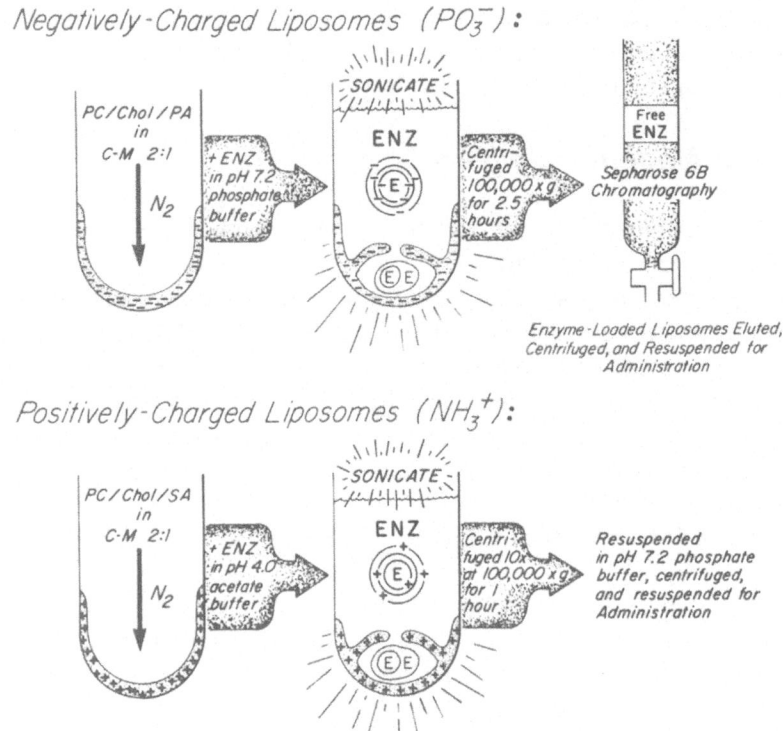

Fig. 6. Liposome entrapment of enzyme.[300]

to the lipid film was routinely entrapped in the negatively or positively charged liposomes.[300] It should be noted that variations in liposome structure (unilamellar versus multilamellar) and composition result in differential *in vivo* stabilities, clearance rates, and tissue distributions.[304–309]

As shown in Fig. 7, the intravenous administration of enzyme in negatively charged liposomes resulted in a rapid clearance of enzymatic activity from the circulation with a concomitant uptake of activity primarily in the liver; in addition, approximately 20% of dose was recovered in the kidney for almost 4 days. Administration of enzyme entrapped in positively charged liposomes was characterized by a rapid hepatic uptake of activity to a maximum level of 75% of dose by 1 hr with a subsequent retention of activity for up to 11 days, 3 days longer than that observed with enzyme entrapped in negatively charged liposomes (Fig. 7). With

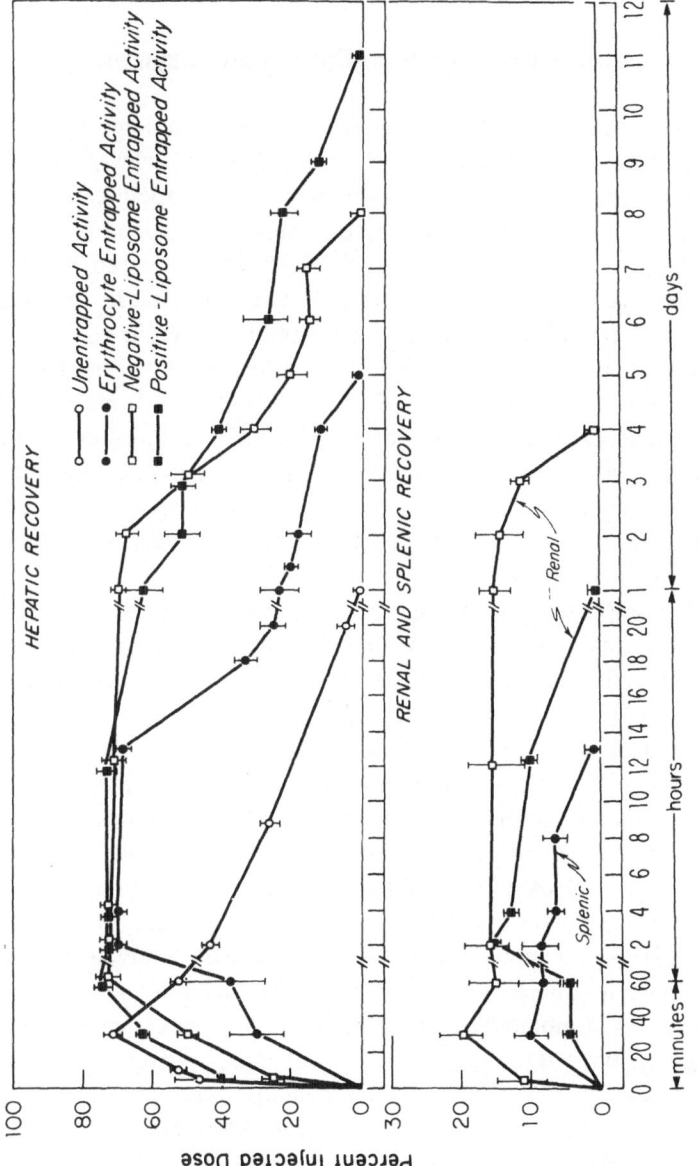

Fig. 7. Comparison of the *in vivo* fates of bovine β-glucuronidase activity unentrapped (O—O) or entrapped in negatively-charged liposomes (□—□), positively-charged liposomes (■—■), or autologous erythrocytes (●—●) in β-glucuronidase-deficient mice.

enzyme entrapped in either negatively or positively charged liposomes, no activity was observed in the bone marrow or brain.[300]

Caveat to Administration of Liposome-Entrapped Enzymes

Hepatic subcellular fractionation studies and immunologic evaluations provided important data concerning the potentially deleterious effects of liposomes as carriers for enzymes or other therapeutic agents. At each time point after injection of β-glucuronidase entrapped in negatively charged liposomes, approximately 70% of the recovered bovine activity was detected in the lysosomally enriched fraction. In marked contrast, however, the administration of enzyme entrapped in positively charged liposomes (containing stearylamine) caused a temporary labilization of the lysosomal membranes of the recipients, resulting in the intracellular release of endogenous, potent lysosomal hydrolases.[310] In addition, cellular and humoral immune responses were elicited by negatively charged liposomes. Liposomes failed to prevent antienzyme antibody-complex formation since the decreased tissue recovery of enzyme in previously sensitized mice was due presumably to degradation of antigen–antibody complexes. Moreover, the liposome vesicle itself elicited a cellular immune response by enhancing phagocytic activity in the reticuloendothelial cells of liver and spleen and the activation of peritoneal macrophages.[311]

Similar findings were reported by Van Rooijen and Van Nieuwmegen[312,313] following the repeated injection of free human serum albumin (HSA), free HSA and buffer-loaded negatively charged liposomes, or liposome-entrapped HSA into rabbits. Although free HSA did not elicit an immune response, HSA entrapped in liposomes resulted in a marked response. An immune response was also detected following the administration of free HSA and empty liposomes confirming the adjuvant nature of the liposome itself. These studies suggested that the adjuvant activity of macrophages was stimulated by the digestion of liposomal membranes. Other investigators have shown that positively charged liposomes activate human complement[314] and negatively charged liposomes interact with plasma constituents causing the release of entrapped markers.[315]

These studies identified the potentially harmful physiologic and immunologic effects that might be associated with the use of liposomes composed of certain lipids and emphasize the need to fully evaluate these carriers in animal models prior to human trials.

Erythrocyte-Entrapment—*In Vivo* Fate

The demonstration that enzymes can be entrapped in human erythrocytes suggested the possibility that these biodegradable vesicles may be useful as *in vivo* carriers of exogenous enzyme. Therefore, we explored the feasibility of enzyme entrapment in autologous murine erythrocytes and evaluated the effectiveness of these enzyme-containing vesicles as a strategy for efficient enzyme delivery in our β-glucuronidase-deficient murine system.

Bovine β-glucuronidase was entrapped in autologous murine erythrocytes essentially by the hypotonic exchange method of Ihler *et al.*,[316] as illustrated in Fig. 8. For erythrocyte entrapment, bovine enzyme was mixed with packed, washed murine erythrocytes. Hypotonic exchange was induced by the rapid addition of 5.0 ml of distilled water and terminated after 60 sec by the addition of NaCl to restore isotonicity. The enzyme-containing erythrocytes were recovered by centrifugation; the cells were then washed repeatedly until no further activity was detected in the supernatant. The entrapped activity remained sedimentable after repeated washing of the erythrocytes with either 0.15 M or 0.3 M NaCl, but was readily released into solution following brief sonication of the cells. Approximately 4% of available bovine β-glucuronidase was routinely entrapped in murine erythrocytes under these conditions.[302] The amount of protein entrapped has been shown to be a function of the hypotonic exchange conditions, including degree and duration of hypotonicity, concentration of various ions, temperature, and the molecular weight or size of the molecule to be entrapped.[316–318] Methods using variations of the hypotonic exchange procedure have been developed to improve entrapment efficiency.[319–323]

As shown in Fig. 7, erythrocyte-entrapped activity was primarily recovered in the liver following intravenous injection. Maximal hepatic

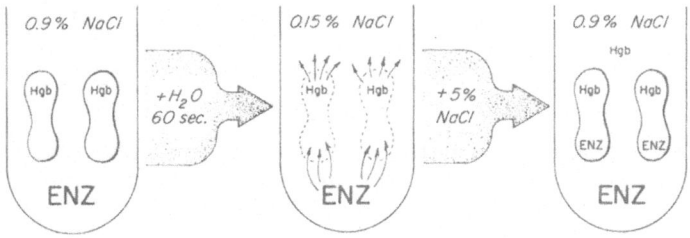

Fig. 8. Erythrocyte entrapment of enzyme by hypotonic exchange.[302]

uptake (70% of dose) occurred at 2 hr; it was maintained at that level up to 13 hr postinjection and retained in hepatic tissues as long as 5 days or five times longer than unentrapped enzyme. Hepatic subcellular fractionation indicated that more than 80% of the recovered exogenous activity was detected in the lysosomally enriched fraction at various time points after administration.

Splenic uptake of erythrocyte-entrapped activity was ~10% of dose; however, as shown in Fig. 9, the chemical and enzymatic modification of loaded erythrocytes significantly increased splenic uptake.[310] In support of these findings, Jancik et al.[324] injected autologous neuraminidase-treated erythrocytes into rats and demonstrated that they were sequestered in the reticuloendothelial cells of the liver and spleen to a significantly greater extent than untreated control erythrocytes.

Importantly, the plasma clearance and tissue delivery of enzyme entrapped in erythrocytes appears to be dependent on the entrapment method.[320] Several techniques for erythrocyte entrapment have been described including hypotonic exchange,[302,310,315,321,322,325] dialysis,[323] and drug-induced endocytosis.[327] These methods have been evaluated recently with respect to factors which may effect the in vivo fate of the administered erythrocytes and/or cause potential immunologic complications.[320] These include the exposure of antigenic determinants normally masked in the erythrocyte membrane, the surface adhesion of en-

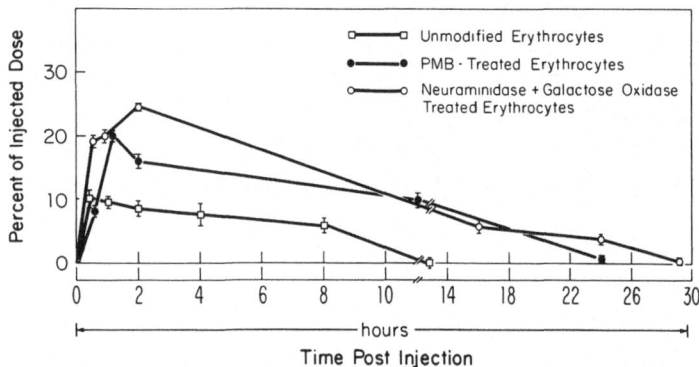

Fig. 9. Splenic recovery of bovine β-glucuronidase activity entrapped in untreated (□—□) and chemically (PMB, p-hydroxymercurobenzoate (●—●)), or enzymatically (neuraminidase and galactose oxidase, ○—○) modified murine erythrocytes following administration to β-glucuronidase-deficient mice.[310]

zyme or other entrapped markers, or alterations of erythrocyte morphology. Although there was little, if any, release of entrapped enzyme, the various entrapment procedures induced different morphologic changes as shown in Fig. 10. The different morphologies might be useful for target delivery. For example, enzyme-loaded cells which become echinocytes or stomatocytes presumably will be cleared rapidly by the reticuloendothelial system, whereas cells which tend to retain their discoid shape will be more deformable and will be retained in the circulation longer. Other modifications of the erythrocyte membrane, prior to loading, such as treatment with sulfhydryl-active reagents,[328] neuraminidase,[329,330] or glutaraldehyde[326] may enhance splenic uptake. Alternatively, coating the erythrocytes with various components such as immunoglobulins may prolong the survival of the erythrocytes in the circulation. In addition, procedures which restore deformed cells to a discoid morphology (e.g., ATP and EDTA[331]) might be useful for the treatment of conditions in which toxic compounds in the plasma could enter the circulating cells and be inactivated by the entrapped agent.

Erythrocyte-Entrapped Enzyme—Immunologic Evaluation

The immune response of repeatedly administered erythrocytes loaded with bovine β-glucuronidase to β-glucuronidase-deficient mice was evaluated.[332] Characterization of the entrapment method demonstrated no detectable β-glucuronidase on the erythrocyte surface by the sensitive hemagglutination test. Repeated intravenous administration to mice of buffer- or enzyme-loaded erythrocytes failed to elicit an immune response to the erythrocyte carrier or entrapped protein; in addition, the fate of entrapped activity in previously sensitized mice was unaltered. Moreover, erythrocyte entrapment protected enzyme from circulating anti-β-glucuronidase antibodies in mice previously sensitized subcutaneously with unentrapped enzyme in Freund's incomplete adjuvant.[302]

These findings indicated that erythrocytes loaded with β-glucuronidase were immunologically indistinguishable from intact erythrocytes *in vivo* and *in vitro*. Thus, entrapment of enzyme preparations in nonimmunogenic, biodegradable autologous erythrocytes (in contrast to liposomes) may effectively protect administered activity from the immune surveillance system, thereby avoiding untoward reactions and maximizing the amount of activity that reaches target subcellular sites of pathology.

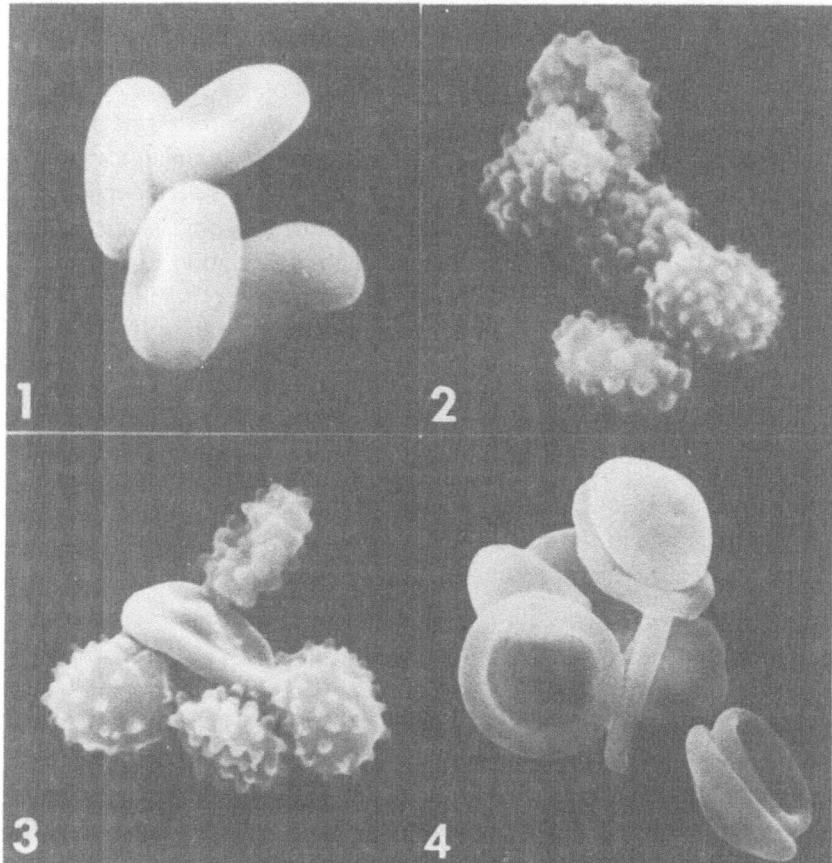

Fig. 10. Scanning-electron-microscopic appearance of normal biconcave human erythrocytes (1) and enzyme-entrapped erythrocytes loaded by various hypotonic exchange methods (2–6), by a dialysis technique (7) and by chlorpromazine-induced endocytosis (8). For details see Fiddler *et al.*[320]

Enzyme Delivery Strategies for Neural Uptake

As noted in Table VII, the target site of many lysosomal storage diseases is the neuron. Previous animal and human replacement trials have demonstrated the inability of intravenously administered enzymes to cross the blood–brain barrier and gain access to neural tissue.[226] Intrathecal administration has been attempted[233]; however, no evidence of enzyme uptake in neural tissue was demonstrated. The inherent limitations and dangers of repeated intrathecal injections warrant the devel-

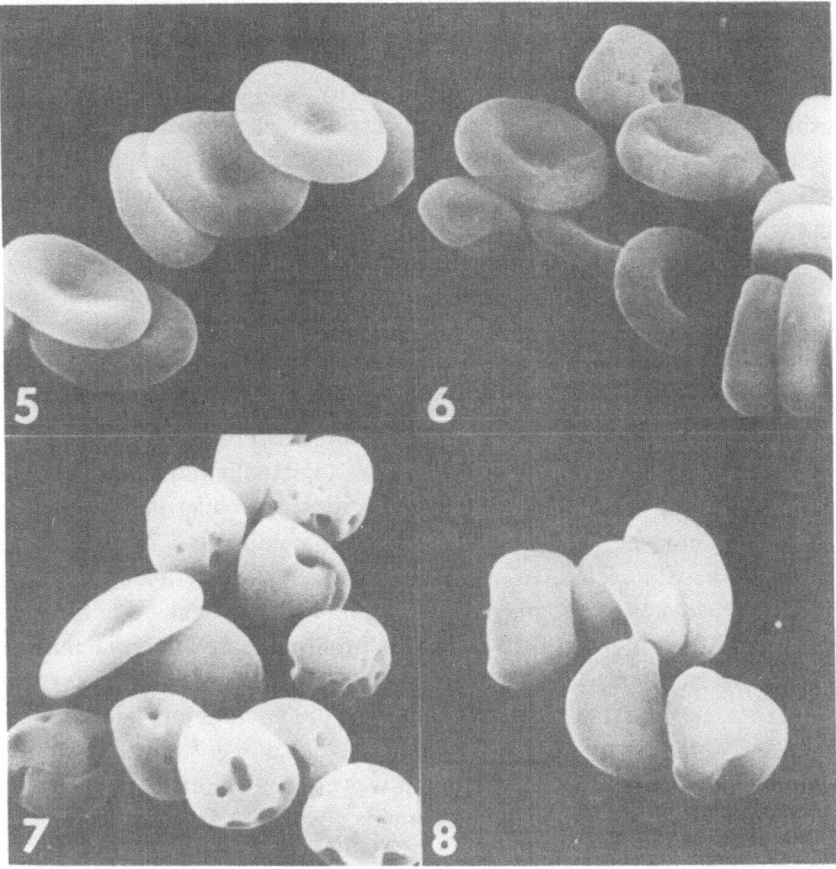

Fig. 10. (*Continued*)

opment of other strategies to achieve uptake of intravenously administered enzyme across the blood–brain barrier.

Effective replacement in these disorders requires uptake of the enzyme not only into brain tissues, but specifically into the lysosomes of the neurons in which the pathologic substrate is accumulated. This requisite presents another major obstacle, since even if a reasonable means were devised to achieve access to neural tissues, it is unlikely that the exogenous enzyme would be taken up by the lysosomes of the relatively nonphagocytic neurons. Instead, the enzyme would presumably be sequestered in the more endocytic neuroglial cells.

However, several investigators have attempted to modify the

blood–brain barrier in animal systems in order to assess neural enzyme uptake. Barringer *et al.*[333] demonstrated that the blood–brain barrier of Sprague–Dawley rats could be temporarily and reversibly opened by osmotic agents for macromolecular access without apparent brain injury. Following intracarotid infusion of 1.6 molal mannitol or arabinose, peroxidase or α-mannosidase was injected. Histochemical staining of brain tissue obtained 72 hours after infusion revealed uptake of peroxidase by neurons, as well as their processes and axonal projections; an occasional glial cell was stained. Following subcellular fractionation, about half of the recovered peroxidase was in the lysosomally enriched fraction.

Rattazzi *et al.*[334,335] evaluated the effectiveness of hyperbaric oxygen (2.5 ATA; 100% O_2 for 90 min) and intracarotid air microembolism to reversibly open the blood–brain barrier of cats. Following exposure to hyperbaric oxygen, β-hexosaminidase was injected into the femoral vein and the neural uptake determined. In order to prolong the half-life of exogenous enzyme in the circulation and maximize its CNS exposure, the rapid hepatic clearance of the enzyme was partially inhibited by infusion of mannans or ovomucoid (i.e., mannose/N-acetylglucosamine system). This inhibition resulted in a 3- to 5-fold increase in neural enzyme uptake over that observed without blockage of the hepatic uptake receptor. It was notable that the mannan infusion did not interfere with enzyme uptake by brain tissue, suggesting the existence of a different recognition marker for neural uptake. However, it was concluded that the hyperbaric oxygen treatment was unlikely to provide significant amounts of enzyme to the brain.

More encouraging results were obtained in preliminary experiments of intracarotid air microembolization. Following the injection of small volumes of air, mannans and then human β-hexosaminidase B were administered intravenously to a cat with Sandhoff disease.[335,336] One hour after enzyme injection, the animal was sacrificed; the exogenous β-hexosaminidase reached 20–30% of the endogenous level in normal feline brain. These results suggest that air microembolization or similar mechanical methods to open the blood–brain barrier—provided they are reversible and do not result in neurologic injury—permit access to neuronal sites of substrate deposition by the nonselective extravasation of plasma into the brain parenchyma.

A third approach[337] involves the investigation of neuronal receptor systems for lysosomal hydrolases, analogous to the receptors for neurotransmitters and neuroregulatory peptides. Kusiak *et al.*[337] have investigated the *in vitro* binding of human placental [^{125}I]-β-hexosaminidase A to isolated rat synaptosomes and synaptic plasma membranes. Binding was saturable, pH-, Ca^{2+}-, and time-dependent, trypsin-sensitive, and inhibited by an excess of unlabeled hexosaminidase. These findings indicated the existence of a receptor system for lysosomal hydrolases in isolated rat brain synaptosomes. Further studies are required to document the occurrence of this system in human nervous tissues and to determine the nature of the recognition marker and membrane receptor. If a recognition system for specific neuronal uptake is documented, and if the mechanisms to open the blood–brain barrier prove reversible and safe, there is a potential basis to overcome the obstacles to enzyme replacement for disorders with primary neurologic involvement.

Clinical Application of Enzyme Delivery Strategies

Table IX summarizes the application of receptor- and carrier-mediated delivery strategies in recent clinical trials of enzyme replacement. For the most part, these studies were designed to evaluate the capability of these delivery strategies for the selective targeting of enzyme to the major sites of pathology in each disorder. Receptor-mediated strategies included the use of the Ashwell recognition signal for prolonged circulation of α-galactosidase A forms in Fabry disease[338] and the LDL receptor-mediated uptake for α-glucosidase replacement in glycogenosis type II.[339]

Carrier-mediated enzyme delivery strategies also have been attempted in humans. Replacement of liposome-entrapped α-glucosidase and β-glucosidase have been attempted in glycogenosis type II[340] and Gaucher type 1 disease,[341] respectively. These clinical trials were conducted prior to the animal studies which demonstrated immune responses to liposome-entrapped antigens and serve to underscore the need for careful evaluation in animal model systems prior to human use. In contrast, the feasibility and safety of erythrocyte-entrapped enzyme delivery, demonstrated first in animal models,[302] was extended for the enhanced reticuloendothelial cell uptake of β-glucosidase in patients with Gaucher disease type 1.[230]

Receptor Mediated Delivery—Administration of α-Galactosidase A Forms in Fabry Disease

In Fabry disease, the preferential deposition of circulating substrate in the vascular endothelium is responsible for the major manifestations

TABLE IX. Human Trials Exploiting Delivery Strategies

Disease	Year	Enzyme administered/source	Delivery strategy	Organ uptake	Substrate catabolism (tissue/substrate)	Reference
Glycogenosis type II	1976	Acid α-glucosidase/A. niger	Liposome entrapment	± liver	± liver/glycogen-muscle	340
Gaucher type 1	1977	β-Glucosidase/placenta	Liposome entrapment	– muscle	± glycogen	341, 347
	1977	β-Glucosidase/placenta	Erythrocyte entrapment		– plasma, RBC + leukocytes, platelets/ GL-1[a]	230
Glycogenosis type II	1979	Acid α-glucosidase/liver	LDL-conjugated	– muscle	± muscle/glycogen	339
Fabry	1979	α-Galactosidase A/spleen (s) or plasma (p)	Receptor-mediated	+ + + liver (s) + liver (p)	(s) + plasma/GL-3[b] (p) + + + plasma/GL-3	338

[a] GL-1, glucosyl ceramide.
[b] GL-3, trihexosyl ceramide.

of the disease. Therefore, α-galactosidase A replacement endeavors must be directed to the depletion of the accumulated substrate in the circulation and vascular endothelium. The recent finding that α-galactosidase A forms purified from plasma and splenic tissues were differentially glycosylated, suggested the intriguing possibility that these enzyme forms may be biologically coded for differential distribution and retention in the plasma by exploitation of the sialic acid recognition process described by Ashwell and Morrell.[242]

In our laboratory, highly purified α-galactosidase A from human plasma[342] and homogeneous enzyme from human spleen[343] have been characterized. Both enzymes bound to Concanavalin A-Sepharose, indicating that they were glycoproteins. In addition, immunodiffusion studies revealed a line of identity for both forms against rabbit anti-human splenic α-galactosidase A antibody. Although the two forms had similar physical and kinetic properties, the splenic form had a single isoelectric point at pH 4.3 whereas the plasma form had a pI of 3.7 and migrated more electronegatively on polyacrylamide disc gel electrophoresis at pH 7.0. After neuraminidase treatment, the pI of the plasma form, its K_m and migration on polyacrylamide gels were essentially the same as those of the treated or untreated splenic enzyme.[343] These results suggested differences in the posttranslational modification of the two forms, the plasma enzyme being more electronegative, consistent with it being uniquely sialylated or having more sialylated carbohydrate moieties than the tissue form. In addition, these findings were consistent with previous studies[344] demonstrating that plasma forms of various lysosomal hydrolases were more electronegative than their respective tissue counterparts.

Based on these *in vitro* findings and the plasma disappearance kinetics of sialylated and asialoglycoproteins in animals,[242] it was hypothesized that the more sialylated plasma enzyme, compared to the splenic form, would have a prolonged retention in the circulation and mediate an increased hydrolysis of circulating substrate.

In order to evaluate this hypothesis, two patients with Fabry disease received multiple intravenous injections of the unentrapped splenic and plasma enzyme forms.[338] The typical disappearance kinetics of equivalent doses of plasma and splenic enzymes from the circulation of the two recipients are shown in Fig. 11. The splenic form was rapidly cleared with a $t_{1/2}$ of approximately 10 min. In marked contrast, the plasma form had a slower plasma disappearance with a $t_{1/2}$ of about 70 min. The metabolic clearance rate of the splenic enzyme was 253 ml/min, which was

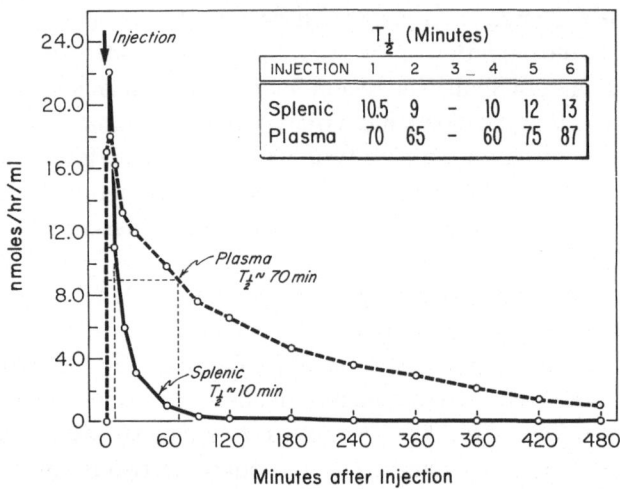

Fig. 11. Disappearance from plasma of splenic and plasma forms of α-galactosidase A activity following intravenous administration to a patient with Fabry disease.[338]

about eight times faster than that of the plasma enzyme. These clearance curves were compatible with distribution of the respective enzymes in at least two compartments, the plasma form being retained longer in the injected compartment. We postulated that the differential clearance kinetics were due to the presence of sialic acid residues on the plasma form of α-galactosidase A. These findings suggested the molecular recognition process for clearance of homologous hydrolases administered to humans was similar to the receptor-mediated uptake system described by Ashwell.[242]

Although the fate of these enzymes could not be determined, their differential metabolic effectiveness was underscored by their remarkably different substrate clearance and reaccumulation kinetics. The effect of the splenic form on the circulating substrate was rapid and transient and paralleled the rapid disappearance of the enzymatic activity from the plasma (Fig. 12). The splenic form was presumably taken up primarily by the lysosomal apparatus of the liver where it may have hydrolyzed accumulated substrate. In contrast, the prolonged retention of the plasma form in the circulation was associated with about a 25-fold greater substrate clearance (Fig. 13).

A metabolic labeling study was designed to gain insight into the source of the reaccumulating substrate in the circulation.[345] Previously, dideuteroglucose was similarly used to determine the turnover of trihexosyl ceramide in a normal individual and a hemizygote with Fabry disease.[234] Based on these studies, the enzymes were administered at a time when the percent deuterium incorporation into the circulating substrate was increasing and was expected to be about half-maximal. If the percent incorporation of deuterium increased after injection of the enzyme, it would reflect the reaccumulation of newly synthesized substrate (presumably by the liver). However, if the percentage of labeled substrate did not increase or was reduced (due to dilution with unlabeled substrate), it would provide evidence that previously synthesized (stored) substrate reaccumulated in the circulation. Gas chromatographic–mass spectrophotometric analyses indicated that the rate of deuterium incorporation into the reaccumulating substrate, trihexosyl ceramide in plasma, was decreased from 2–36 hr following injection of the plasma enzyme, whereas

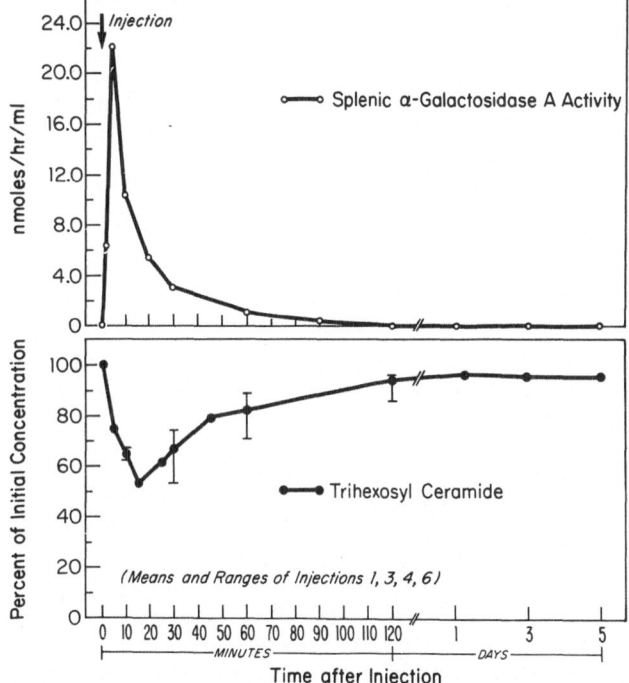

Fig. 12. Effect of intravenously administered splenic α-galactosidase A on plasma concentrations of trihexosyl ceramide (lower) compared to the enzyme's disappearance curve in plasma (upper).[338]

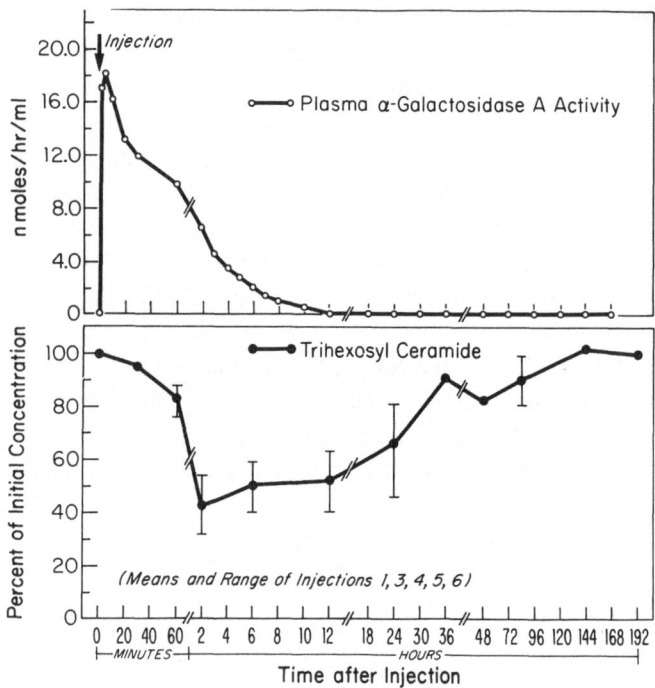

Fig. 13. Effect of intravenously administered plasma α-galactosidase A on plasma concentrations of trihexosyl ceramide (lower) compared to the enzyme's disappearance curve in plasma (upper).[338]

the percent of deuterium incorporation increased with the rapid reaccumulation of substrate following injection of the splenic enzyme.

These preliminary findings suggest that it may be possible to deplete stored substrate by administration of the plasma enzyme. The initial reaccumulation of unlabeled substrate, observed following the injection of the plasma enzyme, indicates that substrate may equilibrate across membranes of cells in which it is stored (e.g., vascular endothelium) to resaturate the depleted plasma pool. If resaturation of the plasma pool by unlabeled substrate can be confirmed in subsequent metabolic labeling experiments with the plasma enzyme, then efforts should be directed to modify the more easily obtained tissue forms of α-galactosidase A for prolonged retention in the circulation. In this way, regular doses of the sialylated plasma or modified tissue enzyme (i.e., using glycoprotein sialyl-transferase[346]) may provide effective treatment for this disease.

Receptor-Mediated Delivery—Administration of LDL-Coupled α-Glucosidase in Glycogenosis Type II

The use of LDL-conjugated enzyme for the target delivery of acid α-glucosidase has been investigated in a terminal patient with glycogenosis type II.[339] Since 95% of LDL is metabolized extrahepatically, it was reasoned that replacement of the deficient enzyme with purified human α-glucosidase coupled to LDL would be preferentially delivered to extrahepatic sites with LDL receptors. In addition, the uptake of the LDL complex by cultured skin fibroblasts from a patient with glycogenosis type II was nine times greater than the uptake of free enzyme.[339] Furthermore, the LDL-complexed activity was maintained in the fibroblasts with a half-life 2.5-fold longer than that of free enzyme.

Two doses of LDL–α-glucosidase were administered to the patient. The enzymatic activity and glycogen content were determined in muscle and hepatic biopsies and postmortem samples. A significant increase in the α-glucosidase activity in muscle, but not in heart or liver, was detected in samples obtained at autopsy 26 days after the second infusion of 13 mg of the complex containing 0.5 μmol/min/mg of α-glucosidase activity. Based on their results, it was estimated that 77% of the administered enzyme was recovered in muscle.[339] The glycogen content was decreased in liver, to a lesser extent in muscle, but not in heart tissue compared to data from previously autopsied patients. In addition, no adverse reactions were observed and antibodies to the enzyme or LDL were not detected. These preliminary results indicate the feasibility of LDL–enzyme conjugates for nonneural, nonreticuloendothelial lysosomal storage diseases. Indeed, this approach may be suited to replacement endeavors in disorders with primary endothelial cell pathology, if the recipients' levels of circulating LDL can be temporarily reduced.

Administration of Liposome-Entrapped α- and β-Glucosidases

Negatively charged liposomes containing *Aspergillus niger* amyloglucosidase were intravenously administered daily over seven days to an 8-month-old patient with glycogenosis type II.[340] The total dose of α-glucosidase administered was 32 μmoles of glucose released from maltose/min. Although her liver size was noted to decrease during the first four days of treatment, there was no other evidence of clinical effect. The patient expired from heart failure on the eighth day; tissues were obtained at autopsy for analysis. Less than 1% of normal α-glucosidase activity was detected in hepatic tissue, and no activity was recovered in skeletal or cardiac muscle. The glycogen content in skeletal muscle was twice that obtained in a preinfusion biopsy. The hepatic glycogen content (5.5%) was lower than the levels usually seen in patients with this disease (8.2% ± 1.6 1SD[281]) and led the investigators to suggest that hepatic catabolism had occurred. However, the design and limited nature of this trial precluded assessment of the biochemical as well as the immunologic aspects of this therapeutic endeavor.

Belchetz *et al.* [341] attempted enzyme replacement in a patient with Gaucher disease type I by the repeated intravenous administration of acid β-glucosidase and [111In]bleomycin entrapped in negatively charged liposomes.[341] Five injections (totaling 14,680 nmol/min hydrolyzed) were given over a period of 13 months. Neither the rate of plasma disappearance of the administered activity nor the glucosyl ceramide concentrations in plasma, erythrocytes, or other tissues was determined. Serial whole body scans over a 9-month period indicated a decreasing hepatic size with sequential doses as well as an increased rate of liposome clearance from the circulation. However, headaches, nausea, difficulty in concentration, sleepiness, and abdominal pain developed in the patient immediately after each administration of liposome-entrapped enzyme. These clinical sequelae presumably represent adverse physiologic and/or immune responses to the liposome vesicle and/or the enzyme preparation. This may be related to the adverse findings of liposome-entrapped enzyme administration in the β-glucuronidase-deficient mice.[311]

Subsequently, the patient received a course of 20 injections over the following 2 years.[347] The amount of lipid and enzyme administered was reduced (i.e., 500–600 nmol/min/injection) which resulted in marked lessening of the side effects. No adverse reactions occurred until the last injection; the patient developed urticaria with vasculitic lesions over the knees, suggesting a type IV delayed hypersensitivity reaction. The enzyme injections were terminated.

Administration of Erythrocyte-Entrapped β-Glucosidase in Gaucher Disease

Beutler *et al.*[230,348] reported the first human trial of an erythrocyte-entrapped enzyme: they administered β-glucosidase in autologous erythrocytes to a patient with Gaucher type 1 disease. Enzyme was entrapped in erythrocytes using a dialysis technique[326] and administered in three doses over a 5-day period. Two single doses of enzyme-entrapped resealed erythrocytes coated with anti-Rh globulin were administered 6 and 7 months later. These enzyme-loaded erythrocytes were cleared from the circulation with a ^{51}Cr half-life of 4.8 hr compared with the slow ^{51}Cr half-life of uncoated erythrocytes (10 days) in an asplenic Gaucher recipient.[259,261] (Gamma-globulin coated, resealed erythrocytes administered intravenously to a Gaucher patient with a spleen had a ^{51}Cr half-life of 22 min).

The erythrocyte-entrapped administrations in the asplenic patient were in series with 11 injections of unentrapped enzyme.[230] These injections resulted in essentially no change in the substrate levels in the recipient's plasma or erythrocytes; however, small decreases were observed in platelet, monocyte, and granulocyte levels, and a significant decrease in the lymphocyte levels.[230]

Subsequently, Beutler et al.[348] infused erythrocyte-entrapped enzyme in a total of six patients with Gaucher disease. One recipient received 10 infusions containing 56.4 nmol/min of enzymatic activity over a period of 7 months. Although these studies were not associated with any significant clinical improvement, perhaps long-term endeavors utilizing large amounts of enzymatic activity will result in clinical benefit. Since there were no significant side effects and no immune responses to the erythrocyte carrier or the entrapped enzyme during these trials, further long-term studies of erythrocyte-entrapped active β-glucosidase should be carried out to determine the efficacy of this carrier for the treatment of Gaucher disease type 1 as well as for the protection of entrapped activity and target delivery of other therapeutic agents.

Prospects for Enzyme Replacement

Recent advances in enzyme technology have already enhanced the practicality of enzyme replacement. The rapid and anticipated progress in cellular engineering to deliver the appropriate highly purified isozyme to target tissue and subcellular sites, while protecting it from bioinactivation and other degradative and immunologic processes, makes the future of this therapeutic modality promising. In addition to chemical modifications for receptor-mediated uptake, various modification procedures can be employed to produce enzyme derivatives designed to maximize their *in vivo* catalytic properties (K_m, V_{max}, and pH optimum), stability, and compartmentalization. For example, when human α-galactosidase A was either complexed with an anti-α-galactosidase antibody or cross-linked with an inorganic bifunctional reagent (hexamethyl-diisocyanate), the activity was significantly more stable than the purified native activity under physiologic conditions *in vitro*.[349] These results suggest that chemical modification of lysosomal hydrolases may be of value to maximize the stability of administered enzymes, at least in the circulation, as well as to potentially stabilize them against intracellular proteolytic destruction, thus prolonging their catalytic lives.

Other chemical modifications of enzymes may promote their differential uptake by target tissues. For example, partial degradation of a high-molecular-weight enzyme, in a manner that permits retention of its catalytic activity, may result in tissue-specific uptake (e.g., by kidney) by virtue of its smaller size. Chemical and/or physical modifications of the membrane surfaces of erythrocyte carriers, similar to those discussed above for unentrapped enzymes, may direct them to target sites; the incorporation of specific carbohydrate, glycolipid, lipopolysaccharide, or glycopeptide "address labels" may provide the appropriate surface marker for specific tissue recognition and selective uptake. Innovative approaches to "educate" erythrocytes by coating them with specific antibodies directed against selected cell or tissue membrane antigens or by chemically modifying erythrocyte surface structures may also "address" these enzyme carriers. These approaches offer exciting possibilities for the delivery of enzymes into the cytoplasm as well as into lysosomes.

An intriguing approach for the treatment of disorders characterized by metabolite accumulation in the blood is the use of enzymes entrapped in erythrocytes modified for prolonged circulation. In this case, circulating toxic substrates of small molecular weight may permeate into erythrocytes loaded with the appropriate enzyme, thus allowing active catalysis and diffusion of the reaction products. Metabolism of the circulating pathogenic substrates presumably would establish equilibrium gradients between the plasma and tissue sites of accumulation, eventually providing a mechanism for the continual clearance and metabolism of the systemic substrate load.

Although the likelihood, frequency, and severity of immunologic reactions to administered exogenous enzymes in recipients receiving long-term replacement therapy are unknown, the potential immunologic complications must be circumvented.[311,350] Chemical modification may provide an important means to render enzymes nonimmunogenic; the co-valent binding of polyethylene glycol to catalase has been reported to reduce its antigenicity in mice.[351] Desired immunologic protection to avoid both immune responses and bioinactivation of enzymes would also be expected when enzymes are camouflaged by entrapment in autologous erythrocytes.

GENE THERAPY

The rapid progress in recombinant DNA technology may have important future implications for the treatment of inherited metabolic diseases. Indeed, the next decade should witness the application of these technologies to human metabolic diseases including (1) the production of large quantities of specific human gene products for replacement endeavors and (2) attempts to transfer isolated normal human genes into the human genome to "cure" the primary genetic defect.

Production of Human Gene Products

The biologic synthesis of human gene products portends an important application of recombinant DNA technology. The initial step, isolation or chemical synthesis of the human gene, has been accomplished for genes whose specific mRNAs can be readily isolated[352] and for genes whose polypeptide products have been sequenced (e.g., see ref. 353). For example, human α- and β-globin messages have been isolated from reticulocytes and liver of patients with various thalassemias.[354] The cDNA for each of these messages can be produced *in vitro* by reverse transcriptase. Alternatively, the genes for several small polypeptide hormones (e.g., somatostatin, growth hormone) have been chemically synthesized based on their amino acid sequences and the genetic code. In these cases, the DNA sequence is chemically synthesized from oligonucleotides and double-stranded DNA generated *in vitro*.[353,355] In fact, Khorana and colleagues[356] have accomplished the remarkable feat of chemically synthesizing a total, biologically functional gene, the *E. coli* suppressor transfer RNA gene (207 base-pairs), including the promoter and distal processing regions.

Having isolated or synthesized the human double-stranded DNA segment, it is then fused to a bacterial plasmid or bacteriophage DNA by recombinant DNA methods for subsequent cloning of the recombinant molecules. In this way, the gene can be produced in quantity for structural and other studies. Specific restriction endonuclease sites and genes conferring drug resistance are usually incorporated into the plasmid or viral vector for selection purposes. In the plasmid vector pBR322, two genes for antibiotic resistance are present, tetracycline and ampicillin resistance. The cloned DNA is usually inserted into one of the antibiotic resistance marker genes rendering the plasmid sensitive to that compound. The recombinant DNA can then be used to transform *E. coli* or other bacteria. The bacteria which contain the plasmid can be detected by plating for the resistant antibiotic marker and the plasmids which carry the recombinant DNA can be selected by their sensitivity to the other resistant marker. Bacterial clones which contain the human DNA sequence can be cultured and large amounts of the human gene can be obtained. The plasmid DNA is separated from bacterial DNA, then restricted and purified on a gel.

The isolation of genes whose gene products have not been sequenced or whose mRNA occurs in small or trace quantities is particularly difficult at present. For these genes, total human DNA is isolated, fragmented with specific restriction endonucleases and then inserted in a plasmid or bacteriophage (lambda) system.[357] In this way, a "library" of clones for the human genome is produced. Following insertion, transformation of *E. coli*, and the initial selection, the bacteria containing plasmids with the specific DNA fragment desired, must be selected. This is the most difficult task and requires a very sensitive means to detect the presence of the particular human DNA fragment or its mRNA-expressed gene product (e.g., RNA or cDNA to DNA hybridization; hybrid-arrested translation and immunoprecipitation of the cell-free translation products[358,359]). Furthermore, gene products which are composed of heteromultimers, whose genes lie on different chromosomes or parts of a chromosome, will require the isolation and cloning of each DNA sequence.

An isolated and cloned human gene (cDNA, not genomic DNA) can be used to biologically produce its gene product in bacteria. For gene expression, the human gene can be linked to a bacterial gene that contains the promoter and control sequences for the enhanced initiation of transcription and translation. Therefore, the *E. coli* β-galactosidase or β-lactamase gene (with appropriate promoter and control sequences) has been linked in phase to the human gene and the recombinant DNA inserted into the plasmid or viral DNA. After transformation, bacterial clones which express the human gene product can be identified and selected for culture. The chimeric mRNA would then be translated into a chimeric protein which can be chemically cleaved into the bacterial and human proteins. Modifications of this strategy have been used to produce human somatostatin, growth hormone, and insulin from the respective human synthesized gene.[353,360,361]

Several problems must be resolved for the bacterial production of these proteins to be practical. First, methods must be developed to improve the yield. Human growth hormone is expressed as only 2% of the bacterial protein. Insertion of several copies of the human gene may result in enhanced yield. Second, the biologic activity of the bacterially produced hormones has not been addressed. Their function *in vivo* must be established. A further obstacle to the bacterial synthesis of certain proteins is the fact that many (e.g., lysosomal hydrolases) undergo extensive posttranslational modifications (e.g., glycosylation, phosphorylation, etc.). If the posttranslational modifications are necessary for biologic function, then chemical or tissue culture systems must be developed so that the protein backbone may be properly processed for *in vivo* function.

With the advent of recombinant DNA technology, it is anticipated that the commercial production of medically useful proteins such as insulin, antihemophilic factor, interferon, selected enzymes, etc., should provide quantities for replacement and other therapeutic endeavors during the 1980s.

Gene Transfer

The ideal treatment for the inherited metabolic defects would be the permanent introduction of new genetic information into the genome of affected individuals. However, therapeutic intervention by gene transfer is currently precluded by the lack of methodology for the precise site-specific insertion of selected DNA segments with the necessary initiation, control and intervening sequences in a manner consistent with normal transcription, translation and regulation of gene expression.

To date, mammalian gene insertion has been accomplished in cultured cells by various methods including (1) the use of viral vectors,[362] (2) cell–cell fusion,[363] and (3) endocytosis.[364,365] Methods are now available for the DNA-mediated transfer of virtually any gene into recipient cells with selectable markers such as thymidine kinase,[366] dihydrofolate reductase,[365] adenosine phosphoribosyltransferase,[367] and hypoxanthine-guanine phosphoribosyltransferase.[368,369] For example, restriction fragments containing the *Herpes* thymidine kinase gene (*tk*), have been isolated, cloned in plasmids, and introducted into tk⁻ recipient cells.[370-373] Recently, a mutant Chinese hamster dihydrofolate reductase gene, *dhfr*, has been used to transform normal mouse cells; following selection of transformants for increased resistance to methotrexate, transformants containing increasing amounts of the transferred gene were obtained. This system also was used to cotransfer the *E. coli* antibiotic resistance plasmid, pBR322 and the *dhfr* gene into animal cells.[365] The exact mechanism of DNA uptake is as yet unknown; however, it appears that the DNA is endocytosed and in some cases incorporated into the genome or remains in very tight association with the cell's chromatin. Use of such dominant drug resistance genes should have important application for the cotransfer of any isolated cloned gene into mammalian cells.

Analogously, techniques for the insertion of drug resistance genes into bone marrow cells of living mice have been reported.[374,375] Mouse bone marrow cells were transformed *in vitro* with the *Herpes simplex* virus thymidine kinase (*tk*) gene[374] or the dihydrofolate reductase (*dhfr*) gene.[375] These genes were used since transformed cells could be detected by their enhanced resistance to methotrexate. The treated cells were then injected into irradiated mice. In both studies, the infused cells (e.g., containing the *tk* gene and a marker chromosome) reconstituted the blood cells of the marrow-depleted mice. This technique can be used for the insertion of other genes linked to the viral *tk* or *dhfr* gene. In fact, the rabbit β-globin gene has been linked and inserted into mouse tissue culture cells, which express at least some sequences of rabbit β-globin mRNA.[376] Extension of this procedure has been suggested for the treatment of sickle cell disease, other hemoglobinopathies, and thalassemias by the replacement of autologous bone marrow cells transformed with drug resistance genes linked to the appropriate human globin gene.[374] Similar gene "transplantation" strategies should be considered for disorders which do not involve bone-marrow-derived pathology. For example, if autologous hepatocytes could be transformed *in vitro* with an attenuated hepatitis virus containing the gene for phenylalanine hydroxylase, the metabolic defect in phenylketonuria might be corrected following transplantation of these cells into the portal vein of the affected patient. Other vectors such as SV-40 and polyoma viruses might also provide the means to introduce new genes into various cells or tissues. However, the major obstacle will be the integration of new genes into the recipient's genome for survival, maintenance, and controlled expression. Until such methodologies are further developed and refined, enzyme replacement therapy and prevention by prenatal diagnosis will continue to be the primary approaches to the treatment of these disorders. Indeed, the significant contributions to the understanding of normal human biochemistry and cell biology made by these inherited diseases mandate our continued efforts to develop specific therapies for the patients born with these severely debilitating disorders.

ACKNOWLEDGMENTS. We are grateful to Ms. Linda Lugo for preparation of the manuscript. The experimental work described in this review was supported in part by grants from The National Foundation–March of Dimes (1-273), the National Institutes of Health (GM 25279), and the Clinical Research Centers Program of the Division of Research Resources, National Institutes of Health (RR 71). R.J.D. is a recipient of a National Institutes of Health Research Career Development Award (1 K04 AM00451).

REFERENCES

1. Stanbury, J. B., Wyngaarden, J. B., and Frederickson, D. S., (eds.), 1978, *The Metabolic Basis of Inherited Disease*, 4th edn., McGraw Hill, New York.
2. McKusick, V. A., 1978, *Mendelian Inheritance in Man*, 5th ed., Johns Hopkins Press, Baltimore.
3. Seakins, J. W. T., Saunders, R. A., and Toothill, C. (eds.), 1973, *Treatment of Inborn Errors of Metabolism*, Churchill Livingstone, Edinburgh.
4. Desnick, R. J., Bernlohr, R. W., and Krivit, W. (eds.), 1973, *Enzyme Therapy in Genetic Diseases*, Williams and Wilkins, Baltimore.
5. Rietra, P. J. G. M., van der Bergh, F. A. J. T. M., and Tager, J. M., 1974, Recent developments in enzyme replacement therapy of lysosomal storage disease, in: *Enzyme Therapy in Lysosomal Storage Diseases* (J. M. Tager, G. J. M. Hooghwinkel, and W. T. Daems, eds.), p. 53, North-Holland, Amsterdam.
6. Desnick, R. J., Thorpe, S. R., and Fiddler, M. B., 1976, Toward enzyme therapy, *Physiol. Rev.* **56:**57–99.
7. Desnick, R. J. (ed.), 1980, *Enzyme Therapy in Genetic Diseases: 2*, Alan R. Liss, New York.
8. Bickel, H., Gerrard, J., and Hickmans, E. M., 1954, The influence of phenylalanine intake on the chemistry and behavior of a phenylketonuric child, *Acta Pediatr.* **43:**64–77.
9. Kaufman, S., Holtzman, N. A., Milstien, S., and Krumbolz, A., 1975, Phenylketonuria due to deficiency of dihydropteridine reductase, *N. Engl. J. Med.* **293:**785–790.

10. Kaufman, S., Berlow, S., Summer, G. K., Milstien, S., Schulman, J. D., Orloff, S., Spielberg, S., and Pueschel, S., 1978, Hyperphenylalaninemia due to a deficiency of biopterin, *N. Engl. J. Med.* **299**:673–679.

11. Tada, K., Personal communication.

12. Schout, J. Daumling, S., Curtius, C. Ch., Niederwieser, A., Bartholome, K., Viscontini, M., Schirchs, B., and Bieuri, J. H., 1978, Tetrahydrobiopterin therapy of atypical phenylketonuria due to defective dihydrobiopterin biosynthesis, *Arch. Dis. Child.* **53**:674–676.

13. Hoskins, J. A., Jack, G., Peiris, R. J. D., Starr, D. J. T., Wade, H. E., Wright, E. C., and Stern, J., 1980, Enzymatic control of phenylalanine intake in phenylketonuria, *Lancet* **1**:392.

14. Nadler, H. L., Inouye, T., and Hsia, D. Y. Y., 1969, Classical galactosemia: A study of fifty cases, in: *Galactosemia* (D. Y. Y. Hsia, ed.), pp. 127–139, C. C Thomas, Springfield.

15. Donnell, G. N., Koch, R., and Bergren, W. R., 1969, Observations on the results of management of galactosemic patients, in: *Galactosemia* (D. Y. Y. Hsia, ed.), pp. 247–268, C. C Thomas, Springfield.

16. Froesch, E. R., 1972, Pentosuria, in: *The Metabolic Basis of Inherited Disease* 3rd edn., (J. G. Stanbury, J. B. Wyngaarden, and D. S. Frederickson, eds.), p. 128, McGraw-Hill, New York.

17. Gray, G. M., 1972, The Hemoglobinopathies, in: *The Metabolic Basis of Inherited Disease,* 3rd edn., (J. G. Stanbury, J. B. Wyngaarden, and D. S. Frederickson, eds.), p. 1457, McGraw-Hill, New York.

18. Komrower, G. M., Lambert, A. M., Cusworth, D. C., and Westall, R. C., 1966, Dietary treatment of homocystinuria, *Arch. Dis. Child.* **41**:666–671.

19. Perry, T. L., Dunn, H. G., Hansen, S., MacDougall, L., and Warrington, P. D., 1966, Early diagnosis and treatment of homocystinuria, *Pediatrics* **37**:502–505.

20. Snyderman, E., Norton, P. M., Roitman, E., and Holt, L. E., Jr., 1964, Maple syrup urine disease, with particular reference to dietotherapy, *Pediatrics* **34**:454–460.

21. Westall, R. G., 1963, Dietary treatment of a child with maple syrup urine disease (branched-chain ketoaciduria), *Arch. Dis. Child.* **38**:485–491.

22. Levy, H. L., Erickson, A. M., Lott, I. T., and Kiertz, D. J., 1973, Isovaleric acidemia: Results of a family study and dietary treatment, *Pediatrics* **52**:83–91.

23. Halverson, S., 1967, Dietary treatment of tyrosinosis, *Am. J. Dis. Child.* **113**:38–40.

24. Kolb, F. O., Earll, J. M., and Harper, H. A., 1967, "Disappearance" of cystinuria in a patient treated with prolonged low methionine diet, *Metabolism* **16**:378–386.

25. Zinneman, H. H., and Jones, J. E., 1966, Dietary methionine and its influence on cystine excretion in cystinuric patients, *Metabolism* **15**:915–921.

26. Bickel, H., Lutz, P., and Schmidt, H., 1973, The treatment of cystinosis with diet or drugs, in: *Cystinosis* (J. D. Schulman, ed.), pp. 199–223, DHEW Publication No. [NIH] 72-249.

27. Corner, B. D., Holton, J. B., Norman, R. M., and Williams, P. M., 1968, A case of histidinemia controlled with a low histidine diet, *Pediatrics* **41**:1074–1080.

28. Gatfield, P. D., Knights, R. M., Devereaux, M., and Pozsonye, J. P., 1969, Histidinemia: A report of four new cases in one family and effect of low histidine diet, *Can. Med. Assoc. J.* **101**:465–469.

29. Shih, V. H., 1978, Urea cycle disorders and other congenital hyperammonemic syndromes, in: *The Metabolic Basis of Inherited Disease,* 4th edn. (J. B. Stanbury, J. B. Wyngaarden, and D. S. Fredrickson, eds.), pp. 362–386, McGraw-Hill, New York.

30. Rosenberg, L. E., 1978, Disorders of propionate, methylmalonate, and cobalamin metabolism, in: *The Metabolic Basis of Inherited Disease*, 4th edn., (J. B. Stanbury, J. B. Wyngaarden, and D. S. Fredrickson, eds.), pp. 411, McGraw-Hill, New York.

31. Batshaw, M., Brusilow, S., and Walser, M., 1975, Treatment of carbamyl phosphate synthetase deficiency with keto analogues of essential amino acids, *N. Engl. J. Med.* **292**:1085–1090.

32. Brusilow, S. W., Batshaw, M. L., and Walser, M., 1979, The use of keto-acids in inborn errors of urea synthesis, in: *Nutritional Management of Genetic Disorders*, Vol. VII (M. Winick, ed.), pp. 65–75, Wiley, New York.

33. Valle, D., Walser, M., Brusilow, S. W., and Kaiser-Kupfer, M., 1980, Gyrate atrophy of the coroid and retina, *J. Clin. Invest.* **65**:371–375.

34. Valle, D., Walser, M., Brusilow, S. W., Kaiser-Kupfer, M., and de Monastero, F., 1980, Long-term results of therapy of gyrate atrophy, *Clin. Res.* **28**:546A.

35. Hsia, Y. E., Coombs, J. T., Hook, L., and Brandt, I. K., 1966, Hepatolenticular degeneration: The comparative effectiveness of D-penicillamine, potassium sulfide and diethyldithiocarbonate as decoppering agents, *J. Pediatr.* **68**:921–926.

36. Richmond, J., Rosenoer, Y. N., Tompsett, S. L., Draper, I., and Simpson, J. A., 1964, Hepatolenticular degeneration (Wilson's disease) treated by penicillamine, *Brain* **87**:619–638.

37. Walshe, J. M., 1979, The management of Wilson's disease with trienthylanetetramine 2HCl (Trien 2HCl), in: *The Management of Genetic Disorders* (C. J. Papadatos, and C. S. Bartsocas, eds.), pp. 271–280, Alan R. Liss, New York.

38. Crawhall, J. C., Scowen, E. F., and Watts, R. W. E., 1964, Further observations on use of D-penicillamine in cystinuria, *Br. Med. J.* **1**:1411–1413.

39. Pippard, M. J., Letsky, E. A., Callender, S. T., and Weatherall, D. J., 1978, Prevention of iron loading in transfusion dependent thalassemia, *Lancet* **1**:1178.

40. Levy, R. I., Fredrickson, D. S., Stone, N. J., Bilheimer, D. W., Brown, W. V., Glueck, C. J., Gotto, A. M., Herbert, P. N., Kwiterovich, P. O., Langer, T., La Rosa, J., Lux, S. E., Rider, A. K., Shulman, R. S., and Sloan, H. R., 1973, Cholestyramine in type II hyperlipoproteinemia, *Ann. Intern. Med.* **79**:51–58.

41. Wyngaarden, J. B., and Kelley, W. N., Disorders of purine and pyrimidine metabolism, in: *The Metabolic Basis of Inherited Disease*, 4th edn. (J. B. Stanbury, J. B. Wyngaarden, and D. S. Fredrickson, eds.), pp. 989, McGraw-Hill, New York.

42. Boss, G. R., and Seegmiller, J. E., 1979, Hyperuricemia and gout, *N. Engl. J. Med.* **300**:1459–1468.

43. Brusilow, S. W., Valle, D. L., and Batshaw, M. L., 1979, New pathways of nitrogen excretion in inborn errors of urea synthesis, *Lancet* **2**:452–454.

44. Brusilow, S. W., and Batshaw, M. L., 1979, Arginine therapy of argininosuccinase deficiency, *Lancet* **1**:124–127.

45. Beaudet, A. L., Michels, V. V., and O'Brien, W. E., 1978, Drug induced orotic aciduria for treatment of ornithine transcarbamylase (OTC) deficiency, *Am. J. Hum. Genet.* **30**:21A.

46. Brusilow, S. W., and Batshaw, M., 1979, Amino acid acylation as supplementary pathways of waste nitrogen excretion (WNE) in urea cycle enzymopathies, *Pediatr. Res.* **13**:417A.

47. Krieger, I., and Tanaka, K., 1976, Therapeutic effects of glycine in isovaleric acidemia, *Pediatr. Res.* **10**:25–29.

48. Hudkoff, M., Cohen, R. M., Puschak, R., Rothman, R., and Segal, S., 1978, Glycine therapy in isovaleric acidemia, *J. Pediatr.* **92**:813–817.

49. Cohn, R. M., Yudkoff, M., Rothman, R., and Segal, S., 1978, Isovaleric acidemia: Use of glycine therapy in neonates, *N. Engl. J. Med.* **299**:996–999.

50. Bartlett, K., and Gompertz, D., 1974, The specificity of glycine *N*-acylase and acylglycine excretion in the organic acidemias, *Biochem. Med.* **10**:15–21.

51. Gibberd, F. B., Page, N. G. R., Billimoria, J. D., and Retsas, S., 1979, Heredopathia atactica polyneuritiformis (Refsum's disease) treated by diet and plasma exchange, *Lancet* **1**:575–576.

52. Hollander, J., and Nusbacher, J. A., Personal communication.

53. Moser, H. W., Braine, H., Pyeritz, R. E., Ullman, D., Murray, C., and Asbury, A., 1980, Therapeutic trial of plasmapheresis in Refsum's disease and Fabry disease, in: *Enzyme Therapy in Genetic Diseases: 2* (R. J. Desnick, ed.), pp. 491–497, Alan R. Liss, New York.

54. King, M. E. E., Breslow, J. L., and Lees, R. S., 1980, Plasma-exchange therapy of homozygous familial hypercholesterolemia, *N. Engl. J. Med.* **302**:1457–1459.

55. Thompson, G. R., Kilpatrick, D., Oakley, C., Steiner, R., and Myant, N., 1978, Reversal of cholesterol accumulation in familial hypercholesterolemia by long-term plasma exchange, *Circulation* **57**:Suppl. 2, 171A.

56. Desnick, R. J., and Schuchman, E., Unpublished results.

57. Johnson, D. L., and Desnick, R. J., 1978, Molecular pathology of Fabry's disease: Physical and kinetic properties of α-galactosidase in cultured endothelial cells, *Biochim. Biophys. Acta* **538**:195–204.

58. Kattlove, H. E., Williams, J. C., Gaynor, E., Spivack, M., Bradley, R. M., and Brady, R. O., 1969, Gaucher cells in chronic myelocytic leukemia: An acquired abnormality, *Blood* **33**:379–390.

59. Lupien, P-J, Moorjani, S., and Awad, J., 1976, A new approach to the management of familial hypercholesterolemias: Removal of plasma-cholesterol based on the principal of affinity chromatography, *Lancet* **1**:1261.

60. Lupien, P-J, Moorjani, S., Lou, M. Brun, D., and Gagne, C., 1980, Removal of cholesterol from blood by affinity binding to heparin-agarose: Evaluation on treatment in homozygous familial hypercholesterolemia, *Pediatr. Res.* **14**:113–117.

61. Burgstaler, E. A., Pineda, A. A., and Ellefson, R. D., 1980, Removal of plasma lipoproteins from circulating blood with a heparin-agarose column, *Mayo Clin. Proc.* **55**:180–184.

62. Buchwald, H., and Varco, R. L. (eds.), 1978, *Metabolic Surgery,* Grune and Stratton, New York.

63. Riddell, A. G., Davies, R. P., and Clark, A. D., 1966, Portacaval transposition in the treatment of glycogen storage disease, *Lancet* **2**:1146–1148.

64. Starzl, T. E., Brown, B. I., Blanchard, H., and Brettschneider, L., 1969, Portal diversion in glycogen storage disease, *Surgery* **65**:504–506.

65. Moore, R. B., Varco, R. L., and Buchwald, H., 1973, Metabolic surgery in the hyperlipoproteinemias, *Am. J. Cardiol.* **31**:148–157.

66. Kelley, W. N., and Wyngaarden, J. B., 1978, The Lesch–Nyhan syndrome, in: *The Metabolic Basis of Inherited Disease,* 4th edn. (J. G. Stanbury, J. B. Wyngaarden, and D. S. Fredrickson, eds.), pp. 1029, McGraw-Hill, New York.

67. Levy, R. I., Fredrickson, D. S., Schulman, R., Bilheimer, D. W., Breslow, J. L., Stone, N. J., Lux, S. E., Sloan, H. R., Krauss, R. M., and Herbert, D. N., 1972, Dietary and drug treatment of primary hyperlipoproteinemia, *Ann. Intern. Med.* **77**:267–294.

68. Melancon, S. B., Dallaire, L., Vincelette, P., Potier, M., and Geoffrey, G., 1979, Early treatment of severe infantile glycine encephalopathy (non-ketotic hyperglycinemia)

with strychnine and sodium benzoate, in: *The Management of Genetic Disorders* (C. J. Papadatos, and C. S. Bartsocas, eds.), pp. 217–229, Alan R. Liss, New York.

69. Bartter, F. C., Pronove, P., Gill, J. R., Jr., and MacCardle, R. C., 1962, Hyperplasia of the juxtaglomerular complex with hyperaldosteronism and hypokalemic alkalosis, *Am. J. Med.* **33**:811–828.

70. Bartter, F. C., Gill, J.R., Jr., Frolich, J. G., Bowden, R. E., Hollifield, J. W., Radfar, N., Keiser, H. R., Oates, J. A., Seyberth, H., and Taylor, A. A., 1976, Prostaglandins are overproduced by the kidneys and mediate hyperreninemia in Bartter's syndrome, *Trans. Assoc. Am. Physicians* **89**:77–91.

71. Gullner, G-H., Bartter, F. C., Cerletti, C., Smith, J. B., and Gill, J. R., Jr., 1979, Prostacyclin overproduction in Bartter's syndrome, *Lancet* **2**:767–770.

72. Smith, W. L., and Lands, W. E. M., 1971, Stimulation and blockade of prostaglandin biosynthesis, *J. Biol. Chem.* **246**:6700–6704.

73. Gullner, G-H, Gill, J. R., Jr., Bartter, F. C., and Smith, J. B., 1980, Correction of increased prostacyclin production in Bartter's syndrome by indomethacin, *Clin. Res.* **28**:559A.

74. Gill, J. R., Jr., and Bartter, F. C., 1978, Evidence for a prostaglandin defect in chloride absorption in the loop of Henle as a proximal cause of Bartter's syndrome, *Am. J. Med.* **65**:766–772.

75. Strauss, R. G., 1974, Failure of methyldopa therapy in Bartter's syndrome, *J. Pediatr.* **85**:101–103.

76. Brook, C. G. D., Zachmann, M., Prader, A., and Murset, G., 1974, Experience with long-term therapy in congenital adrenal hyperplasia, *J. Pediatr.* **85**:12–19.

77. Klein, A. H., Meltzer, S., and Kenny, F., 1972, Improved prognosis in congenital hypothyroidism treated before age three months, *J. Pediatr.* **81**:912–915.

78. Tanner, J. M., Whitehouse, R. J., Hughes, P. C. R., and Vince, F. P., 1971, Effect of human growth hormone treatment for 1 to 7 years on growth of 100 children with growth hormone deficiency, low birth weight, inherited smallness, Turner's syndrome and other complaints, *Arch. Dis. Child.* **46**:745–782.

79. Janeway, C. A., and Rosen, F. S., 1966, The gamma globulins, *N. Engl. J. Med.* **275**:826–831.

80. Danks, D. M., Campbell, M. B., Stevens, B. J., Mayne, V., and Cartwright, E., 1972, Menkes' kinky hair syndrome. An inherited defect in copper apsorption with widespread effects, *Pediatrics* **50**:188.

81. Bucknall, W. E., Haslan, R. H., and Holtzman, N. A., 1973, Kinky hair syndrome: Response to copper therapy, *Pediatrics* **52**:653–657.

82. Menkes, J. H., Alter, M., Steiglider, G. K., Weakly, D. R., and Sung, J. H., 1962, A sex-linked recessive disorder with retardation of growth, peculiar hair and focal cerebral and cerebellar degeneration, *Pediatrics* **29**:764.

83. Dekaban, A. S., and Steusing, J. K., 1974, Letter: Menkes kinky hair disease treated with subcutaneous copper sulphate, *Lancet* **2**:1523.

84. Grover, W. D., and Scrutton, M. C., 1975, Copper infusion therapy in trichopoliodystrophy, *J. Pediatr.* **86**:216–220.

85. Danks, D. M., Cartwright, E., and Stevens, B. J., 1973, Menkes steely hair (kinky hair) disease, *Lancet* **1**:891.

86. Beratis, N. G., Price, P., LaBadie, G. U., and Hirschhorn, K., 1978, ^{64}Cu metabolism in Menkes and normal cultured skin fibroblasts, *Pediatr. Res.* **12**:699–702.

87. Beratis, N. G., Price, P., LaBadie, G. U., and Hirschhorn, K., 1979, Copper metabolism in Menkes disease, *Pediatr. Res.* **13**:206–210.

88. LaBadie, G. U., Hirschhorn, K., Katz, S., and Beratis, N. G., 1981, Increased copper metallothionein in Menkes cultured skin fibroblasts, *Pediatr. Res.* (in press).

89. LaBadie, G. U., Beratis, N. G., Price, P., and Hirschhorn, K., 1981, Studies of the copper-binding proteins in Menkes and normal cultured skin fibroblast lysates, *J. Cell Physiol.* (in press).

90. Evans, G. W., and Johnson, P. E., 1976, Zinc binding factor in acrodermatitis enteropathica (letter) *Lancet* 2:1310.

91. Dillaha, C. J., Lorincz, A. L., and Aavik, O. R., 1953, Acrodermatitis enteropathica. Review of the literature and report of a case successfully treated with diodoquin, *J. Am. Med. Assn.* 152:509–512.

92. Moynahan, E. J., Johnson, F. R., and McMinn, R. M. H., 1963, Acrodermatitis enteropathica: Demonstration of possible intestinal enzyme defect, *Proc. Roy. Soc. Med.* 56:300–301.

93. Cash, R., and Berger, C. K., 1969, Acrodermatitis enteropathica: Defective metabolism of unsaturated fatty acids, *J. Pediatr.* 74:717–729.

94. Moynahan, E. J., 1974, Acrodermatitis enteropathica: A lethal inherited zinc deficiency disorder, *Lancet* 2:399–400.

95. Neldner, K. H., and Hambridge, K. M., 1975, Zinc therapy of acrodermatitis enteropathica, *New Engl. J. Med.* 292:879–882.

96. Costa, T., Reade, T. M., Cole, D. E. C., Nogrady, B., Scriver, C. R., Marais, P., and Glorieux, F. H., 1980, Renal handling of phosphate (Pi) and bone mineralization in X-linked hypophosphatemia (XLH) during treatment with Pi and 1,25-$(OH)_2D_3$, *Pediatr. Res.* 14:521A.

97. Fleischer, L. D., and Gaull, G. E., 1980, Enzyme manipulation by specific megavitamin therapy, in: *Enzyme Therapy in Genetic Diseases:2* (R. J. Desnick, ed.), pp. 239–267, Alan R. Liss, New York.

98. Hsia, Y. E., 1975, Treatment in genetic diseases, in: *The Prevention of Genetic Disease and Mental Retardation* (A. Milunsky, ed.), pp. 277–305, Saunders, Philadelphia.

99. Rosenberg, L. E., 1976, Vitamin responsive inherited metabolic disorders, in: *Advances in Human Genetics,* Vol. 6 (H. Harris and K. Hirschhorn, eds.), pp. 1–74, Plenum Press, New York.

100. Fenton, W. A., and Rosenberg, L. E., 1978, Genetic and biochemical analysis of human cobalamin mutants in cell culture, *Annu. Rev. Genet* 12:223–248.

101. Pascal, T. A., Gaull, G. E., Beratis, N. G., Gillam, G. M., and Tallan, H. H., 1978, Cystathionase deficiency: Evidence for genetic heterogeneity in primary cystathioninuria, *Pediatr. Res.* 12:125–133.

102. Jolly, R. D., Van de Water, N. S., Janmaat, A., Slack, P. M., and McKenzie, R. G., 1980, Zinc therapy in the bovine mannosidosis model, in: *Enzyme Therapy in Genetic Diseases: 2* (R. J. Desnick, ed.), pp. 305–318, Alan R. Liss, New York.

103. Grabowski, G. A., Walling, L., and Desnick, R. J., 1980, Human mannosidosis: *In vitro* and *in vivo* studies of cofactor supplementation, in: *Enzyme Therapy in Genetic Diseases: 2* (R. J. Desnick, ed.), pp. 319–334, Alan R. Liss, New York.

104. Snaith, S. M., 1975, Characterization of jack-bean α-D-mannosidase as a zinc metalloenzyme, *Biochem. J.* 143:83–89.

105. Phillips, N. C., Robinson, D., and Winchester, B. G., 1974, Human liver α-mannosidase activity, *Clin. Chim. Acta* 55:11–21.

106. Grabowski, G. A., Ikonne, J. U., and Desnick, R. J., 1980, Comparative physical, kinetic and immunologic properties of acidic and neutral α-D-mannosidase isozymes from human liver, *Enzyme* 25:13–25.

107. Stonard, M. D., and Webb, M., 1976, Influence of dietary cadmium on the distribution of the essential metals, copper, zinc and iron in the tissues of the rat, *Chem. Biol. Interact.* 15:349–360.

108. Elsas, L. J., Hollins, B., and Pinnell, S. R., 1974, Hydroxylysine-deficient collagen disease: Effect of ascorbic acid, *Am. J. Hum. Genet.* **26**:28A.
109. Elsas, L. J., Miller, R. L., and Pinnell, S. R., 1976, Inherited human collagen lysyl hydroxylase deficiency: Ascorbic acid response, *Clin. Res.* **24**:294A.
110. Ampola, M. G., Mahoney, M. J., Nakamura, F., and Tanaka, K., 1975, Prenatal therapy of a patient with vitamin B_{12}-responsive methylmalonic acidemia, *N. Engl. J. Med.* **293**:313–317.
111. Thompson, R. P. H., 1973, The use and abuse of treatment in the hyperbilirubinemia syndromes, in: *Treatment of Inborn Errors of Metabolism* (J. W. T. Seakins, R. A. Saunders, and C. Toothill, eds.), pp. 215–225, Livingstone, Edinburgh.
112. Rosen, F. S., and Beyler, A., 1980, Hereditary angioneurotic edema and its correction with androgen therapy, in: *Enzyme Therapy in Genetic Diseases: 2* (R. J. Desnick, ed.), pp. 499–508, Alan R. Liss, New York.
113. Gadek, J. E., Fulmer, J. D., Gelfand, J. A., Frank, M. M., Petty, T. L., and Crystal, R. G., 1980, Danazol-induced augmentation of serum α_1-antitrypsin levels in individuals with marked deficiency of this antiprotease, *J. Clin. Invest.* **66**:82–87.
114. Gadek, J. E., Hosea, S. W., Gelfand, J. A., Santaella, M., Wickerhauser, M., Triantaphyllopoulos, D. C., and Frank, M. M., Replacement therapy in hereditary angioedema with partly purified C1 inhibitor, *N. Engl. J. Med.* **302**:542–546.
115. Lascelles, J., 1960, The synthesis of enzymes concerned in bacteriochlorophyll formation in growing cultures of *Rhodopseudomonas spheroides*, *J. Gen. Microbiol.* **23**:487–498.
116. Burnham, B., and Lascelles, J., 1963, Control of porphyrin biosynthesis through a negative-feedback mechanism. Studies with preparations of delta-aminolaevulate synthetase and delta-aminolaevulate dehydratase from *Rhodopseudomonas spheroides*, *Biochem. J.* **87**:462–472.
117. Granick, S., 1962, Porphyrin biosynthesis, porphyrin diseases, and induced enzyme synthesis in chemical porphyria, *Trans. N.Y. Acad. Sci.* **25**:53–65.
118. Granick, S., 1966, The induction *in vitro* of the synthesis of δ-aminolevulinic acid synthetase in chemical porphyria. A response to certain drugs, sex hormones and foreign chemicals, *J. Biol. Chem.* **241**:1359–1375.
119. Waxman, A. D., Collins, A., and Tschudy, D. P., 1966, Oscillations of hepatic δ-aminolevulinic acid synthetase produced *in vivo* by heme, *Biochem. Biophys. Res. Commun.* **24**:675–683.
120. Bonkowsky, H. L., Tschudy, D. P., Collins, A., Doherty, J., Bossenmaier, I., Cardinal, R., and Watson, C. J., 1971, Repression of the overproduction of porphyrin precursors in acute intermittent porphyria by intravenous infusions of hematin, *Proc. Natl. Acad. Sci. USA* **68**:2725–2729.
121. Watson, C. J., Pierach, C. A., Bossenmaier, I., and Cardinal, R., 1977, Postulated deficiency of hepatic heme and repair by hematin infusions in the "inducible" hepatic porphyrias, *Proc. Natl. Acad. Sci. USA* **74**:2118–2120.
122. Watson, C. J., Pierach, C. A., Bossenmaier, I., and Cardinal, R., 1978, Use of hematin in the acute attack of the "inducible" hepatic porphyrias, *Adv. Intern. Med.* **23**:265–286.
123. Welland, F. H., Hellman, E. S., Gaddis, E. M., Collins, A., Hunter, G. W., Jr., and Tschudy, D. P., 1964, Factors affecting the excretion of porphyrin precursors by patients with acute intermittent porphyria. I. The effect of diet, *Metabolism* **13**:232–250.
124. Brodie, M. J., Moore, M. R., Thompson, G. G., and Goldberg, A., 1977, The treatment of acute intermittent porphyria with levulose, *Clin. Sci. Mol. Med.* **53**:365–371.

125. Matas, A. J., Simmons, R. L., and Desnick, R. J., 1978, Transplantation in metabolic disease, in: *Metabolic Surgery* (H. Buchwald, and R. L. Varco, eds.), pp. 177–227, Grune and Stratton, New York.

126. Matas, A. J., Desnick, R. J., Najarian, J. S., and Simmons, R. L., 1978, Clinical and experimental transplantation in enzymatic deficiency disease, *Surg. Gyn. Ob.* **146:**975–986.

127. Hirschhorn, R., 1980, Treatment of genetic diseases by allotransplantation, in: *Enzyme Therapy in Genetic Diseases: 2* (R. J. Desnick, ed.), pp. 429–444, Alan R. Liss, New York.

128. Congdon, C. C., 1971, Bone marrow transplantation, *Science* **116:**171–182.

129. Stiehm, E. R., Jr., Lawlor, G. J., Kaplan, M. S., Greenwald, H. L., Neerhout, R. C., Sengar, D. P. S., and Terasaki, P. I., 1972, Immunologic reconstitution in severe combined immunodeficiency without bone marrow chromosomal chimerism, *N. Engl. J. Med.* **286:**797–803.

130. Bach, F. H., Albertini, R. J., and Joo, P., 1968, Bone marrow transplantation in a patient with the Wiskott–Aldrich syndrome, *Lancet* **2:**1364–1368.

131. Parkman, R., Rappaport, J., Geha, R., Belli, J., Cassady, R., Levey, R., Nathan, D. G., and Rosen, F. S., 1978, Complete correction of the Wiskott–Aldrich syndrome by allograft bone-marrow transplantation, *N. Engl. J. Med.* **298:**921–927.

132. Meuwissen, H. J., Keiserman, M., Taft, E., Pollara, B., and Pickering, R. J., 1978, Marrow transplantation (MTP) in Wiskott–Aldrich Syndrome (WAS): T-cell engraftment with cyclophosphamide (CY), complete engraftment with total body irradiation, *Pediatr. Res.* **12:**483A.

133. Coccia, P. F., Cervenka, J., Teitelbaum, S. L., Kahn, A., Clawson, C. C., and Brown, D. M., 1979, Reversal of human osteopetrosis by bone marrow transplantation: Etiologic implications, *Clin. Res.* **27:**483A.

134. Sorell, M., Rosen, J. F., Kapoor, N., Kirkpatrick, D., Chaganti, R. S. K., Pollack, M. S., Dupont, B., Goossen, C., Good, R. A., and O'Reilly, R. J., 1979, Bone marrow transplant for osteopetrosis in a 10-year-old boy, *Pediatr. Res.* **13:**481A.

135. Coccia, P. F., Krivit, W., Cervenka, J., Clawson, C., Kersey, J. H., Kim, T. H., Nesbit, M. E., Ramsay, N. K. C., Warkentin, P. I., Teitelbaum, S. L., Kahn, A. J., and Brown, D. M., 1980, Successful bone marrow transplantation for infantile malignant osteopetrosis, *N. Engl. J. Med.* **302:**701–708.

136. Gale, R. P., Sparkes, R. S., and Golde, D. W., 1978, Bone marrow origin of hepatic macrophages (Kupffer cells) in humans, *Science* **201:**937–939.

137. Santos, G. W., Elfenbein, G. J., and Tutschka, P. J., 1979, Bone marrow transplantation—present status, *Transplant. Proc.* **11:**182–190.

138. Buckley, R. H., Whisnant, J. K., Schiff, R. I., Gilbertsen, R. B., Huang, A. T., and Platt, M. S., 1976, Correction of severe combined immunodeficiency by fetal liver cells, *N. Engl. J. Med.* **294:**1076–1081.

139. August, C. S., Rosen, F. S., Filler, R. M., Janeway, C. A., Markowski, B., and Kay, H. E. M., 1968, Implantation of a fetal thymus, restoring immunological competence in a patient with thymic aplasia (Di George's syndrome), *Lancet* **2:**1210–1213.

140. Tubergen, D. G., 1974, Thymus transplant in lymphopenic immunodeficiency (Nezelof's syndrome), *J. Pediatr.* **84:**915–920.

141. Dubois, R. S., Giles, G., Rodgerson, D. O., Lilly, J., Martineau, G., Halgrimson, C. G., Shroter, G., Starzl. T. E., Sternleib, I., and Scheinberg, I. H., 1971, Orthotopic liver transplantation for Wilson's disease, *Lancet* **1:**505–508.

142. Groth, C. G., Dubois, R. S., Corman, J., Gustafsson, A., Iwatsuki, S., Royerson, D. O., Halgrimson, C. G., and Starzl, T. E., 1973, Metabolic effects of hepatic replacement in Wilson's disease, *Transplant. Proc.* **5:**829–838.

143. Beart, R. W., Jr., Putnam, C. W., Porter, K. A., and Starzl, T. E., 1975, Liver transplantation for Wilson's disease, *Lancet* 2:176.
144. Sharp. H. L., and Najarian, J. S., Unpublished results.
145. Putnam, C. W., Porter, K. A., Peters, R. L., Aschcavai, M., Redeker, A. G., and Starzl, T. E., 1977, Liver replacement for alpha₁-antitrypsin deficiency, *Surgery* 81:258-292.
146. Hood, J. M., Koep, L. J., Peters, R. L., Schroter, G. P. J., Weil, R., Redeker, A. G., and Starzl, T. E., 1980, Liver transplantation for advanced liver disease with alpha-1-antitrypsin deficiency, *N. Engl. J. Med.* 302:272-274.
147. Macdougall, B. R. D., McMaster, P., Calne, R. Y. and Williams, R., 1980, Survival and rehabilitation after orthotopic liver transplantation, *Lancet* 1:1326-1328.
148. Daloze, P., Delvin, E. E., Glorieux, F. H., Corman, J. L., Bettez, P. and Toussi, T., 1977, Replacement therapy for inherited enzyme deficiency: Liver orthotopic transplantation in Niemann–Pick Type A, *Am. J. Med. Genet.* 1:229-239.
149. DiMagno, E. P., Hermon-Taylor, J., Go, V. L. W., Lillehei, R. C., and Summerskill, W. H. J., 1971, Functions of a pancreaticoduodenal allograft in man, *Gastroenterology* 61:363-370.
150. Jonasson, I., 1979, Transplantation of the pancreas, *Transplant. Proc.* 11:325-330.
151. Hathaway, W. E., Mull, M. M., Githens, J. H., Groth, C. G., Marchioro, T. L., and Starzl, T. E., 1969, Attempted spleen transplant in classical hemophilia, *Transplant* 7:73-80.
152. Groth, C. G., Hagenfeldt, L., Dreborg, S., Lofstrom, B., Ockerman, P. A., Samuelson, L., Svennerholm, L., Werner, B., and Westberg, G., 1971, Splenic transplantation in a case of Gaucher disease, *Lancet* 1:1260-1262.
153. Groth, C. G., Collste, H., Dreborg, S., Hakansson, G., Lundgren, G., and Svennerholm, L., 1980, Attempt at enzyme replacement in Gaucher disease by renal transplantation, in: *Enzyme Therapy in Genetic Diseases: 2* (R. J. Desnick, ed.), pp. 475-490, Alan R. Liss, New York.
154. Hoyer, J. R., Kjellstrand, C. M., Simmons, R. L., Najarian, J. S., Mauer, S. M., Buselmeier, T. J., Michael, A. F., and Vernier, R. L., 1973, Successful renal transplantation in 3 children with congenital nephrotic syndrome, *Lancet* 1:1410-1412.
155. Sorensen, L. B., 1966, Suppression of the shunt pathway in primary gout by azathioprine, *Proc. Natl. Acad. Sci. USA* 55:571-575.
156. Flatmark, A. L., Hovig, T. Myhre, E., and Gjone, E., 1977, Renal transplantation in patients with familial lecithin: cholesterol acyltransferase deficiency, *Transplant. Proc.* 9:1665-1669.
157. Deodhar, S. D., Tung, K. S. K., Zuhlke, V., and Nakamoto, S., 1969, Renal homotransplantation in a patient with primary familial oxalosis, *Arch. Pathol.* 87:118-124.
158. Mahoney, C. P., Striker, G. E., Fetterman, G. H., Hickman, R. O., Schneider, J., and Marchioro, T. L., 1973, Renal transplantation in childhood cystinosis: Effects of the metabolic disease and renal allografts on each other, in: *Enzyme Therapy in Genetic Diseases* (R. J. Desnick, R. W. Bernlohr, and W. Krivit, eds.), pp. 141-148, Williams and Wilkins, Baltimore.
159. Desnick, R. J., Simmons, R. L., Allen, K. Y., Najarian, J. S., and Krivit, W., 1972, Fabry's disease: Correction of enzymatic deficiencies by renal transplantation, *Surgery* 72:203-210.
160. Philippart, M., Franklin, S. S., and Gordon, A., 1972, Reversal of an inborn sphingolipidosis (Fabry's disease) by kidney transplantation, *Ann. Intern. Med.* 77:195-200.
161. Desnick, R. J., Allen, K. Y., Simmons, R. L., Woods, J. E., Anderson, C. F., Najarian, J. S., and Krivit, W., 1973, Fabry disease: correction of the enzymatic deficiency by

renal transplantation, in: *Enzyme Therapy in Genetic Disease*, (R. J., Desnick, R. W. Bernlohr, and W. Krivit, eds.), pp. 88–96, Williams and Wilkins, Baltimore.

162. Jacky, E., 1976, Fabrysche Erkrankung (Angiokeratoma corporis diffusum universale): günstiger Verlauf nach Nierentransplantation, *Schweiz. Med. Wochenschr.* **106**:703–709.

163. Clarke, J. T. R., Guttmann, R. D., Wolfe, L. S., Beaudoin, J. G., and Morehouse, D. D., 1972, Enzyme replacement by renal allotransplantation in Fabry's disease, *N. Engl. J. Med.* **287**:1215–1218.

164. Spense, M. W., Mackinnon, K. E., Burgess, J. K., d'Entremont, D. M., Belitsky, P., Lannon, S. G., and MacDonald, A. S., 1976, Failure to correct the metabolic defect by renal allotransplantation in Fabry's disease, *Ann. Intern. Med.* **84**:13–20.

165. Grunfeld, J. P., LePorrier, M., Droz, D., Bensaude, I., Hinglais, N., and Crosnier, J., 1975, La transplantation rénale chez les sujets atteints de maladie de Fabry, *Nouv. Presse. Méd.* **4**:2081–2086.

166. Gibbs, D. A., Spellacy, E., Roberts, A. E., and Watts, R. W. E., 1980, The treatment of lysosomal storage diseases by fibroblast transplantation: Some preliminary observations, in: *Enzyme Therapy in Genetic Diseases: 2* (R. J. Desnick, ed.), pp. 457–474, Alan R. Liss, New York.

167. Dean, M. F., Muir, H., Benson. P. F., Button, L. R., Boylston, A., and Mowbray, J., 1976, Enzyme replacement by fibroblast transplantation in a case of Hunter disease, *Nature* **261**:323–325.

168. Dean, M. F., Stevens, R. L., Muir, H., Benson, P. F., Button, L. R., Anderson, R. L., Boylston, A., and Mowbray, J., 1979, Enzyme replacement therapy by fibroblast transplantation, *J. Clin. Invest.* **63**:138–145.

169. Dean, M. F., Muir, H., Benson, P., and Button, L., 1980, Enzyme replacement in the mucopolysaccharidoses by fibroblast transplantation, in: *Enzyme Therapy in Genetic Diseases:2* (R. J. Desnick, ed.), pp. 445–456, Alan R. Liss, New York.

170. Willner, J. P., Matalon, R., Beratis, N., Ritch, R., Rose, J., Hirschhorn, K., and Desnick, R. J., 1978, Failure of fibroblast allograft for enzyme replacement in mucopolysaccharidosis Type VI, *Pediatr. Res.* **12**:459.

171. Barton, R. W., and Neufeld, E. F., 1971, The Hurler corrective factor, *J. Biol. Chem.*, **246**:7773–7779.

172. Cantz, M., Chrambach, A., Bach, G., and Neufeld, E. F., 1972, The Hunter corrective factor, *J. Biol. Chem.* **247**:5456–5462.

173. Kresse, H., and Neufeld, E. F., 1972, The Sanfilippo A corrective factor, *J. Biol. Chem.* **247**:2164–2170.

174. Haskins, M. E., Jezyk, P. F., Desnick, R. J., McDonough, S. K., and Patterson, D. F., 1979, Alpha-L-iduronidase deficiency in a cat: A model of mucopolysaccharidosis I, *Pediatr. Res.* **13**:1294–1297.

175. Jezek, P. F., Haskins, M. E., Patterson, D. F., Mellman, W. J., and Greenstein, M., 1977, Mucopolysaccharidosis in a cat with arylsulfatase B deficiency: A model of Maroteaux–Lamy syndrome, *Science* **198**:834–840.

176. Haskins, M. E., Jezyk, P. F., Desnick, R. J., McDonough, S. K., and Patterson, D. F., 1979, Mucopolysaccharidosis in a domestic short-haired cat—a disease distinct from that seen in the Siamese cat, *J. Am. Vet. Med. Assn.* **175**:384–387.

177. Leonard, R. J., Lazarow, A., and Hegre, O. D., 1973, Pancreatic islet transplantation in the rat, *Diabetes* **22**:413–428.

178. Sutherland, D. E. R., Hong, C., and Najarian, J. S., 1980, Cellular transplantation for enzymatic and metabolic deficiencies, in: *Enzyme Therapy in Genetic Disease: 2* (R. J. Desnick, ed.), pp. 207–217, Alan R. Liss, New York.

179. Mukherjee, A. B., and Krasner, J., 1973, Induction of an enzyme in genetically deficient rats after grafting of normal liver, *Science* **182:**68–71.

180. Bodmer, W. F., Jones, E. A., Barnstable, C. J., and Bodmer, J. G., 1978, Genetics of HLA: The major human histocompatibility system, *Proc. Roy. Soc. Lond.* **202:**93–102.

181. Thorsby, E., Albrechtsen, D., Bergholtz, B. O., Hirshberg, H., and Solheim, B. G., 1978, Identification and significance of products of the HLA-D region, *Transplant. Proc.* **10:**313–323.

182. Hirschhorn, R., and Martin, D. W., Jr., 1978, Enzyme defects in immunodeficiency diseases, *Springer Seminars Immunopathol.* **1:**299–320.

183. Yulish, B. S., Stern, R. C., and Polmar, S. H., 1980, Partial resolution of bone lesions in a child with severe combined immunodeficiency disease and adenosine deaminase deficiency after enzyme-replacement therapy, *Am. J. Dis. Child.* **134:**61–63.

184. Staal, G. E. J., Stoop, J. W., Zegers, B. J. M., Siegenbeek van Heukelom, L. H., van der Vlist, M. J. M., Wadman, S. K., and Martin, D. W., 1980, Erythrocyte metabolism in purine nucleoside phosphorylase deficiency after enzyme replacement therapy by infusion of erythrocytes, *J. Clin. Invest.* **65:**103–108.

185. Aceto, J., Jr., Frasier, S. D., Hayles, A. B., Meyer-Bahlburg, H. F. L., Parker, M. L., Munschauer, R., and DiChiro, G., 1972, Collaborative study of the effects of human growth hormone in growth hormone deficiency. I. First year of therapy, *J. Clin. Endocrinol. Metab.* **35:**483–496.

186. Preece, M. A., Tanner, J. M., Whitehouse, R. N., and Cameron, N., 1976, Dose dependence of growth response to human growth hormone in growth hormone deficiency, *J. Clin. Endocrinol. Metab.* **42:**477–483.

187. Brook, C. G. D., Zachmann, M., Prader, A., and Murset, G., 1974, Experience with long-term therapy in congenital adrenal hyperplasia, *J. Pediatr.* **85:**12–19.

188. Rappaport, R., Bouthreuil, E., Marti-Henneberg, C., and Basmaciogullari, A., 1973, Linear growth rate, bone maturation and growth hormone secretion in prepubertal children with congenital adrenal hyperplasia, *Acta Paediatr. Scand.* **62:**513–519.

189. Fisher, D. A., Dussault, J. H., Foley, T. P., Jr., Klein, A. H., LaFranchi, S., Larsen, P. R., Mitchell, M. L., Murphey, W. H., and Walfish, P. G., 1979, Screening for congenital hypothyroidism: Results of screening one million North American infants, *J. Pediatr.* **94:**700–705.

190. Ginsberg-Fellner, F., 1980, Insulin-dependent diabetes mellitus, *Pediatr. Annu.* **9:**142–153.

191. Hilgartner, M. W. (ed.), 1976, *Hemophilia in Children*, P. S. G., New York.

192. Biggs, R., 1976, *Hemostasis and Thrombosis*, 2nd edn., Blackwell Scientific, London.

193. Hitzig, W. H., and Muntener, U., 1975, Conventional immunoglobulin therapy, in: *Immunodeficiency in Man and Animals* (D. Bergsma, R. A. Good, and J. Finstad, eds.), *Birth Defects Orig. Art. Ser.* **11:**339–342.

194. Janeway, C. A., and Rosen, F. S., 1966, The gamma globulins. IV. Therapeutic uses of gamma globulin, *N. Engl. J. Med.* **275:**826–831.

195. DeDuve, C., 1964, From cytases to lysosomes, *Fed. Proc.* **23**(2):1045–1049.

196. O'Brien, J. S., Miller, A. L., Loverde, A. W., and Veath, M. L., 1973, Sanfilippo disease type B: Enzyme replacement and metabolic correction in cultured fibroblasts, *Science* **181:**753–755.

197. Lagunoff, D., Nicol, D. M., and Pritzl, P., 1973, Uptake of β-glucuronidase by deficient human fibroblasts, *Lab. Invest.* **29:**449–453.

198. Porter, M. T., Fluharty, A. L., and Kihara, H., 1971, Correction of abnormal cerebroside sulfate metabolism in cultured metachromatic leukodystrophy fibroblasts, *Science* **172**:1263-1265.

199. Dawson, G., Matalon, R., and Li, Y. T., 1973, Correction of the enzymatic defect in cultured fibroblasts from patients with Fabry's disease: Treatment with purified α-galactosidase from ficin, *Pediatr. Res.* **7**:684-690.

200. Cantz, M., and Kresse, H., 1974, Sandhoff's disease: Defective glycosaminoglycan catabolism in cultured fibroblasts and its correction by β-*N*-acetylhexosaminidase, *Eur. J. Biochem.* **47**:581-590.

201. Neufeld, E. F., and Fratantoni, J. C., 1970, Inborn errors of mucopolysaccharide metabolism, *Science* **169**:141-146.

202. Bach, G., Friedman, R., Weissman, B., and Neufeld, E. F., 1972, The defect in the Hurler and Scheie syndromes: Deficiency of α-L-iduronidase, *Proc. Natl. Acad. Sci. USA* **69**:2048-2051.

203. Brot, F. E., Glaser, J. H., Roozen, K. J., Sly, W. S., and Stahl, P. D., 1974, *In vitro* correction of deficient human fibroblasts by β-glucuronidase from different human sources, *Biochem. Biophys. Res. Commun.* **57**:1-8.

204. DeBarsy, T., Jacquemin, P., Van Hoof, F., and Hers, H. G., 1973, Enzyme replacement in Pompe disease: An attempt with purified human acid α-glucosidase, in: *Enzyme Therapy in Genetic Diseases* (R. J. Desnick, R. W. Bernlohr, and W. Krivit, eds.), pp. 184-190, Williams and Wilkins, Baltimore.

205. Mersmann, G., von Figura, K., and Buddecke, E., 1976, Storage of mannose containing material in cultured human mannosidosis cells and metabolic correction by pig kidney α-mannosidase, *Hoppe-Slylers Z. Physiol. Chem.* **357**:641-648.

206. Baudhuin, P. H. G., Hers, H. G., and Loeb, H., 1964, An electron microscopic and biochemical study of type II glycogenosis, *Lab Invest.* **13**:1139-1152.

207. Hug, G., and Schubert, W. K., 1967, Lysosomes in type II glycogenosis, *J. Cell Biol.* **35**:C1-C6.

208. Lauer, R. M., Mascarinas, T., Racela, A. S., Diehl, A. M., and Brown, B. I., 1968, Administration of a mixture of fungal glucosidases to a patient with type II glycogenosis (Pompe's disease), *Pediatrics* **42**:672-678.

209. Greene, H. L., Hug, G., and Schubert, W. K., 1969, Metachromatic leukodystrophy: treatment with arylsulfatase A, *Arch. Neurol.* **20**:147-153.

210. Nadler, H. B., Mourao, P. A. S., Toledo, S. P. A., and Dietrich, C. P., 1974, Mucopolysaccharide degradation and excretion after hyluronidase injection in patients with Hunter's and Hurler's syndrome, *Clin. Chim. Acta* **50**:245-255.

211. Austin, J. H., 1967, Some recent findings in leukodystrophies and in gargoylism, in: *Inborn Errors of Sphingolipid Metabolism* (S. M. Aronson, and B. W., Volk, eds.), pp. 359-387, Pergamon, Oxford.

212. Di Ferrante, N., Nichols, B. L., Donnelly, P. V., Neri, G. Hrgovcic, R., and Berglund, R. K., 1971, Induced degradation of glycosaminoglycans in Hurler's and Hunter's syndromes by plasma infusions, *Proc. Natl. Acad. Sci. USA* **68**:303-307.

213. Dekaban, A. S., Holden, K. R., and Constantopoulos, G., 1972, Effects of fresh plasma or whole blood transfusions on patients with various types of mucopolysaccharidosis, *Pediatrics* **50**:688-692.

214. Di Ferrante, N., Nichols, B. L., Knudson, A. G., McCredie, K. B., Singh, J., and Donnelly, P. V., 1973, Mucopolysaccharide-storage diseases: Corrective activity of normal human serum and lymphocyte extracts, in: *Enzyme Therapy in Genetic Diseases* (R. J. Desnick, R. W., Bernlohr, and W. Krivit, eds.), pp. 31-40, Williams and Wilkins, Baltimore.

215. Knudson, A. G., Jr., Di Ferrante, N., and Curtis, J. F., 1971, Effect of leukocyte

transfusion in a child with type II mucopolysaccharidosis, *Proc. Natl. Acad. Sci. USA* **68:**1738–1741.

216. Erickson, R. P., Sandman, R., Robertson, W., Van, B., and Epstein, C. J., 1972, Inefficiency of fresh frozen plasma therapy of mucopolysaccharidosis II, *Pediatrics* **50:**693–701.

217. Dean, M. F., Muir, H., and Benson, P. F., 1973, Mobilization of glycosaminoglycans by plasma infusion in mucopolysaccharidosis Type II—two types of response, *Nature New Biol* **243:**143–146.

218. Kolodny, E. H., 1973, discussion, in: *Enzyme Therapy in Genetic Diseases* (R. J. Desnick, R. W. Bernlohr, and W. Krivit, eds.), p. 40, Williams and Wilkins, Baltimore.

219. Sly, W. S., Glaser, J. G., Roozen, K. Brot, F., and Stahl, P., 1974, Enzyme replacement studies with β-glucuronidase in β-glucuronidase deficiency, in: *Enzyme Therapy in Lysosomal Storage Diseases* (J. M. Tager, G. J. M. Hooghwinkel, and W. T. Daems, eds.), pp. 288–289, North-Holland, Amsterdam.

220. Desnick, R. J., Krivit, W., Snyder, P. D., Desnick, S. J., and Sharp, H. L., 1972, Sandhoff's disease: Ultrastructural and biochemical studies, in: *Sphingolipids Sphingolipidoses and Allied Disorders* (S. M. Aronson, and B. V. Volk, eds.), pp. 351–371, Plenum, New York.

221. Mapes, C. A., Anderson, R. L., Sweeley, C. C., Desnick, R. J., and Krivit, W., 1970, Enzyme replacement in Fabry's disease, an inborn error of metabolism, *Science* **169:**987–989.

222. Van den Bergh, F. A. J. T. M., 1978, Ph.D. Thesis, University of Amsterdam.

223. Brady, R. O., Tallman, J. F., Johnson, W. G., Gal, A. E., Leahy, W. R., Quirk, J. M., and Dekaban, A. S., 1973, Replacement therapy for inherited enzyme deficiency: Use of purified ceramidetrihexosidase in Fabry's disease, *N. Engl. J. Med.* **289:**9–14.

224. Brady, R. O., Pentchev, P. G., and Gal, A. E., 1975, Investigations in enzyme replacement therapy in lipid storage diseases, *Fed. Proc.* **34:**1310–1315.

225. Thorpe, S. R., Fiddler, M, B., and Desnick, R. J., 1974, Enzyme therapy IV: A method for determining the *in vivo* fate of bovine β-glucuronidase in β-glucuronidase deficient mice, *Biochem. Biophys. Res. Commun.* **61:**1464–1470.

226. Johnson, W. G., Desnick, R. J., Long, D. M., Sharp, H. L., Krivit, W., Brady, B., and Brady, R. O., 1973, Intravenous injection of purified hexosaminidase A into a patient with Tay–Sachs disease, in: *Enzyme Therapy in Genetic Diseases* (R. J. Desnick, R. Bernlohr, and W. Krivit, eds.), pp. 120–124, Williams and Wilkins, Baltimore.

227. De Barsy, T., Jacquemin, P., Van Hoof, F., and Hers, H. G., 1973, Enzyme replacement in Pompe's disease: An attempt with purified human α-glucosidase, in: *Enzyme Therapy in Genetic Diseases* (R. J. Desnick, R. Bernlohr, and W. Krivit, eds.), pp. 184–190, Williams and Wilkins, Baltimore.

228. Brady, R. O., Pentchev, P. G., Gal, A. E., Hibbert, S. R., and Dekaban, A. S., 1974, Replacement therapy for inherited enzymatic deficiency: Use of purified glucocerebrosidase in Gaucher's disease, *N. Engl. J. Med.* **291:**989–993.

229. Beutler, E., Dale, G. L., and Kuhl, W., 1977, Enzyme replacement with red cells, *N. Engl. J. Med.* **296:**942–943.

230. Beutler, E., Dale, G. L., Guinto, E., and Kuhl, W., 1977, Enzyme replacement therapy in Gaucher's disease: Preliminary clinical trial of a new enzyme preparation, *Proc. Natl. Acad. Sci. USA* **74:**4620–4623.

231. Brady, R. O., Barringer, J. A., Gal, A. E., Pentchev, P. G., and Furbish, R. S., 1980, Status of enzyme replacement therapy for Gaucher disease, in: *Enzyme Therapy in Genetic Diseases:2* (R. J. Desnick, ed.), pp. 361–368, Alan R. Liss, New York.

232. Brady, R. O., 1977, Heritable catabolic and anabolic disorders of lipid metabolism, *Metabolism* **26:**329–345.

233. Von Specht, B.U., Geiger, G., Arnon, R., Passwell, J., Keren, G., Goldman, B., and Padeh, B., 1979, Enzyme replacement in Tay–Sachs disease, *Neurology* **29**:858–864.
234. Vance, D. E., Krivit, W., and Sweeley, C. C., 1975, Metabolism of neutral glycosphingolipids in plasma of a normal and a patient with Fabry's disease, *J. Biol. Chem.* **250**:8119–8124.
235. Dawson, G. L., and Sweeley, C. C., 1970, *In vivo* studies on glycosphingolipid metabolism in porcine blood, *J. Biol. Chem.* **245**:410–416.
236. Goldstein, J. L., and Brown, M. S., 1977, The low-density lipoprotein pathway and its relation to atherosclerosis, *Ann. Rev. Biochem.* **46**:897–930.
237. Desnick, R. J., Klionsky, B., and Sweeley, C. C., 1978, Fabry disease, in: *The Metabolic Basis of Inherited Disease*, 4th edn. (J. B. Stanbury, J. B. Wyngaarden, and D. S. Frederickson, eds.), pp. 810–840, McGraw Hill, New York.
238. Clarke, J. T. R., Stoltz, J. M., and Mulcahy, M. R., 1976, Neutral glycosphingolipids of serum lipoproteins in Fabry's disease, *Biochim. Biophys. Acta* **431**:317–322.
239. Van Den Bergh, F. A. J. T. M., and Tager, J. M., 1976, Localization of neutral glycosphingolipids in human plasma, *Biochim. Biophys. Acta* **441**:391–402.
240. Vance, D. E., Krivit, W., and Sweeley, C. C., 1969, Concentrations of glycosylceramides in plasma and red cells in Fabry's disease—a glycolipid lipidosis, *J. Lipid Res.* **10**:188–194.
241. Dawson, G. L., Kruski, A., and Scanu, A. M., 1976, Distribution of glycosphingolipids in the serum lipoproteins of normal subjects and patients with hypo- and hyper-lipidemias, *J. Lipid Res.* **17**:125–131.
242. Ashwell, G., and Morell, A. G., 1974, The role of surface carbohydrate in the hepatic recognition of circulating glycoproteins, in: *Advances in Enzymology*, Vol. 41 (A. Meister, ed.), pp. 99–128, John Wiley, New York.
243. Teichberg, V. I., Silman, I., Beitsch, D. D., and Resheff, G., 1975, A β-D-galactoside binding protein from electric organ tissue of *Electrophorus electrus*, *Proc. Natl. Acad. Sci. USA* **72**:1383–1387.
244. Hill, R. L., Pizzo, S. V., Imber, M., Lehrman, M., Prieels, J. P., Glasgow, L. R., Guthrow, C. E., and Paulson, J. C., 1980, Receptors on hepatocytes that bind ligands containing fucosyl, α1,3N-acetylglucosamine linkages, in: *Enzyme Therapy in Genetic Diseases:2* (R. J. Desnick, ed.), pp. 85–91, Alan R. Liss, New York.
245. Baynes, J. W., and Wold, F., 1976, Effect of glycosylation on the *in vivo* circulating half-life of ribonuclease, *J. Biol. Chem.* **251**:6016–6024.
246. Brown, T. L. Henderson, L. A., Thorpe, S. R., and Baynes, J. W., 1978, The effect of α-mannose terminal oligosaccharides on the survival of glycoproteins in the circulation, *Arch. Biochem. Biophys.* **188**:418–428.
247. Stahl, P., Six, H., Rodman, J. S., Schlesinger, P. H., Tulsiani, D. R. P., and Touster, O., 1976, Evidence for specific recognition sites of mediated clearance of lysosomal enzymes *in vivo*, *Proc. Natl. Acad. Sci. USA* **73**:4045–4059.
248. Stahl, P., Rodman, J. S., Doebber, T., Miller, M. J., and Schlesinger, P., 1978, Specific recognition and uptake of lysosomal enzymes and modified glycoproteins by rat tissue, in: *Protein Turnover and Lysosomal Function* (H. L. Segal and D. J. Doyle, eds.), pp. 479–496, Academic Press, New York.
249. Stahl, P., Rodman, J. S., Miller, M. J., and Schlesinger, P. H., 1978, Evidence for receptor-mediated binding of glycoproteins, glycoconjugates, and lysosomal glycosidases by alveolar macrophages, *Proc. Natl. Acad. Sci. USA* **75**:1399–1403.
250. Hudgin, R. L., Pricer, W. F., Ashwell, G. L., Stockert, F. J., and Morell, A. G., 1974, The isolation and properties of a rabbit liver binding protein specific for asialoglycoproteins, *J. Biol. Chem.* **249**:5536–5543.
251. Morell, A. G., Gregoriadis, G., Schlenberg, F. H., Hickman, J., and Ashwell, G.,

1971, The role of sialic acid in determining the survival of glycoproteins in the circulation, *J. Biol. Chem.* **246**:1461–1467.

252. Stahl, P., Rodman, J. S., and Schesinger, P., 1978, Clearance of lysosomal hydrolases following intravenous infusion, *Arch. Biochem. Biophys.* **177**:594–605.

253. Brown, M. S., and Goldstein, J. L., 1976, Receptor-mediated control of cholesterol metabolism, *Science* **191**:150–151.

254. Goldstein, J. L., and Brown, M. S., 1976, The LDL-pathway in human fibroblasts: A receptor mediated mechanism for the regulation of cholesterol metabolism, *Curr. Topics Cell. Reg.* **11**:147–170.

255. Youngdahl-Turner, P., Allen, R. H., and Rosenberg, L. E., 1978, Binding and uptake of transcobalamin II by human fibroblasts, *J. Clin. Invest.* **61**:133–141.

256. Neufeld, E. F., 1980, The uptake of enzymes into lysosomes: An overview, in: *Enzyme Therapy in Genetic Diseases: 2* (R. J. Desnick, ed.), pp. 77–84, Alan R. Liss, New York.

257. Wold, R., 1980, Enzyme recognition and modification, in: *Enzyme Therapy in Genetic Diseases: 2* (R. J. Desnick, ed.), pp. 129–139, Alan R. Liss, New York.

258. Kawasaki, T., and Ashwell, G., 1976, Chemical and physical properties of a hepatic membrane protein that specifically binds asialoglycoproteins, *J. Biol. Chem.* **251**:1296–1300.

259. Kawasaki, T., and Ashwell, G., 1976, Carbohydrate structure of glycopeptides isolated from a hepatic membrane-binding protein specific for asialoglycoproteins, *J. Biol. Chem.* **251**:5292–5297.

260. Stowell, C. P., and Lee, Y. C., 1978, The binding of D-glucosyl-neoglycoproteins to the hepatic asialoglycoprotein receptor, *J. Biol. Chem.* **253**:6107–6111.

261. Rogers, J. C., and Kornfeld, S., 1971, Hepatic uptake of proteins coupled to fetuin glycopeptide, *Biochem. Biophys. Res. Commun.* **45**:622–632.

262. Lee, Y. C., Stowell C. P., and Krantz, M. J., 1976, 2-Imino-2-methoxyethyl-1-thioglycosides: New reagents for attaching sugars to proteins, *Biochemistry* **15**:3956–3960.

263. Krantz, M. J., Holtzman, N. A., Stowell, C. P., and Lee, Y. C., 1976, Attachment of thioglycosides to proteins: Enhancement of liver membrane binding, *Biochemistry* **15**:3963–3967.

264. Marsh, J. W., Denis, J., and Wriston, J. C., Jr., 1977, Glycosylation of *Escherichia coli* L-asparaginase, *J. Biol. Chem.* **252**:7678–7684.

265. Sando, G. N., 1978, Synthetic inhibitors of receptor-mediated endocytosis of a lysosomal enzyme by cultured fibroblasts, *Fed. Proc.* **37**:1502–1512.

266. Wilson, G., 1978, Effect of reductive lactose amination on the hepatic uptake of bovine pancreatic ribonuclease A dimer, *J. Biol. Chem.* **253**:2070–2077.

267. DeWaard, A., Hickman, S., and Kornfeld, S., 1976, Isolation and properties of β-galactoside binding lectins of calf heart and lung, *J. Biol. Chem.* **251**:7581–7590.

268. Stahl, P., Schlesinger, P. H., Rodman, J. S., and Doebber, T., 1976, Recognition of lysosomal glycosidases *in vivo* inhibited by modified glycoproteins, *Nature* **264**:86–88.

269. Achord, D. T., Brot, F. E., and Sly, W. S., 1977, Inhibition of the rat clearance system for agalacto-orosomucoid by yeast mannans and mannose, *Biochem. Biophys. Res. Commun.* **77**:409–415.

270. Kawasaki, T., Etoh, R., and Yamashina, I., 1978, Isolation and characterization of mannan-binding protein from rabbit liver, *Biochem. Biophys. Res. Commun.* **81**:1018–1025.

271. Stahl, P., and Schlesinger, P., 1978, Mannose/N-acetylglucosamine receptor: Plasma clearance and macrophage uptake of glycoconjugates and lysosomal glycosidases, *Fed. Proc.* **38**:467.

272. Hickman, S., Shapiro, L. J., and Neufeld, E. F., 1974, A recognition marker required

for uptake of a lysosomal enzyme by cultured fibroblasts, *Biochem. Biophys. Res. Commun.* **57**:55–61.

273. Shapiro, L. J., Hall, C. W., Leder, I. G., and Neufeld, E. F., 1976, The relationship of α-L-iduronidase and Hurler correction factor, *Arch. Biochem. Biophys.* **172**:156–161.

274. Glaser, J. H., Roozen, K. L., Brot, F. E., and Sly, W. S., 1975, Multiple isoelectric and recognition forms of human β-glucuronidase activity, *Arch. Biochem. Biophys.* **166**:536–542.

275. Kaplan, A., Fischer, D., Achord, D. T., and Sly, W. S., 1977, Phosphohexosyl recognition is a general characteristic of pinocytosis of lysosomal glycosidases by human fibroblasts, *J. Clin. Invest.* **60**:1088–1093.

276. Sando, G. N., and Neufeld, E. F., 1977, Recognition and receptor-mediated uptake of a lysosomal enzyme, α-L-iduronidase, by cultured human fibroblasts, *Cell* **12**:619–627.

277. Ullrich, K., Mersmann, G., Weber, E., and von Figura, K., 1978, Evidence for lysosomal enzyme recognition by human fibroblasts via a phosphorylated carbohydrate moiety, *Biochem. J.* **170**:643–649.

278. Brown, M. S., and Goldstein, J. L., 1975, Lipoprotein receptors and the genetic control of cholesterol metabolism in cultured human cells, *Naturwissenschaften* **62**:385–389.

279. Goldstein, J. L., and Brown, M. S., 1974, Binding and degradation of low density lipoproteins by cultured human fibroblasts, *J. Biol. Chem.* **249**:5153–5162.

280. Goldstein, J. L., Dana, S. E., Faust, J. R., Beaudet, A. L., and Brown, M. S., 1975, Role of lysosomal acid lipase in the metabolism of plasma low density lipoprotein, *J. Biol. Chem.* **250**:8487–8495.

281. Brown, M. S., and Goldstein, J. L., 1975, Regulation of the activity of the low density lipoprotein receptor in human fibroblasts, *Cell* **6**:307–316.

282. Brown, M. S., and Goldstein, J. L., 1974, Familial hypercholesterolemia: Defective binding of lipoproteins to cultured fibroblasts associated with impaired regulation of 3-hydroxy-3-methylglutaryl Coenzyme A reductase, *Proc. Natl. Acad. Sci. USA* **71**:788–792.

283. Ho, Y. K., Brown, M. S., Kayden, H. J., and Goldstein, J. L., 1976, Binding, internalization and hydrolysis of low density lipoprotein in long term lymphoid cell lines from a normal subject and a patient with homozygous familial hypercholesterolemia, *J. Exp. Med.* **144**:444–455.

284. Goldstein, J. L., and Brown, M. S., 1975, Lipoprotein receptors, cholesterol metabolism and atherosclerosis, *Arch. Pathol.* **99**:181–184.

285. Albers, J. J., and Bierman, E. L., 1976, The effect of hypoxia on uptake and degradation of low density lipoproteins by cultured human arterial smooth muscle cells, *Biochim. Biophys. Acta* **424**:422–429.

286. Stein, O., and Stein, Y., 1976, High density lipoproteins reduce the uptake of low-density lipoproteins by human endothelial cells in culture, *Biochim. Biophys. Acta* **431**:363–368.

287. Hamilton, R. L., 1972, Synthesis and secretion of plasma lipoproteins, *Adv. Exp. Med. Biol.* **26**:7–24.

288. Frederickson, D. S., Goldstein, J. L., and Brown, M. S., 1978, The familial hyperlipoproteinemias, in: *The Metabolic Basis of Inherited Disease,* 4th edn. (J. B. Stanbury, J. B. Wyngaarden, and D. S. Fredrickson, eds.), pp. 604–655, McGraw Hill, New York.

289. Brown, M. S., Ho, Y. K., and Goldstein, J. L., 1976, The low density lipoprotein pathway in human fibroblasts: Relation between cell surface receptor binding and endocytosis of low density lipoproteins, *Ann. N.Y. Acad. Sci.* **275**:244–257.

290. Brown, M. S., Brannan, P. G., Bohmfalk, H. A., Brunschede, G. Y., Dana, S. E.,

Helgeson, J., and Goldstein, J. L., 1975, Use of mutant fibroblasts in the analysis of the regulation of cholesterol metabolism in human cells, *J. Cell Physiol.* **85**:425–436.

291. Mahley, R. W., Weisgraber, K. H., Melchior, G. W., Innerarity, T. L., and Holcombe, K. S., 1980, Inhibition of receptor-mediated clearance of lysine and arginine-modified lipoproteins from plasma of rats and monkeys, *Proc. Natl. Acad. Sci. USA* **77**:225–229.

292. Goldstein, J. L., Basu, S. K., Brunschede, G. Y., and Brown, M. S., 1976, Release of low density lipoprotein from its cell surface receptor by sulfated glycosaminoglycans, *Cell* **7**:85–95.

293. Anderson, R. G. W., Goldstein, J. L., and Brown, M. S., 1976, Localization of low density lipoprotein receptors on plasma membrane of normal human fibroblasts and their absence in cells from a familial hypercholesterolemia homozygote, *Proc. Natl. Acad. Sci. USA* **73**:2434–2438.

294. Furbish, S. F., Steer, C. J., Barringer, J. A., Jones, E. A., and Brady, R. O., 1978, The uptake of native and desialylated glucocerebrosidase by rat hepatocytes and Kupffer cells, *Biochem. Biophys. Res. Commun.* **81**:1047–1053.

295. Rattazzi, M. C., McCullough, R. A., Downing, C. J., and Kung, M-P., 1979, Towards enzyme therapy in G_{M2} gangliosidosis: β-hexosaminidase infusion in normal cats, *Pediatr. Res.* **13**:916.

296. Rattazzi, M. C., Baker, H. J., Cork, L. C., Cox, N. R., Lanse, S. B., McCullough, R. A., and Munnell, J. F., 1979, The domestic cat as a model for human G_{M2} gangliosidosis: Pathogenetic and therapeutic aspects, in: *Models for the Study of Inborn Errors of Metabolism* (F. A. Hommes, ed.), p. 57, American Elsevier, Amsterdam.

297. Schlesinger, P., Rodman, J. S., Frey, M., Lang, S., and Stahl, P., 1976, Clearance of lysosomal hydrolases following intravenous infusion, *Arch. Biochem. Biophys.* **177**:606–614.

298. Fiddler, M. B., and Desnick, R. J., 1977, Enzyme therapy: Differential *in vivo* retention of bovine hepatic, renal and splenic β-glucuronidases and evidence for enzyme stabilization by intermolecular exchange, *Arch. Biochem. Biophys.* **179**:397–408.

299. Gregoriadis, G., and Ryman, B. E., 1971, Liposomes as carriers of enzymes or drugs: A new approach to the treatment of storage diseases, *Biochem. J.* **124**:58P–63P.

300. Steger, L. D., and Desnick, R. J., 1977, Enzyme therapy VI: Comparative *in vivo* fates and effects on lysosomal integrity of enzyme entrapped in negatively and positively charged liposome, *Biochim. Biophys. Acta* **464**:530–546.

301. Weissman, G., Bloomgarden, D., Kaplan, R., Cohen, C., Hoffstein, S., Collins, T., Gotlieb, A., and Nagle, D., 1975, A general method for the introduction of enzymes, by means of immuno-globulin-coated liposomes into lysosomes of deficient cells, *Proc. Natl. Acad. Sci. USA* **72**:88–92.

302. Thorpe, S. R., Fiddler, M. B., and Desnick, R. J., 1975, Enzyme therapy V. *In vivo* fate of erythrocyte-entrapped β-glucuronidase in β-glucuronidase-deficient mice, *Pediatr. Res.* **9**:918–923.

303. Gregoriadis, G., and Leathwood, R. D., 1971, Enzyme entrapment in lysosomes, *FEBS Lett.* **14**:95–97.

304. Gregoriadis, G., and Davis, C., 1979, Stability of liposomes *in vivo* and *in vitro* is promoted by their cholesterol content and the presence of blood cells, *Biochem. Biophys. Res. Commun.* **89**:1287–1293.

305. Deshmukh, D. S., Bear, W. D., Wisniewski, H. M., and Brocherhoff, H., 1978, Long-living liposomes as potential blood carriers, *Biochem. Biophys. Res. Commun.* **82**:328–334.

306. Milsmann, M. H. W., Schwendener, R. A., and Weder, H. G., 1978, The preparation

of large single bilayer liposomes by a fast and controlled dialysis, *Biochem. Biophys. Acta* **512**:147–155.

307. Daemer, D., and Bangham, A. E., 1976, Large volume liposomes by an ether vaporization method, *Biochim. Biophys. Acta* **512**:147–155.

308. Szoka, F., and Papahadjopoulos, D., 1978, Procedure for preparation of liposomes with large internal aqueous space and high capture by reverse-phase evaporation, *Proc. Natl. Acad. Sci. USA* **75**:4194–4198.

309. Schuster, B. G., Neidig, M. Alving, B. M., and Alving, C. R., 1979, Production of antibodies against phosphocholine, phosphatidylcholine, sphingomyelin, and lipid A by injection of liposomes containing lipid A, *J. Immunol.* **122**:900–905.

310. Desnick, R. J., Fiddler, M. B., Thorpe, S. R., and Steger, L. D., 1977, Enzyme entrapment in erythrocytes and liposomes for the treatment of lysosomal storage diseases, in: *Biomedical Applications of Immobilized Enzymes and Proteins* (T. M. S. Chang, ed.), pp. 227–244, Academic Press, New York.

311. Hudson, L. D. S., Fiddler, M. B., and Desnick, R. J. J., 1979, Enzyme therapy X: Immune response induced by enzyme- and buffer-loaded liposomes in C3H/HeJ *Gus*[h] mice, *J. Pharmacol. Exp. Ther.* **208**:507–514.

312. Van Rooijen, N., and van Nieuwmegen, R., 1977, Liposomes in immunology: The immune response against antigen-containing liposomes, *Immunol. Commun.* **6**:489–498.

313. Van Rooijen, N., and van Niewmegen, R., 1979, Liposomes in immunology: Impairment of the adjuvant effects of liposomes by incorporation of the adjuvant lysolecithin and the role of macrophages, *Immunol. Commun.* **8**:381–396.

314. Cunningham, C. M., Kingzette, M., Richards, R. L., Alving, C. R., Lint, T. F., and Gewurz, H. J., 1979, Activation of human complement by liposomes: A model for membrane activation of the alternative pathway, *Immunology* **122**:1237–1242.

315. Finkelstein, M. C., and Weismann, G., 1979, Enzyme replacement via liposomes. Variations in lipid composition determine liposomal integrity in biological fluids, *Biochim. Biophys. Acta* **587**:202–216.

316. Ihler, G. M., Glew, R. H., and Schnure, F. W., 1973, Enzyme loading of erythrocytes, *Proc. Natl. Acad. Sci. USA* **70**:2663–2668.

317. Baker, R. F., 1967, Entry of ferritin into human red cells during hypotonic haemolysis, *Nature* **215**:424–426.

318. Bodeman, H., and Passow, H., 1972, Factors controlling the resealing of the membrane of human erythrocyte ghosts after hypotonic hemolysis, *J. Membrane Biol.* **8**:1–4.

319. Dale, G. L., Villacorte, D. F., and Beutler, E., 1977, High-yield entrapment of proteins into erythrocytes, *Biochem. Med.* **18**:220–225.

320. Fiddler, M. B., Hudson, L. D. S., White, J. G., and Desnick, R. J., 1980, Enzyme therapy XIV: Comparison of methods for enzyme entrapment in human erythrocytes, *J. Lab. Clin. Med.* **96**:307–317.

321. Ihler, G., Alan, L., Purpura, J., and Glad, R. H., 1975, Enzymatic degradation of uric acid by uricase-loaded erythrocytes, *J. Clin. Invest.* **56**:595–602.

322. Updike, S., Wakamaya, R. T., and Lightfoot, E. N., 1976, Asparaginase trapped in red blood cells: Action and survival, *Science* **193**:681–687.

323. Deloach, J., and Ihler, G., 1977, A dialysis procedure for loading erythrocytes with enzymes and drugs, *Biochim. Biophys. Acta* **496**:136–145.

324. Jancik, J. M., Schquer, R., Andres, K. H., and von During, M., 1978, Sequestration of neuraminidase-treated erythrocytes, *Cell Tiss. Res.* **186**:209–226.

325. Tyrrell, D. A., and Ryman, B. E., 1976, The entrapment of therapeutic agents in resealed erythrocyte "ghosts" and their fate *in vivo*, *Biochem. Soc. Trans.* **4**:677–681.

326. De Loach, J., Peters, S., Pinkard, I., Glew, R., and Ihler, G., 1977, Effect of glutaraldehyde treatment on enzyme-loaded erythrocytes, *Biochim. Biophys. Acta* **496**:136–145.

327. Ben-Bassat, I., Bensch, D. G., and Shrien, S. L., 1972, Drug-induced erythrocyte membrane internalization, *J. Clin. Invest.* **51**:1833–1840.
328. Jacob, H. S., and Jandl, J. H., 1962, Effects of sulfhydral inhibition on red blood cells II. Studies *in vivo*, *J. Clin. Invest.* **41**:1514–1520.
329. Durocher, J. R., Payne, R. C., and Conrad, M. E., 1975, Role of sialic acid in erythrocyte survival, *Blood* **45**:11.
330. Seaman, G. V. F., and Uhlenbruck, G., 1963, The surface structure of erythrocytes from some animal sources, *Arch. Biochem. Biophys.* **100**:493–501.
331. Weed, R. I., La Celle, P. L., and Merrill, E. W., 1969, Metabolic dependence on red cell deformability, *J. Clin. Invest.* **48**:795–800.
332. Fiddler, M. B., Hudson, L. D. S., and Desnick, R. J., 1977, Enzyme therapy VIII: Immunologic evaluation of repeated administration of erythrocyte-entrapped β-glucuronidase to β-glucuronidase deficient mice, *Biochem. J.* **168**:141–145.
333. Barringer, J. A., Rappoport, S. I., and Brady, R. O., 1980, Access of enzymes to brain following osmotic alteration of the blood-brain barrier, in: *Enzyme Therapy in Genetic Diseases: 2* (R. J. Desnick, ed.), pp. 195–205, Alan Liss, New York.
334. Rattazzi, M. C., Lanse, S. B., McCullough, R. A., Nester, J. A., and Jacobs, E. A., 1980, Towards enzyme replacement in G_{M2} gangliosidosis: Organ deposition and induced central nervous system uptake of human β-hexosaminidase in the cat, in: *Enzyme Therapy in Genetic Diseases: 2* (R. J. Desnick, ed.), pp. 179–193, Alan R. Liss, New York.
335. Rattazzi, M. C., Appel, A. M., and Nester, J. A., 1979, Towards enzyme therapy in G_{M2} gangliosidosis: Visceral organ and CNS uptake of human β-hexosaminidase in normal cats, *Am. J. Hum. Genet.* **30**:59A.
336. Ratazzi, M. C., Appel, A. M., Baker, H. J., and Nester, J. A., 1981, Toward enzyme replacement in G_{M2} gangliosidosis; inhibition of hepatic uptake and induction of CNS uptake of human β-hexosaminidase in the cat, in: *Lysosomes and Lysosomal Storage Diseases* (J. W. Calletian and J. A. Lowden, eds.), pp. 405–424, Raven Press, New York.
337. Kusiak, J. W., Toney, J. H., Quirk, J. M., and Brady, R. O., 1979, Specific binding of ^{125}I-labeled β-hexosaminidase A to rat brain synaptosomes, *Proc. Natl. Acad. Sci. USA* **76**:982–985.
338. Desnick, R. J., Dean, K. J., Grabowski, G. A., Bishop, D. F., and Sweeley, C. C., 1979, Enzyme therapy in Fabry disease: Differential *in vivo* plasma clearance and metabolic effectiveness of plasma and splenic α-galactosidase, *Proc. Natl. Acad. Sci. USA* **76**:5326–5330.
339. Williams, J. C. and Murray, A. K., 1980, Enzyme replacement in Pompe disease with an α-glucosidase low density lipoprotein complex, in: *Enzyme Therapy in Genetic Diseases: 2* (R. J. Desnick, ed.), pp. 415–423, Alan R. Liss, New York.
340. Tyrell, D. A., Ryman, B. E., Kieton, B. R., and Dubovitz, V., 1976, Use of liposomes in treating type II glycogenosis, *Br. Med. J.* **12**:88.
341. Belchetz, P. E., Braidman, I. P., Crawley, J. C. W., and Gregoriadis, G., 1977, Treatment of Gaucher's disease with liposome-entrapped glucocerebroside:β-glucosidase, *Lancet* **2**:116–117.
342. Bishop, D. F., and Sweeley, C. C., 1978, Plasma α-galactosidase A: Properties and comparisons with tissue α-galactosidase, *Biochim. Biophys. Acta* **525**:399–409.
343. Bishop, D. F., and Desnick, R. J., Unpublished observation.
344. Swallow, D. M., Stokes, D. C., Corney, G., and Harris, H., 1974, Differences between the *N*-acetylhexosaminidase isozymes in serum and tissues, *Ann. Hum. Genet.* **37**:287–302.
345. Desnick, R. J., Dean, K. J., Grabowski, G. A., Bishop, D. F., and Sweeley, C. C.,

1980, Enzyme therapy XVII: Metabolic and immunologic evaluation of α-galactosidase A replacement in Fabry disease, in: *Enzyme Therapy in Genetic Diseases: 2* (R. J. Desnick, ed.), pp. 393–414, Alan R. Liss, New York.

346. Paulson, J. C., Beranek, W. E., and Hill, R. L., 1977, Purification of sialyltransferase from bovine colostrum by affinity chromatography on CDP-agarose, *J. Biol. Chem.* **252:**2356–2362.

347. Gregoriadis, G., Neerunjun, D., Meade, T. W., Goolamali, S. K., Weereantne, H., and Bull, G., 1980, Experiences after long-term treatment of a type I Gaucher disease patient with liposome entrapped glucocerebrosidase:β-glucosidase, in: *Enzyme Therapy in Genetic Diseases: 2* (R. J. Desnick, ed.), pp. 383–392, Alan R. Liss, New York.

348. Beutler, E. L., Dale, G. L., and Kuhl, W., 1980, Replacement therapy in Gaucher disease, in: *Enzyme Therapy in Genetic Diseases: 2* (R. J. Desnick, ed.), pp. 369–381, Alan R. Liss, New York.

349. Snyder, P. D., Wold, F., Bernlohr, R. W., Dullum, C., Desnick, R. J., Krivit, W., and Condie, R. M., 1974, Enzyme therapy II: Purified human α-galactosidase A: Stabilization to heat and protease degradation by complexing with antibody and by chemical modification, *Biochim. Biophys. Acta* **350:**432–436.

350. Boyer, S. H., Siggers, D. C., and Krueger, L. J., 1973, Caveat to protein replacement therapy for genetic disease. Immunological implications of accurate molecular diagnosis, *Lancet* **2:**654–659.

351. Abuchawski, A. T., Es, V., Palczud, N. C., and Davis, F. R., 1974, Preparation and properties of non-immunogenic catalase, *Fed. Proc.* **33:**1317–1320.

352. Schmickel, R., Wilson, G., and Waterson, J., 1980, Techniques for the isolation of specific and functional templates, in: *Enzyme Therapy in Genetic Diseases: 2* (R. J. Desnick, ed.), pp. 43–51, Alan R. Liss, New York.

353. Itakura, K., Hirose, T., Crea, R., Riggs, A. D., Heyneker, H. L., Bolivar, F., and Boyer, H. W., 1977, Expression in *Escherichia coli* of a chemically synthesized gene for the hormone somatostatin, *Science* **198:**1056–1063.

354. Forget, B. G., Housman, D., Benz, E. J., and McCaffrey, R. P., 1975, Synthesis of DNA complementary to separated alpha and beta globin messenger RNAs, *Proc. Natl. Acad. Sci. USA* **72:**984–988.

355. Martial, J. A., Hallewall, R. A., Baxter, J. D., and Goodman, H. M., 1979, Human growth hormone: Complementary DNA cloning and expression in bacteria, *Science* **205:**602–606.

356. Khorana, H. G., 1979, Total synthesis of a gene, *Science* **203:**614.

357. Nagata, S., Taira, H., Hall, A., Johnrud, L., Streuli, M., Ecsodi, J., Boll, W., Cantell, K., and Weissman, C., 1980, Synthesis in *E. coli* of a polypeptide with human leukocyte interferon activity, *Nature* **284:**316–318.

358. Villa-Komaroff, L., Broome, S., Naber, S. P., Efstratiadis, A., Lomedies, P., Tizard, R., Chick, W. L., and Gilbert, W., 1980, The synthesis of insulin in bacteria: A model for the production of medically useful proteins in prokaryotic cells, in: *Enzyme Therapy in Genetic Diseases: 2* (R. J. Desnick, ed.), pp. 53–68, Alan R. Liss, New York.

359. Villa-Komaroff, L., Efstratiadis, A., Broome, S., Lomedius, P., Tizard, R., Naber, S. P., Chick, W. L., and Gilbert, W., 1978, A bacterial clone synthesizing pro-insulin, *Proc. Natl. Acad. Sci. USA* **75:**3727–3731.

360. Goeddel, D. V., Kleid, D. G., Bolivar, F., Heyneker, H. L., Yansura, D. G., Crea R., Hirose, T., Kraszewski, A., Itakura, K., and Riggs, A. D., 1979, Expression in *Escherichia coli* of chemically synthesized genes for human insulin, *Proc. Natl. Acad. Sci. USA* **76:**106–110.

361. Baxter, J. D., 1980, Recombinant DNA and medical progress, *Hosp. Pract.* **15:**57–67.

362. Munyon, W., Kraiserburd, E., Davies, D., and Mann, J., 1971, Transfer of thymidine kinase to thymidine kinase-less L-cells by infection with ultraviolet-irradiated *Herpes simplex* virus, *J. Virol.* **7**:813–821.

363. Fournier, R. E. K., and Ruddle, F. H., 1977, Microcell-mediated transfer of murine chromosomes into mouse, Chinese hamster, and human somatic cells, *Proc. Natl. Acad. Sci. USA* **74**:319–323.

364. Bacchetti, S., and Graham, F. I., 1977, Transfer of the gene for thymidine kinase to thymidine kinase deficient human cells by purified *Herpes simplex* viral DNA, *Proc. Natl. Acad. Sci. USA* **74**:1590–1594.

365. Wigler, M., Pellicer, A., Silverstein, S., and Axel, R., 1978, Biochemical transfer of single-copy eucaryotic genes using total cellular DNA as donor, *Cell* **14**:725–731.

366. Wigler, M., Perucho, M., Kurtz, D., Dana, S., Pellicer, A., Axel, R., and Silverstein, S., 1980, Transformation of mammalian cells with an amplifiable dominant acting gene, *Proc. Natl. Acad. Sci. USA* **77**:3567–3570.

367. Wigler, M., Pellicer, A., Silverstein, S., Axel, R., Urlaub, G., and Chasin, L., 1979, DNA-mediated transfer of the adenine phosphoribosyltransferase into mammalian cells, *Proc. Natl. Acad. Sci. USA* **76**:1373–1376.

368. Willecke, K., Klomfose, M., Miraw, R., and Dohmer, J., 1979, Intraspecies transfer via total cellular DNA of the gene for hypoxanthinine phosphoribosyl-transferase into cultured mouse cells, *Mol. Gen. Genet.* **170**:179–185.

369. Graf, L. A., Jr., Urlaub, G., and Chasin, L., 1979, Transformation of the gene for hypoxanthinine phosphoribosyltransferase, *Somat. Cell Genet.* **5**:1031–1044.

370. Wigler, M. H., Silverstein, S., Lee, L. S., Pellicer, A., Ching, V. and Axel, R., 1977, Transfer of purified *Herpes* virus thymidine kinase gene to cultured mouse cells, *Cell* **11**:223–232.

371. Colbere-Garapin, F., Chousterman, S., Horodniceaunu, F., Kourilsky, P. and Garapin, A. C., 1979, Cloning of the active thymidine kinase gene of *Herpes simplex* virus type 1 in *Escherichia coli* K-12, *Proc. Natl. Acad. Sci. USA* **76**:3755–3759.

372. Mantei, N., Boll, W., and Weissman, C., 1979, Rabbit β-globin mRNA production in mouse L cells transformed with cloned rabbit β-globin chromosomal DNA, *Nature* **281**:40–46.

373. Wigler, M., Sweet, R., Sim, G. K., Wold, B., Pellicer, A., Lacy, E., Maniatis, T., Silverstein, S., and Axel, R., 1979, Transformation of mammalian cells with genes from procaryotes and eucaryotes, *Cell* **16**:777–785.

374. Mercola, K. E., Stang, H. D., Browne, J., Salser, W., and Cline, M. J., 1980, Insertion of a new gene of viral origin into bone marrow cells of mice, *Science* **208**:1033–1036.

375. Cline, M. J., Stang, H., Mercola, K., Morse, L., Ruprecht, R., Browne, J., and Salser, W., 1980, Gene transfer in intact animals, *Nature* **284**:422–427.

376. Mantei, N., Boll, W., and Weissman, C., 1979, Rabbit β-globin mRNA production in mouse L-cells transformed with cloned rabbit β-globin chromosomal DNA, *Nature* **281**:40–43.

Addenda

CHAPTER 1: THE PI POLYMORPHISM
Genetic, Biochemical, and Clinical Aspects of Human α₁-Antitrypsin

Magne K. Fagerhol and Diane Wilson Cox

NOMENCLATURE

Guidelines for human gene nomenclature were presented at the 5th International Workshop on Human Gene Mapping, Edinburgh Conference (1979). According to these guidelines, the locus for α_1-antitrypsin would be *PI*; allele *PI∗M*, etc.; phenotype PI MZ, etc.; genotype *PI∗M/PI∗Z*. The null allele would be *PI∗QO*. Other changes are currently under consideration.

CHROMOSOME MAPPING OF Gm-Pi

Several other chromosome locations for the Gm-Pi linkage group have been suggested by recent studies using hybrid cells. These studies show inconsistent results, and the location of this linkage group cannot yet be considered firmly established. Using human hepatoma–mouse RAG cell hybrids, Turner and Turner (1980) found no correlation between α_1AT production and chromosome 6 markers. The larger number of chromosomes present in these hybrids has made analysis difficult. In another study using somatic cell hybrids between mouse myeloma cells and human lymphocytes or established cell lines, human chromosome 14 was the only chromosome present in all independent hybrids producing gamma heavy chains (Croce *et al.*, 1979).

REFERENCES

Fifth International Workshop on Human Gene Mapping, Edinburgh Conference, 1979, in: *Human Gene Mapping 5* (H. J. Evans, J. L. Hamerton, H. P. Klinger, and V. A. McKusick, eds.), p.96, S. Karger AG, New York.

Turner, B. M., and Turner, V. S., 1980, Secretion of alpha-1-antitrypsin by an established human hepatoma cell line and by human/mouse hybrids, *Somatic Cell Genet.* **6:**1–14.

Croce, C. M., Shander, M., Martinis, J., Cicurel, L., D'Ancona, G. G., Dolby, T. W., and Koprowski, H., 1979, Chromosomal location of the genes for human immunoglobulin heavy chains, *Proc. Natl. Acad. Sci.* **76:**3416–3419.

CHAPTER 2: SEGREGATION ANALYSIS

R. C. Elston

Recently, two computer programs for pedigree segregation analysis have become available: PAP (Hasstedt and Cartwright, 1979) and POINTER (Lalouel and Yee, 1980). The latter calculates the likelihood for nuclear families under the Morton–MacLean mixed model, reparametrized and modified so as to approximate the joint likelihood of a set of nuclear families that make up a single multigenerational pedigree. Both programs have been implemented using GEMINI (Lalouel, 1979) to maximize the likelihood.

Graepel (1981) has investigated several numerical approximations for calculating the mixed model (34), some of which can be easily extended to more than two-generation data. These approximations are quite different from that used by Lalouel (1978) and offer the prospect of providing an efficient algorithm to calculate a mixed-model likelihood for pedigrees. Graepel also reanalyzed the immunoglobulin E data of Gerrard *et al.* (1978) and concluded, in agreement with Rao *et al.* (1980), that it is most likely that there is a major gene segregating in these data. He noted, however, that the one large family eliminated for Ott's analysis is responsible for a disproportionately large share of the evidence for the presence of a major gene, and that the currently available computer programs sometimes yield results that are different enough to be a cause for concern about their numerical accuracy.

REFERENCES

Gerrard, J. W., Rao, D. C., and Morton, N. E., 1978, A genetic study of immunoglobulin E, *Am. J. Hum. Genet.* **30:**46–58.

Graepel, J., 1981, Multifactorial models and likelihoods for the segregation analysis of quantitative traits, unpublished Ph.D. thesis, University of North Carolina at Chapel Hill.

Hasstedt, S., and Cartwright, P., 1979, PAP pedigree analysis package, Technical Report No. 13, Department of Medical Biophysics and Computing, University of Utah Medical Center.

Lalouel, J. M., 1978, Recurrence risks as an outcome of segregation analysis, in: *Genetic Epidemiology* (N. E. Morton and C. S. Chung, eds.), pp. 255–284, Academic Press, New York.

Lalouel, J. M., 1979, GEMINI—A computer program for optimization of general nonlinear functions, Technical Report No. 14, Department of Medical Biophysics and Computing, University of Utah Medical Center.

Lalouel, J. M., and Yee, S., 1980, POINTER: Computer programs for complex segregation analysis of nuclear families with pointers, Technical Report No. 3, Population Genetics Laboratory, University of Hawaii.

Rao, D. C., Lalouel, J. M., Morton, N. E., and Gerrard, J. W., 1980, Immunoglobulin E revisited, *Am. J. Hum. Genet.* **32:**620–625.

Index